THE

MACADAMIA

AUSTRALIA'S GIFT TO THE WORLD

This is the front cover of the 1893 Department of Agriculture booklet, *The Cultivation of the 'Australian Nut'* by Fred Turner, which detailed all aspects of the nut and its culture to stimulate interest in the planting of macadamias. It is the first book published on the cultivation of the macadamia.

THE
MACADAMIA

AUSTRALIA'S GIFT TO THE WORLD

A history of the finest nut in the world
and the industry it supports

Ian McConachie AM

Australian Scholarly Publishing

First published 2024 by
Australian Scholarly Publishing Pty Ltd
7 Lt Lothian Street North
North Melbourne, Victoria 3051
www.scholarly.info / contact@scholarly.info

ISBN 9781923267107

Author: Ian McConachie
Text © Wild Macadamia Conservation, 2024 (previously Macadamia Conservation Trust)
Images © Photographs, maps, illustrations and diagrams as credited

www.wildmacadamias.org.au wild@macadamias.org

Proceeds from the sale of this book will support the conservation of wild macadamia trees in Australia.

The author respectfully acknowledges the Aboriginal Traditional Custodians on whose Country the wild macadamia flourishes, and their continuing connection to Country. Respects are paid to Elders past, to Elders present and to emerging leaders.

All images featuring Aboriginal people, or created by them, as well as insights gained from discussion with Aboriginal people are reproduced with their permission. Aboriginal readers are advised that some images may include people who have now died.

Printed in Australia by Print Strategy Management, Melbourne

Cover and content designer: David Morgan

Cover image: Young macadamia orchard framed by the Glass House Mountains in Queensland, c. 2011, owned by Peter Harvey and now farmed by Michael Harvey and Donna Strong. Credit: Michael Harvey.

Dedicated to my most supportive and loving wife, Janet Kathleen,
who has generously tolerated my eccentricities
and to all the people who in the past, present or future have
added, are adding or will add to the macadamia story.

Contents

Maps and Tables

Acronyms and More

Acronyms

ACIAR Australian Centre for International Agricultural Research
AHC Australian Horticultural Corporation
AMHA Australian Macadamia Handlers Association
AMS Australian Macadamia Society Ltd
ANFIC Australian Nurserymen's Fruit Improvement Company
ANIC Australian Nut Industry Council
AVG Abnormal Vertical Growth (Tall Tree Syndrome)
CMSYB California Macadamia Society Yearbook
COSOP Code of Sound Orchard Practices
CREEC Caboolture Regional Environment and Education Centre
CSIRO Commonwealth Scientific and Industrial Research Organisation
DAF (Queensland) Department of Agriculture and Fisheries
DIS dry-in-shell (nuts)
HACCP Hazard Analysis and Quality Control Points
HAES Hawaii Agricultural Experiment Station
HAL Horticulture Australia Limited (later HIA)
HIA Horticultural Innovation Australia
HMNA Hawaii Macadamia Nut Association
HRDC Horticultural Research and Development Corporation
IMSC International Macadamia Symposium Committee (researchers)
INC International Nut and Dried Fruit Council
IOM Integrated Orchard Management (concept)
IP Intellectual Property
IPM Integrated Pest Management (concept)
ITC (US) International Trade Commission
IVBGR International Board for Plant Genetic Studies (Italy)
KR kernel recovery (see also SKR)
LIDAR Light Detection and Ranging (imaging)
MC moisture content
MCT Macadamia Conservation Trust
MFM Macadamia Farm Management
MHRS Maroochy Horticultural Research Station (QDPI Nambour)
MIA Macadamia Industries Australia
MPA Macadamia Plantations of Australia
MPC Macadamia Processing Company Ltd

Continued >

mya	million years ago
NIH	nut-in-husk
NIS	nut-in-shell
NSW	New South Wales
QA	Quality Assurance
QAAFI	Queensland Alliance for Agriculture and Food Innovation
QDAF	Queensland Department of Agriculture and Fisheries (current in 2024)
QDPI	Queensland Department of Primary Industries (to 2009)
Qld	Queensland
R&D	research and development
SAMAC	Southern African Macadamias Growers Association
SIAP	Strategic Investment Advisory Panel
SKR	sound kernel recovery (percentage)
USITC	see ITC
WIS	wet-in-shell (nuts)
WMO	World Macadamia Organisation

Measurements

1 metre [m]	39.37 inches or 3.3 feet
1 hectare [ha]	2.47 acres
28.35 grams [g]	1 ounce
1 kilogram [kg]	2.2 pounds [lbs]
1 tonne [t]	1,000 kilograms = 0.984 UK tons
20° Celsius [C]	68° Fahrenheit [F]
100° C	212° F

Australian Currency Equivalences (over time)

1 penny	1 cent
12 pennies [pence]	1 shilling
20 shillings	1 pound
1 pound [£1.00]	2 dollars [$2.00]

Foreword

Australians are a nation of story-tellers. First Nations peoples filled their world with stories of the creation of the land, waters, beings and the sky. The First Fleet brought the Irish with their tall tales and wild songs, then there was Ned Kelly and later Phar Lap. We know the stories of the stump-jump plough and the merino, the Sydney Harbour Bridge and Opera House, the winged keel and Warnie. May Gibbs even told stories of Banksia Men and gumnuts.

But one story, the story of an Australian icon, our 'gift to the world', has yet to be told. This book attempts to right that wrong, fill that gap, by telling the story of macadamias.

Ian McConachie has dedicated a large part of his life to not just recording the story of *Macadamia* trees and nuts, but contributing to it, enriching it, playing a starring role. So who better to tell the tale?

While not a young man, Ian came bounding into my office shortly after I had taken on the role of CEO of the Australian Macadamia Society. My days were filled with supporting the commercial macadamia industry and its growers, processors and marketers. The origins of the amazing plant that was the basis of all this activity had not occupied much of my time. The beautiful natural habitat, millennia-long relationship with First Nations peoples, and the chaotic, frustrating but ultimately rewarding story of its development as Australia's premier native food plant export was largely unknown to me.

As Ian began to tell his story in my Lismore office, I felt that I was looking back in time to fifty million years ago and links to the continent of Gondwana. I heard of the sharing of a precious nut, and trade between an ancient culture with a delicious food and an invader with new ways and ideas.

The scene shifts to enthusiastic nut 'missionaries', convinced that they had found the next sugar, chocolate or nutmeg. Orchards planted, nut-cracking machines invented and factories built. Beginning to build a dream, only to find that the Hawaiians had got there before them. Disappointment, indifference and distractions followed. But through it all was an unwavering belief in the unique quality of this Australian nut.

You may think that I have gotten a little carried away in the recounting. But this is a story with it all: heroes, villains, wrong turns and lost ways, new inventions, David and Goliath battles over international trade, and colourful characters – some with a utopian vision, others perhaps driven by the dream of a quick profit.

But this is an unfinished story. There is still time to right some wrongs and work with Indigenous peoples for the next chapters, developing more sustainable ways of growing and new uses based on traditional medicines, novel components and their application. Macadamias make up less than 2 per cent of the world trade in tree nuts – a trade that is growing strongly on the back of new dietary and lifestyle trends. Whether vegetarian, vegan, paleo or gluten free, nuts are part of the plan.

Continued >

The Australian macadamia industry is the only agricultural industry in Australia, possibly in the world, to invest its own funds in understanding the plight of the original species and developing a conservation plan for its rescue and future. The Macadamia Conservation Trust has driven this extraordinary effort, despite public indifference and bureaucratic inaction, to the point where every wild macadamia population has now been mapped. Your purchase of this book will assist the dedicated volunteers of the Trust (now Wild Macadamia Conservation) to protect these remaining wild macadamias.

I hope that, after reading this book, you will be inspired to play a small part in the continuing drama, whether as a consumer, grower, new product developer or, most importantly, by supporting the work of Wild Macadamia Conservation. I can guarantee that, should you take on the role, your life will be enriched.

Enjoy!

Jolyon Burnett
Former CEO Australian Macadamia Society

Preface

This book is intended primarily as a history of the macadamia nut, its evolution, its use and traditions with Aboriginal people, the slow start of an industry which spread to many countries and its possible future. It is believed to be the first detailed history of the macadamia. It is written both as a reference source for the Australian and global macadamia industry and to tell a story that may appeal to all who have an interest in the macadamia and Australia's heritage.

The research and preparation of this book has been a labour of love extending over fifty years. This is the second version of the resulting writing. The first version commenced in the 1970s and it was intended then to release it as *The Macadamia Story*; however, publication was frozen due to the liquidation of the publisher. Then by the time it was released as part of the *California Macadamia Society Yearbook* in 1983, it had limited circulation and parts of it were already out of date. The only parts used in this new book are the first and last sentences.

Research continues to uncover more information, more pioneers and more stories. Too often these conflict in minor or major ways with existing knowledge or assumptions. Exploring these new reports often opens previously unknown history. The volume of information available to me is beyond the scope of this book. My quandary has been how much detail to include to ensure completeness and a sound record and not put the reader to sleep. For the professional historian, my files hold much more detail and may be available. It is planned to have these archived at the Queensland Museum. There is no doubt that this book contains errors for which I am responsible, important omissions and conclusions that will be controversial. Historians will be dismayed at my lack of discipline and, by attempting to cover so much, readers and researchers will dispute, add to and correct some of what I've written. My hope is that the following pages will be a record, however imperfect, that tells the story of the finest nut in the world. This book attempts to consolidate much of the history as well as a summary of the current era from the writer's perspective until the start of 2024.

Therefore, I tender my apologies to all the people who have played a role in our history or have been passionate about macadamias and who deserved to be recorded but were not acknowledged. The major sources of this history are written primary documents, including newspapers and personal communications. There were also many people who verbally passed on their knowledge or opinions, and these were used where I was confident that they were sound. For many years, I did not adequately record the sources of the information I obtained. References are not listed if the information is considered readily available. Being an old amateur historian, the references in this book will vary from sound to poor, be sometimes incomplete and confusing. Any errors and omissions in this book will, if possible, be corrected in future printings. Almost every day I find something new which adds to or is different from what I've written. This book, largely completed in late 2023, is intended to record much of the early history but rapid ongoing industry changes from 2024 will show my folly of interpreting current events and speculating on the future.

As I reside in and am familiar with Australia, the ancestral home of the macadamias, I have focused mainly on its local history. It is presumptuous of me to think that I could write fairly about macadamias in other countries, about the experiences of other people and get their story in perspective. Yet I have been foolish and tried. There are many hard-to-believe reports, and stories exaggerated with many tellings which largely resulted from the very special regard that was held for macadamias by our forebears. Enjoy the lore of the macadamia!

Prologue: A Life with Macadamias

One day, Queensland nut trees will be famous.

(Madge Morrow to her young nephew, Ian McConachie, in 1946)

Born in Brisbane in 1936, I grew up in macadamia country. Seedling macadamia trees grew in backyards, and all the kids knew which ones produced reliable crops of thin-shelled nuts. We collected nuts from under the trees when permitted, and climbed fences to steal a few when we were not.

My grandparents showed me how to crack nuts with a hammer or vice, or by finding a hole in the concrete where the nut would not roll. The nuts were hard to crack and the kernels were often damaged, but macadamia enthusiasts were said to have 'bruised fingers and a satisfied smile'.

From about ten years of age, we roamed the bush, playing games and searching for fool's gold. A favourite spot was the small patch of rainforest between Pine Mountain and Sankeys Mountain, near the suburb of Holland Park and Mount Gravatt in Brisbane. This area was surrounded by harsh eucalypt forest, and the strip of remnant rainforest along the creek has endured through many climate changes. We did not notice any macadamia trees, but fifty years later I heard that wild macadamias were growing there and found them in the very area where we had played. Now only two wild trees remain and the population is threatened with extinction.

My Aunt Madge, who lived in the foothills of Mount Coot-tha in Brisbane's western suburbs, played a major role in my early life. An intellectual, she introduced me to the great books of the past, developing my imagination and sparking my interest in exploring the world. In her backyard, she had four large, seedling macadamia trees that were likely to have been planted in the 1920s or earlier. She enthused about each tree's characteristics, the potential of the macadamia, and their importance as a native crop. As a lad, I remember cracking these nuts with her, roasting them in butter and eating them with her, delighting in their superb flavour and texture.

I was a scrawny, painfully shy child, but I had a normal, fun-filled upbringing, enjoying the bounties of boyhood. I was fortunate to have supportive, loving parents and a best friend in my brother Garth. My imagination was unlimited, which possibly assisted me later to have some vision for opportunities. I was inquisitive, forever asking questions. I loved to debate the great questions of life. I was, and still am, fascinated by the complexities, contradictions, subtleties, grandness and madness of humankind. I wanted to be a great scientist and undertake research that would benefit the world. Despite my introversion I was, with trepidation, able to force myself to do the things I feared. Slowly I developed confidence in what I could do well.

My parents had high standards, a strong work ethic, and narrow views on religion and politics. At school, I was a generalist and moderately successful academically. Since my parents were unable to fund my study at university, I obtained a position as laboratory assistant at Barnes Milling Ltd, and planned to become a qualified industrial chemist by attending evening classes at the Brisbane Technical College, now the Queensland University of Technology. This study extended to seven long years of classes, mainly four nights a week, punctuated by youthful distractions, loss of part of a year through National Service military training and extensions to the course. After graduation, I completed a four-year course in Business Management. Qualifying in both science and management created employment opportunities, and by specialising in the food industry I became a food technologist. I'd like to think that my decision to become a chemist was not the easy access to 100 per cent ethanol!

The Barnes Milling factory, producing flour and stockfeed, was located at what is now Brisbane's South Bank. It was surrounded by seedy hotels, a brothel and a hairdresser's shop. I was occasionally enticed into the hotels, but never the brothel. I got to know the hairdresser well. He had been born in the 1890s at the inner suburb of Spring Hill, and, like many barbers, was a great story-teller. He had grown up and played with Aboriginal people, and his knowledge of local history seemed to me to be immense.

After I had worked for one month in the flourmill's laboratory, the chief chemist had a dispute with management and resigned. His assistant resigned in support, and I found myself acting chief chemist at sixteen years of age. I threw myself into the position with enthusiasm, immaturity and inexperience, and was advised that my elevated position would be short term. However, the replacement chief chemist could not cope, and at seventeen I became chief chemist. This involved responsibility for all aspects of quality in two factories and a 50 per cent increase in salary to £5 ($10) per week. I learnt valuable lessons: to make quick decisions and to worry but not to worry. I thrived on the responsibility, but I suspect that I was an eager beaver who was a pain in the rear end to all the staff.

A job change led me to Tristram's, a soft drink and fruit juice manufacturer, where the discipline of working with a competent food chemist was a necessary learning experience. While at Tristram's, I met George Orford, who owned a large soft drink factory in Toowoomba. Visiting his factory one weekend, I observed him doing odd jobs, and I asked why he

was doing this work when he had a large staff to do it for him. He said he enjoyed doing it and had learnt that, to be a success, you had to make your work your hobby. That philosophy appealed to me and was consciously adopted.

Ian McConachie as an eager young food scientist, 1956. Credit: T. Tristram.

From 1961 to 1977 I was employed at Nutta Products, a Queensland subsidiary of the now Goodman Fielder Group, where I was responsible for quality and later Operations Manager at a company which produced margarine, vegetable oils and a wide range of table nuts.

The Managing Director of Nutta Products was Neil Furness, a disciplinarian of the old school, who was a wonderful mentor to me. Unbeknown to him, he planted in me the desire to establish my own business. An entrepreneur with outstanding business skills, Neil had taken over a struggling family business and developed it into a large, successful company, which was then acquired by the Goodman Fielder Group.

Neil was a macadamia enthusiast. When expanding the company in 1948, he had designed and constructed a macadamia cracker which utilised hacksaw blades to cut through the shell. This machine was ineffective due to its extreme slowness and his lack of awareness that the nuts needed to be dried to shrink the kernel. (The concept of making a cut in the shell so as to gently open it was adopted fifty years later.) I studied this machine and it was obviously never going to work economically.

About 1965, a newspaper article invited people to attend a meeting to learn about the potential of the macadamia nut. The presentation was made by Doug Winterton, a food technologist who worked for the Queensland Department of Primary Industries. Doug told the thirty people attending that the grafting of macadamia trees had been mastered, tree culture was understood, superior Hawaiian varieties were becoming available, and research had been undertaken to process efficiently and maximise quality. He advised that if enterprising people planted an orchard, they would be rich in eight years.

To become rich quickly had been my dream. I was hooked. Through Doug, I updated my knowledge of growing macadamias. I then began searching for suitable land but gradually realised that it was not going to be as easy as Doug had suggested. A lack of funds and a smidgen of common sense caused me to defer becoming a farmer, but a lifetime passion for macadamias had been generated.

In the Government Food Laboratories, where the technology for the macadamia industry was developed, Doug Winterton was responsible to respected food scientists Dr S.A. (Sandy) Trout and Rowland Leverington. Sandy took a personal interest in me because, thirty years earlier, he had courted my Aunt Madge. I commenced a life-long friendship with Rowland, who was responsible for implementing the Queensland Government's support for the emerging industry. He discussed with me his research into macadamia quality and processing. After visiting Hawaii, he informed me of the industry

there. These men were to guide and support me, although neither lived into old age.

The Queensland Government and New South Wales Government were keen to promote the growing of macadamias, due to their potential as a food crop and an export earner. In the 1960s, with State Government support, Steve Angus, the only processor in Australia, was attracted from Murwillumbah in New South Wales to build a 'modern' factory near Brisbane. At the same time, Colonial Sugar Refining Co. Ltd (CSR) entered the industry, spent several million dollars in research, and planted 90,000 trees. Over the same period, a number of small commercial orchards were planted, and it was obvious that the Australian crop was set to expand during the 1970s.

I studied macadamias wherever I could find backyard trees or one of the few small commercial orchards. Every time I went for a drive in the country, I studied the soils and environment and tried to assess whether macadamias could be grown there. My daughter Amanda gradually became my strong supporter and later would search for residential trees to add to my list. Many of the trees planted at the back of homes were surrounded by a chicken run, where they thrived and produced large crops – an observation that led me and others to over-estimate the potential crop yield per tree when orchards were actually planted.

My growing interest led to a love of the land and a dream to become a farmer. I began to realise the enormous destruction that had taken place during the years of European settlement and our need to conserve our flora and fauna and to better understand the balances of nature. Also, during the 1960s, I became convinced that macadamias would become an important Australian industry and I could have an opportunity to play a part in it.

To this stage, my career at Nutta Products had been a satisfying learning experience. It gave me the opportunity to study all aspects of a wide range of tree nuts, particularly their processing and quality. To add to the company's promotion of nuts, I planted at the factory site at Zillmere a small display orchard, which included most of the nuts they processed. Although macadamias were then a distant dream, I secured two grafted trees and I germinated a number of seedling trees.

I undertook regular training at the head office in Sydney, becoming an applied lipid chemist, avidly interested in the technology of fats and oils. My long-term interest in human nutrition helped when poly-unsaturated margarine was introduced. This was cholesterol free and claimed to reduce the risk of heart disease. I became the company's Queensland spokesperson on cholesterol and health benefits, which required me to continually update my knowledge. A lesson in applied psychology was that margarine had a low-priced stigma to it, so marketing the new margarine at a higher price than butter made it respectable to offer to guests.

In the late 1960s, I was corresponding with Cathy Cavaletto from the University of Hawaii, who was the acknowledged world authority on macadamia technology and later in her career was conferred as Associate Professor. I also commenced corresponding with Len Hobson in South Africa, who was promoting the growing of macadamias there and had developed new techniques for propagation. Norm Greber, whose contribution is described separately in this book, worked for and was loyal to CSR at Beerwah, but his passion for the industry led to our meeting and enjoying many stimulating discussions. My introduction to growing and processing macadamias was at Gympie, where I met Bernie Mason, whose 1,000-tree orchard had been planted by Italian prisoners just after World War II. The small factory he had built under his house gave me some understanding of macadamia processing.

When my interest became known, a few enterprising growers, largely through Norm Greber, asked me to provide an independent assessment of the quality and suitability of local varieties they had selected. This led to my learning of growers' requirements and developing laboratory techniques to assess the percentage and quality of sound kernel and the level and type of unsound kernel.

The largest independent growers in Australia at that time were Jack Gowen and his brother Beau, who owned 'Sahara Farms' at the Glass House Mountains. When I phoned Jack, he invited me to his home, showed me his orchards, and talked about the industry. He was a wonderful visionary, and became a great friend. Jack encouraged Nutta Products to become the second processor in Australia as an alternative to CSR. Many of the independent new growers were concerned that CSR would dominate the industry and control the price paid to them.

Jack also promoted the formation of the Australian Macadamia Society (AMS), and I eagerly put up my hand to join. At first, my application was refused because some members strongly believed that only growers should be accepted. However, Col Heselwood, the first President, supported open membership, and I joined the AMS a few months after its formation in 1974. Some members of the AMS Committee, believing that the processors exploited them, were upset at the thought of accepting a processor's representative, but by late 1974 I was invited to join the Committee and remained on the Board for thirty years. Committee meetings were held in the homes of Col and Miriam Heselwood, Eric and Beth Cottam, and Jack and Ruby Gowen. Most General Meetings were held at the Beerwah Hall, which was devilish cold in winter.

The Committee at that time was naturally all male! Some of the wives would attend to provide supper, but they, of course, did not interrupt the hard-working men! We all contributed, and I would prepare well-roasted kernels ground into macadamia butter and spread it on cracker biscuits. Several times I provided macadamia oil as a cosmetic and recommended that the ladies rub it on one elbow before retiring each night, and after a week look at both elbows in the mirror. One of the ladies saw the oil as a way to eternal youth and promoted it vigorously.

During the 1970s, my dream of developing and owning my own business expanded, wavered and changed. I recorded my thoughts and options, and listed the advantages and disadvantages of many types of business. Broadly, my plans were to establish an analytical laboratory, offering consultancies that would include macadamias.

In 1972, the Mary Kathleen uranium mine in north-western Queensland had closed, and many assets, including laboratory equipment, were put up to auction. I successfully tendered for some of the equipment, which would form the basics of a very suitable laboratory. A truck I dispatched to pick it up returned empty. The equipment had been stolen and there was no redress. This set me back psychologically more than financially, but may have been a blessing in disguise.

At this time, only about five people had mastered a grafting technique to propagate macadamias, and there was some belief that the secret was likely to be only passed down from father to son. However, through the efforts of pioneer innovator Reg Young, I was taught the patch budding technique by expert propagator Eric Cottam. I had prepared for this by growing seedlings and quickly budded twenty seedling trees. To my dismay, not one bud took. Eric then told me he had budded over 500 at the same time, and only one or two had taken. I had a lot to learn.

Norm Greber showed Eric the Greber graft, which we both successfully adopted, and as a result I established a nursery of about 500 trees in the small backyard of my home in Brisbane. This nursery was the first part of my business plan, and the business was named Macs Macs. I later moved the nursery to some unused land at Nutta Products, where I would put in an hour's work most mornings before commencing at the company.

The parent company of Nutta Products was Allied Mills Limited, which became Goodman Fielder Limited. Annual Christmas functions were held in Sydney, where I would meet people who were to assist me in different ways. The first was Phil Henry, an olive guru, who managed the company's purchase of Olive Home at Robinvale in Victoria. The largest olive grove in Australia, it contained 530 acres of irrigated olives, planted from 1946. Phil

and I had a common interest in discussing the development problems of pioneer horticultural industries. With Berry Spooner, another employee, I discussed the potential, difficulties and opportunities for the emerging macadamia industry. Berry had purchased the 'Alamo Plantation' at Dunoon (see Map 1 for location of this and other main towns, districts and features of macadamia country in Australia), which was the second grafted, commercial macadamia orchard in New South Wales. Berry later moved to the Northern Rivers and developed a substantial orchard management business for macadamia investors.

With the growth in plantings, Nutta Products had the opportunity to enter the processing and marketing of macadamias, and Neil Furness gave me total support. This resulted in a trip to Hawaii in 1974. At the University of Hawaii, Cathy Cavaletto and her associate Dr Harry Yamamoto gave me every assistance and even arranged for the annual classes on macadamia technology and processing to be held while I was there.

I also met Dr R.A. (Dick) Hamilton, who played a major role in the development of the Hawaiian industry. He became a close friend, and we later worked together on the taxonomy of the macadamia species and varietal selections. Ignorant of Hawaiian customs, I wore my brand-new business suit to the University and found that I was the only person not wearing an iconic Hawaiian shirt. I also met Dr Frank Scott, an able agricultural economist and statistician, who gave me a lot of background.

Then we flew to the Big Island and visited the processors. Royal Hawaiian were in the process of changing their name to Mauna Loa. They gave me an excellent reception, which led to a close association for over twenty years.

I had an appointment to visit Hawaiian Holiday at Honokaa, but found the area picketed by sugar-cane workers on strike. These big-chested, intimidating Hawaiians became most friendly after learning I was a macadamia enthusiast from Australia. I was refused entry to the factory, and subsequently learned that I was a pawn in their confidential attempts to sell the business. Possibly I was suspected of arranging my visit so I could report on the plant to prospective buyers. Eventually one of the staff told me that they were instructed not to talk to me, which they felt was unjust after my travelling from Australia. They quietly gave me a set of their work clothes and said they would pretend I was one of them for twenty minutes. They even allowed me to take my camera, provided they did not see me use it.

I also visited the co-operative factory on the Kona Coast, which was a converted coffee plant of a low standard. I felt I related well to the manager, and we spent a lot of time talking. On leaving, I met a Japanese-Hawaiian gardener, who turned out to be one of their macadamia pioneers and had a perceptive insight into the industry. Several months after arriving back in Australia, I received a legal notice, subpoenaing me to appear in a Hawaiian Court. Apparently, someone had a long-term contract for purchase of the factory's kernel, and some of that kernel had been sold elsewhere. The interpretation of my time with the manager was that I had secretly purchased the kernel and had it shipped to Australia. I refused to return to Hawaii and possibly avoided knowing the inside of a Hawaiian jail.

Back in Brisbane, my proposal to establish a macadamia processing factory for Nutta Products was approved, and by 1975 the design and construction was complete. We received strong support from independent growers, and, as Australia's second processor, cracked about 20 tonnes of nut-in-shell (NIS) in the first year. The kernel was roasted in coconut oil, which was then considered the ideal cooking medium due to its neutral flavour, then the nuts were salted and packed in 50g nitrogen-flushed, high-grade, pink pouches. In a stroke of marketing brilliance, the company announced that these exclusive packs would be available to current clients only, on a quota system. Demand exceeded supply, and the year's production was quickly exhausted. This marketing approach, which lasted several years, added to the prestige of the macadamias.

At this time, my disillusionment with large corporations intensified. I was ill-equipped to handle the consumption of alcohol required to succeed in a network of favouritism. I was too naive to play company politics by climbing the ladder of promotion and kicking someone else down – but it was done to me. I had expected that on the retirement of the managing director I would become general manager, which would have enabled me to continue my involvement with macadamias. When I declined a vague invitation to move to head office, I was told I would be considered only as a valued employee with little or no prospects for advancement. This did not impress me and was the trigger to start a full life with macadamias.

I had formed a friendly business relationship with managed macadamia orchards entrepreneurs Mel Braham and Tom Hoult of Macadamia Plantations of Australia at Dunoon, in the Northern Rivers. They were considering their next venture, and were looking for a manager for their overall operations and someone to construct their processing plant. Several times they had invited me to join them, but desiring to be my own master I had declined. I prepared a new business plan with three separate opportunities. First, if I could secure a contract with Macadamia Plantations of Australia, this would keep bread on the table. Second, a food science consultancy would offer services to the food industry. Third, I would offer a full service in Queensland to investor-owners for the selection, development and management of macadamia orchards. Braham and Hoult readily entered into a two-year contract for me to design, construct and commission their factory, and so began my first business, Mac Food Consultants.

My departure from Nutta Products in 1977 was traumatic. Neil Furness was aware of my dissatisfaction and the probability of my resignation. I had arranged to give him early advice but when I offered up to three months' notice and he informed head office, they reacted aggressively. I was on holidays at the time, and was ordered to immediately return the company car, which I refused to do as I was away. When I returned from holidays, I was met by the Company Secretary, supervised for every second while collecting my personal but not macadamia papers, and escorted off the property without being allowed to say goodbye to my friends. Although staff members were forbidden to speak to me, my true friends ignored this and contacted me at home. Several years later, when I phoned Neil Furness to wish him well in retirement, he admitted that the manner of my treatment was one of his personal regrets.

In 1978, building the factory for Mel Braham and Tom Hoult at Dunoon was an introduction to the politics and personalities of the rapidly expanding industry in northern New South Wales. Shortly after commencing, I found that some of the staff were growing marijuana in the orchards and there was significant theft of grafted trees and materials. When the person responsible pleaded guilty, I became interim manager of the orchards. I was on a rapid learning curve and most anxious to find a permanent manager.

Braham and Hoult had a controversial image as they were unashamedly promoters and businessmen before they were farmers, but they treated me well. Jointly, we engaged Keith Ainsbury from CSR as their manager, and Keith was to play a larger-than-life role in the development of the industry until his early death. Each fortnight, I spent two to four days at Dunoon and often invited prominent people to come with me on the long drive to Lismore. One of these was Australia's first commercial processor, Steve Angus, who told fascinating tales of the industry from the 1930s on.

The small Macadamia Plantations of Australia factory was as technically up-to-date as was possible at that time. Due to funding constraints, it was not possible to purchase a modern vacuum foil pouch sealer, so we gas-flushed with a stainless-steel wand using carbon dioxide. This gas was absorbed by the moisture in the kernels, resulting in an apparently vacuumed pack. The next year, a retail packing operation was introduced to offer high-quality roasted and salted nuts in metal cans.

My Mac Food Consultants provided consultancy in applied food technology and also developed a brochure on investment in macadamia growing. Dr George Gray from Melbourne, who was establishing an orchard at Clunes in the Northern Rivers, invited me to dinner, not for my social charms but to determine whether he should recommend my services to his friends. The business of establishing macadamia orchards then boomed, limited only by the availability of suitable land.

I closed Mac Food Consultants and opened Macadamia Consultants Pty Ltd. A few years later, Australian Macadamia Management Pty Ltd was established. From 1980 to 1985, fourteen macadamia orchards were developed from rough, bare land in the Gympie district on behalf of absentee investor-owners.

In the Northern Rivers and south-east Queensland, orchards were being planted at a rapid rate, and many of the growers wanted to control their future by owning their processing and marketing operations. In the strictest confidence, I developed a proposal for eight large macadamia growers to build and operate a modern factory. The intent was to invite more growers as shareholders later. One of these growers considered he had a moral obligation to inform the three major processors about the plan. The growers and I were then subjected to intimidation, so the proposal was, unfortunately, dropped. Had it proceeded, it might have changed the structure of macadamia processing in Australia.

After this disappointment, I designed and developed The House of Macadamias, a small Melbourne-based processing plant for George Gray and his partners.

An event which disturbed me concerned a nut-cracker invented by Paul Shaw from California. I became close friends with Paul and Pauline, stayed at their home in California, and spent time with them when they came to Australia. Paul was a man of integrity but, with limited means, could not afford to patent his invention in macadamia-producing countries. He appointed me his agent in Australia and I sold one Shaw Cracker on the basis that, as it was not patented, a condition applied that no one could take drawings from it or reproduce it. Two years later, it was being manufactured in Australia and sold under another brand name.

During the 1980s, many entrepreneurs were attracted to the emerging macadamia industry. I became involved with some of them and occasionally had my fingers burnt. Unable to fund my proposal to establish a prestigious retail tourist outlet on the Gold Coast, I sought partners to develop it. My detailed plan was accepted by a prominent businessman, who told me there were delays in securing finance. Then one day I opened a newspaper to read that my proposed partners were proceeding without me. This was a harsh lesson about trust. Their venture failed!

Developing and managing macadamia orchards for investor-owners at that time was not for the faint hearted. Many of the investors lacked an understanding of business realities and were only seeking a legal tax shelter. Sometimes their requirements were determined over a glass or three of port with their colleagues at the Melbourne Club.

Some of the dramatic situations I encountered while managing orchards are amusing in retrospect. One day, for example, I discovered to my embarrassment that two young ladies who were renting a farmhouse expected to pay the rent in kind. On another property bought for development, an employee had a fierce dog which I later found was to guard his marijuana crop. On another large farm, after dismissing an employee for trading in marijuana, I found most of the rest of the staff had disappeared. They were all involved and feared the police being called.

Before mechanical harvesters were widely used, we would have up to eighty people at any time picking up macadamias by hand on a contract basis. These people exposed me to Australia's different social classes, and made me aware of many personal tragedies. We negotiated the pay rate based on the number of nuts on the ground. One season, a group of militant women demanded and secured ever-higher rates. Then, one weekend, a group of stranded female English backpackers sought work, and on the Monday morning it was found that they had picked up over three times the average daily rate of our current pickers. I had been conned!

For many years, a group of local mothers handpicked the nuts on a piece-rate basis. Aware that they were not earning a reasonable wage, I went to dismiss them. They explained that nut picking was the highlight of their year. They brought their young children and a grandma or two, and paid the grandmas to mind the babies. They admitted to talking and listening to music while they picked up the nuts. They had picnic lunches under the trees, savouring their cask wine before taking a quick nap, then picking up a few more nuts.

One day I made the mistake of sending a male and a female employee to pick up nuts in an isolated section of an orchard. Driving to check on them, I found them hastily putting on their clothes. Another time, the police phoned, asking if we had a particular person working for us. We had a man with a different name but a similar description and later the police arrived at the orchard and arrested him for murder.

One of my farm managers was a fine person but became increasingly lazy. I checked workers in part by assessing tasks against the number of recorded tractor hours. Then someone who had been to the orchard at night found the tractor idling in the shed, which explained the apparent increase in hours worked. Instead of sacking the manager on the spot, I gave him one last chance and set rigorous controls. But he was a friend of the orchard owner and reported that I was treating him unfairly so I was the one to get sacked. Such is life!

The orchards we were managing commenced cropping, and in 1985 we formed a company, Suncoast Gold Macadamias, to handle and market the crop. In 1987, we built a factory, which in time became the second largest macadamia factory in Australia and the third largest in the world. It was based on co-operative principles and in time over 140 macadamia growers became shareholders.

Prior to 1987, Suncoast Gold produced very small quantities of high-quality roasted and salted jar packs solely for the orchard owners and for gifts. The nuts were dried in a 200-litre drum with a hole cut in the side and a ladies hair-drier inserted to blow cool and then heated air. The kernels were then roasted in coconut oil in a domestic deep-fryer, salted in a plastic bucket, and hand-filled into jars with a sliver of dry ice added to create a vacuum. They were superb. Coconut oil later became an unacceptable frying medium, due to its high level of saturated fatty acids.

From the early 1980s, I visited Bundaberg, 200 kilometres north of Gympie, whose economy was largely based on sugar-cane. There I admired the flat ground, high land utilisation, water availability and obvious farming efficiencies. Whenever I inquired about purchasing farmland around Bundaberg, there was a real risk that the cane farmers would want to string me up to the nearest tree. However, during a downturn in the sugar industry, some farmers began to grow vegetable crops, and the attitude to outsiders growing other crops softened. I could see that this district had potentially the most efficient macadamia farmland in Australia, and continued to seek opportunities to develop orchards there.

In 1987, Lincoln Doggrell and I, along with my wife Jan, formed Macadamia Farm Management Pty Ltd, purchased sugar-cane farms with irrigation and machinery, and commenced developing macadamia orchards on behalf of investor-owners. We were the pioneers of the Bundaberg industry. About a year later, we saw two people walking through an orchard, looking at our newly planted trees. They were Phil Zadro and John Wilkie from New South Wales, who had told us we were crazy going so far north but decided to come up to assess us. During the next thirty years, Phil Zadro planted several million macadamia trees in the Bundaberg district, which now is the largest macadamia growing and

producing area in Australia with over five million trees.

In 1981 my wife and I bought a small, rundown dairy farm in Wolvi, near Gympie, and found ourselves milking cows for a few weeks in order to retain the milk quota entitlement. We named our property 'Warrawee', which was a misnomer because that is an Aboriginal word meaning 'a place to rest'. We planted macadamias as well as mangoes, lychees and stone fruit, before realising the error of our ways and becoming just macadamia farmers. Over eight years, we developed our paradise, and no one had a better life style. We became custodians of 6,000 trees and enjoyed our 34 hectares of flora, fauna, rainforest, creek and welcome solitude.

Being a farmer fulfilled a bitter-sweet dream. It is enormously satisfying to commune with nature, though not to deal with its forces – hail, frosts, cyclonic winds, bushfires, pests, floods and droughts.

Ian McConachie picking up macadamias with a Bag-A-Nut machine. Credit: J. McConachie.

Being interested in macadamias in the wild (see Map 2 for location of wild macadamia trees in Australia), I was intrigued by a report of a few *Macadamia integrifolia* being found in a rainforest gully off Granite Creek north-west of Bundaberg. Only twenty-three trees were originally found in 1982 by local field naturalists, Ray Jansen, Keith Sarnadsky and Eric Zillman. Keith generously gave the credit to Ray and we arranged to meet him. The day started disastrously when Ray showed us a small tree he had grown and I told him it was not *M. integrifolia*. Ray accused us of being city-slickers and of trying to claim the credit. It took all of my good wife Jan's skills and a carton of beer for Ray to continue to talk to us. Eventually, with Ray and Jan, we trekked to see the trees in an isolated rainforest and realised that they were different to any macadamia we had ever seen. Ray had the observational skills to notice and find the trees and we had the ability to liaise with the Royal Botanic Garden in Sydney to have them described and named. They were the first new macadamia species to be found in 130 years. Named *Macadamia jansenii* after Ray Jansen, they could well change the commercial industry with its breeding potential and possible ability to handle global warming. Mentoring this new species has been a highlight.

Janet and Ian McConachie, 2018. Credit: I. McConachie.

By the mid-1990s, my body was slowly losing the battle to keep pace with my mind, and I was feeling the pressure of business and the demands of being executive chairman of Suncoast Gold. In 1996, we sold Australian Macadamia Management Pty Ltd, then our share in Macadamia Farm Management Pty Ltd and retired from Suncoast Gold.

Semi-retired in my seventies, I became the message boy on our orchard, played at farming, and began to write this book. Having the support of my wife, Janet Kathleen, in encouraging me to go off on tangents and indulging in my hobbyhorses has made my life complete. I'm still active in the industry and like to believe I can tell others what the industry needs to do to advance.

An unexpected reward was the outcome of our amateur varietal assessment and selection trials, resulting in the release of an elite new variety. In 1982, there was an opportunity to be given 220 small seedlings which were claimed to be derived from 'good parents'. The trees were assessed annually, culled, culled again and again, and one tree was selected which was of moderate size, hardy, could produce a commercial crop in Year 3, was a high yielder, with a high kernel recovery and good quality characteristics. It was named MCT1 and we gifted it to the Macadamia Conservation Trust, who protected it by securing Plant Breeder Rights. Royalties from the sale of grafted trees are now providing most considerable funds to support conservation (see Map 3 for location of remaining Australian macadamia habitat) and, from 2024, it will be commercialised in Africa and elsewhere. In November 2022, one of these trees was planted by the Governor of Queensland as the Queen Elizabeth II Memorial Tree. It overlooks the city of Brisbane and might be the macadamia tree with the best view in the world. By 2023, over half a million MCT1 trees had been sold and royalties will fund conservation possibly for ever.

Over the years, I have had the privilege of working in the Australian and international macadamia industries, in Hawaii, Costa Rica, Ecuador, Malawi, India and South Africa (see Map 4 for location of these and other commercial macadamia-producing countries). On Australia Day 2006, I was honoured to become a Member of the Order of Australia for my services to the macadamia industry. In 2014, I was made an Honorary Senior Fellow of the University of the Sunshine Coast for science-based advancements to macadamias and in 2017 received the AMS Norm Greber Award for contributions to macadamias. 1n 2018, I received a Lifetime Achievement Award from the Eighth International Macadamia Symposium in China. How privileged and fortunate I've been for having fun indulging in my hobby.

In 2013, after thirty-three years on 'Warrawee', we sold to a delightful couple from north Queensland and now live on one acre outside Gympie, with six special macadamias trees for company. My passion remains and this book was completed. Conservation of the wild macadamias, becoming the industry conscience on the importance of quality, talking too much and probably being tolerated as the industry dinosaur is a great way to fill each day!

Seeing the Australian industry grow from 60 tonnes to 60,000 tonnes in sixty years and certain to reach 100,000 tonnes is more than satisfying. Macadamias have been so good to me; I hope I've contributed a little and I've certainly had a ball.

Ian McConachie receiving Australian Institute of Food Science and Technology Award 1997 for contribution to a science-based industry. Credit: Australian Institute of Food Science and Technology.

Map 1: Main towns, districts and features of macadamia country mentioned in this book

Map 2: All wild macadamias come from a small area of Australia's east coast

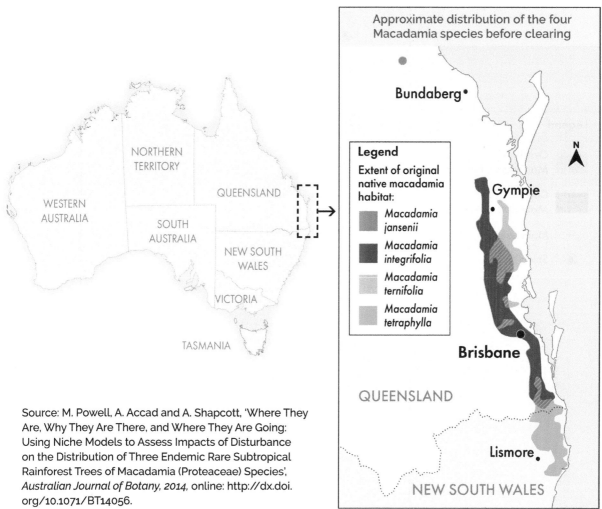

Source: M. Powell, A. Accad and A. Shapcott, 'Where They Are, Why They Are There, and Where They Are Going: Using Niche Models to Assess Impacts of Disturbance on the Distribution of Three Endemic Rare Subtropical Rainforest Trees of Macadamia (Proteaceae) Species', *Australian Journal of Botany, 2014*, online: http://dx.doi.org/10.1071/BT14056.

Map 3: Approximate extent of remaining Macadamia habitat in 2014

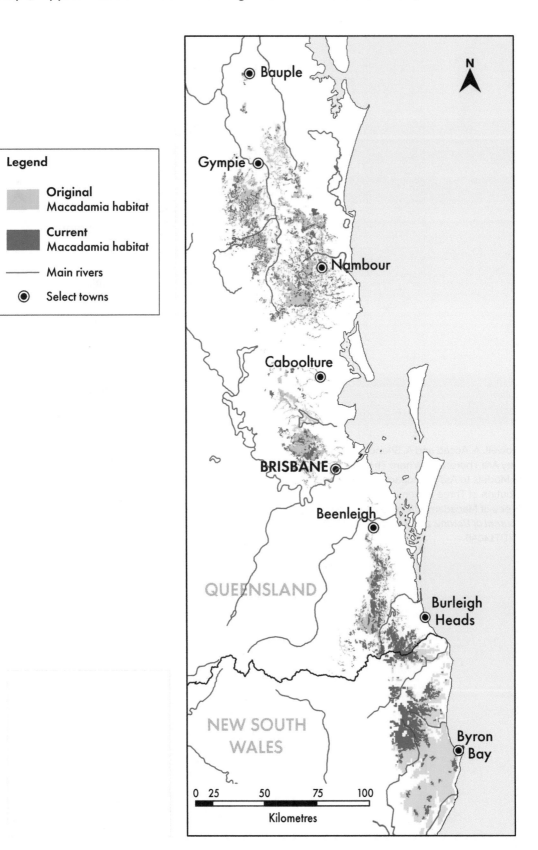

Part 1

IN THE BEGINNING

Chapter 1

MACADAMIA COUNTRY

The evolution of macadamias, their natural habitat
and how they came close to extinction

*Fossilised macadamia-like pollen has been found across Australia in
stratified rock from the Late Cretaceous around 85 million years ago.*

(Summarised from Mary Dettman and David Jarzen, 1998) [1]

Evolution

The macadamia has a very long history, almost all of it unseen by humans.

Way back in the mists of prehistory, a flowering plant evolved which today is called the macadamia.[2] An attractive, evergreen tree, it flowers in spring, then bears nuts, which mature in autumn and fall to the ground over a period of months. The nuts are protected by fibrous, green husks, and the creamy round kernels are enclosed in hard brown shells. When dried, the species with edible kernels are crisp and crunchy, with a distinctive, delicate flavour.

The evolution of the macadamia commenced many millions of years ago. Over millennia, mutations or natural variations occurred and those that had an ecological advantage in the changing environmental conditions survived to become the four separate species we know today.

Macadamias belong to the family Proteaceae, which is the third largest botanical family in Australia. This family of plants originated at least 90 million years ago (mya) in the southern hemisphere, in the prehistoric mother continent of Gondwana.[3]

About 336 mya, a supercontinent, Pangaea, covered much of the world's landmass. Over the next 200 mya, monumental rifts in Pangaea created Gondwana, which embraced present-day South Africa, India, Antarctica, South America, New Zealand and Australia. About 150 mya, Gondwana began to break up, and by 45 mya the resultant continents and islands had closely assumed their present form and position. Australia then developed in isolation.[4]

Angiosperms, the world's flowering plants, appear suddenly in the fossil record in the Cretaceous era, around 140 mya, and in the Palaeocene, from 65 mya, they started to assume more modern forms, diversifying into the genera we know today.[5]

Under some conditions, pollen from flowers is virtually indestructible. If it becomes fossilised, it can later be extracted, analysed and interpreted to give a cautious indication of plant type. Fossilised pollen shows that by 65 mya, ancestors of the macadamia were growing in eastern Australia[6] and New Zealand.[7] The fossilised fruit of ancestral forms of macadamia have been recorded on the eastern coast of Australia.[8]

From 39 mya and for the next 20 million years, sea levels in Australia were high, and a carbon dioxide-rich environment produced a greenhouse ecosystem in which plant life flourished.[9]

The Proteaceae family diversified, adapting to the changing climate and environment. Rainforests in the mountain ranges created by volcanic activity in northern New South Wales and southern Queensland became the enduring home of the macadamia (see Map 2). Over millions of years, the climate became drier. Massive environmental changes, together with one-in-a-thousand-year events – such as droughts, fires, storms, floods, and extremes of temperature and wind – resulted in the loss of rainforest. By 29 mya, the ancestors of macadamias had evolved, and probably existed in a form we would recognise today.[10] According to Mast et al. (2008), trees that also evolved from this common ancestor managed to get to both South Amer-

ica and South Africa, possibly by floating along the Antarctic Circumpolar Current after a massive flood event.[11] These ancestors evolved into the lacewood trees of Central and Southern America (the *Panopsis* genus with at least twenty-five species) and the wild almond or *ghoeboontjie* of South Africa, the single species of the *Brabejum* genus.

Through millennia, isolation, natural selection and adaptation to the environment resulted in plants taking their separate evolutionary paths. Macadamias retreated to isolated pockets in the foothills of Australia's eastern ranges, where they now exist in remnants of subtropical rainforest.[12] The evolution of the various macadamia species occurred through generations of genetic mutations and recombining DNA, resulting in beneficial traits with a survival advantage. The four species of macadamia had a common ancestor about 7 mya.[13] Sometime since then, the two edible species, *M. integrifolia* and *M. tetra-*

A large, wild *M. integrifolia* in the foothills of Mount Tamborine, first recorded in the 1870s. Credit: I. McConachie.

phylla, diverged from *M. ternifolia* and *M. jansenii,* the two species with bitter nuts. Despite *M. ternifolia* and *M. jansenii* now being in widely separated geographic locations, it is possible that one may be an ancestor of the other.[14]

By the time humans arrived in Australia, the Gondwana remnants had retreated into rainforest niches. Where low numbers of macadamias remained, they were vulnerable to loss from inbreeding, and their limited capacity to reproduce in thick rainforest meant that they came close to extinction.

In summary, four species of macadamia are distributed within and on the edges of some south-east Queensland and northern New South Wales rainforests. The fourth species, *M. jansenii,* has only been located in a single rainforest valley 160 kilometres further north.

Fortunately, two of the species just happen to have outstanding qualities of flavour and taste.

Macadamia country today

In 1963, the pioneer macadamia grower Norm Greber described the narrow strip of coastal country that today is the natural habitat of the macadamia:

> Throughout this area, in a terrain of mountainous country, with its many broken valleys and fast-running, winding streams, grew the macadamia, along with many other scrub trees.[15]

To the layperson, macadamia country (see Map 3) is found in both wet and dry subtropical rainforests (known to European settlers as 'scrub') from the Richmond River in New South Wales to Mount Bauple in Queensland. *M. jansenii* exists in a small outlier further north. Macadamia country includes rainforest as well as some tall wet eucalypt forest with rainforest understorey. Only a small percentage of rainforest is macadamia country, and wild trees exist in isolated pockets.[16]

Rainforests are lush, rich, enchanting worlds. Large trees dominate, while smaller species struggle to reach the light or have adapted to almost perpetual shade. Amid the unseen life of the flora and fauna, birds twitter and sometimes a rocky stream tin-

Janet McConachie finding a wild *M. integrifolia* in rainforest at Amamoor. Credit: I. McConachie.

Typical wild macadamia rainforest heartland. Cascade Creek, Amamoor. Credit: I. McConachie.

kles, and there is a subtle, organic aroma. Alone in a rainforest, you could be the only person left in the world.

Few of the people who explore rainforests would recognise a wild macadamia tree, because the trees do not stand out from neighbouring species. An experienced searcher will look along roadsides, at the edge of the rainforest or in places where light allows the trees to expand. Distinctive leaves may emerge at the top of the canopy, where they have reached sunlight, or drooping, low branches may show typical leaves. Rarely, a few rat-eaten nuts lying on the forest floor will tell of a nearby, adult tree. Trunk lenticels (breathing pores) are another means of identification. Sometimes the trunks are multi-stemmed as a result of damage from fire, stress or wild animals. Although macadamias are difficult to locate and photograph, once one tree is spotted, others are easier to find.

With enough light and adequate rainfall, an adult tree may produce an annual crop, although numerous native pests and diseases will attack the flowers and nuts. Unless trees are in relatively open areas, they may produce only in favourable seasons. Small seedlings may be found under the canopy. Larger seedlings may establish where nuts washed away from the parent tree have found a suitable growing environment.

Within the rainforest, growth can be extremely slow. First observed in the 1960s, one wild macadamia in the Amamoor Creek Valley was, in 1979, about 70 centimetres tall, with a spindly trunk and about eighteen leaves. During the next forty-five years, this healthy tree did not grow any taller. It is still spindly, with about twenty-two leaves.[17]

When examining a macadamia tree, it is not always possible to tell if it is wild or introduced. The age of wild trees is difficult to determine, because the annual growth of macadamias varies widely, depending on its access to resources, so growth rings are imprecise and difficult to see. Techniques such as carbon dating for trees less than 500 years old are only now being applied and give the age to about forty years. The Wild Macadamia Hunt, a citizen science project, has encouraged landowners to report very old residential trees or even wild trees. DNA analysis has indicated that some of the residential trees originated from wild populations that have since been lost.[18]

From the 1860s, when settlers first appreciated the edible qualities of macadamias, trees were often planted near the family home. Many of these homes

Wild *M. integrifolia* tree at Amamoor that has remained dormant for at least sixty years. With Uncle Russell Bennet (Jirimir) from the Gubbi Gubbi. Credit: I. McConachie.

Trevor Miles with wild *M. integrifolia* at Hotham Creek, Gold Coast hinterland, that his grandfather rested under in the early twentieth century. Credit: I. McConachie.

are long gone, but the trees may remain, and nuts washed away may have germinated downstream. Most of the small orchards that were planted from the 1890s onward have disappeared, sometimes leaving isolated trees.

Two hundred years of European settlement have effected massive changes on macadamia habitat. Believing the bush to be an endless resource, settlers cleared the rainforest for farming and urban development. In northern New South Wales, for example, macadamias were distributed in the Big Scrub, which consisted of 74,000 hectares of almost continuous rainforest from Lismore to Ballina to Mul-lumbimby. Within fifty years of settlement, fewer than 700 hectares remained, resulting in the probable loss of 99 per cent of the wild macadamias in that area.

By 2010, wild macadamias had been recorded in only 196 localities in Queensland, which indicates their sparse distribution. Botanists from the Queensland Herbarium have studied 2,400 sites of remnant vegetation, finding wild macadamias in only nineteen of them. Without conservation, wild macadamias will become extinct in many existing habitats and will eventually be confined to a few remnants that have viable populations.

Wild *M. integrifolia* at Mooloo in the Mary Valley, that might well be the tree from which seed nuts were sent to the USA, including its then territory of Hawaii, in the early 1900s by Herbert J. Rumsey. This would make it the ancestor of most Hawaiian varieties. Credit: I. McConachie.

Macadamia integrifolia

Chapter 2

BOTANY: MEET THE MACADAMIAS

Our changing understanding of the macadamia genus and how it fits into the tree of life

... a tree of oriental subtropical Australia, with leaves three in a whorl or rarely opposite ...
A beautiful genus, allied to Adenostephanus, Orites and Xylomelum.

(Ferdinand Mueller, 1857)

An overview of the macadamia species

Botanically, macadamias are classified as Family Proteaceae, Sub-Family Grevilleoideae, Tribe Macadamieae, with the macadamia being one of eighteen genera in the Macadamieae Tribe.[1] The Proteaceae are a family of flowering plants that originated in the ancient continent of Gondwana at least 90 million years ago. The family name derives from Proteus, the Greek Sun God who could change his shape at will, and was first used to describe a South African genus by scientist Carl Linnaeus in 1735. Taxonomy is continually being refined and the Proteaceae family now has over fifty genera and about 1,600 species, nearly all in the southern hemisphere.

Many Proteaceae such as Proteas, Banksias, Dryandra, Grevilleas, Waratah and Hakea produce attractive flowers, but only two species, both Macadamias, have been commercialised for food. Unlike most other domesticated plants with a selection history of thousands of years, macadamias are new to practices of selective breeding. Commercial macadamia cultivars are only a few generations distant from their rainforest ancestors. Other food crops such as olives were domesticated some 5,000 years ago, grapes around 3,500 years ago and cashews possibly 2,000 years ago.[2] Macadamias were domesticated only in the last 1,000 years.

The rich and varied plant life of Australia was a revelation to European botanists. In 1857, the Victorian Government Botanist, Dr Ferdinand Mueller (later Baron von Mueller), named the macadamia after Dr John Macadam, Secretary of the Philosophical Institute of Victoria. Naming a specimen of leaves and flowers *Macadamia ternifolia*, Mueller described it as:

> ... a tree of oriental subtropical Australia, with leaves three in a whorl or rarely opposite, lanceolate or oblong flat, or above the base entire, net-veined, with stomata on the lower side; racemes terminal pendunculate; flowers twine, with a solitary bract. A beautiful genus, allied to Adenostephanus, Orites and Xylomelum.[3]

For all their skills, early Australian botanists often failed to spot the salient characteristics of a species. Some of the early specimens came from habitats where there were natural hybrids with overlapping morphology. Advances in taxonomy and the ability to define species by genetic markers have provided greater certainty.[4]

Until 2008, most botanists accepted that there were thirteen species of macadamia: three endemic to New Caledonia, one to both north Australia and New Caledonia, at least one to the Celebes, four in north Queensland and another four, described as the 'southern clade' from northern New South Wales to southern Queensland.

In 2008, a team led by Dr Peter Weston of the Royal Botanic Garden in Sydney and Professor Austin Mast of Florida State University reclassified all but the four southern macadamia species.[5] They placed the north Queensland and Indonesian species, which had been variously named *M. heyana, youngiana, verticillata, whelani, hildebrandt, claudiensis, erecta* and *grandis*, into two new genera, *Lasjia* (*L. claudiensis, grandis, whelanii, hildebrandii* and *erecta*) and *Nothorites* (*N. megacarpus*).

The team also found that two species assumed to be close relatives of macadamias probably diverged from macadamias at least 70 mya. These are now called *Floydia prealta* (the Possum, Ball or Mullumbimby nut), once *Macadamia prealta*, named after the late Alex Floyd, a passionate botanist of northern New South Wales who studied the macadamia in the 1960s and 1970s, and *Hicksbeachia pinnatifolia* (Red Bopple nut). Both species grow in or near macadamia habitat.

One hundred years of species confusion

The species confusion commenced with Dr Ferdinand Mueller who together with Walter Hill found both *M. ternifolia* and *M. integrifolia* trees in the Pine River Valley in 1857 but did not distinguish between the two. Mueller recorded a single new species that year, naming it *Macadamia ternifolia*. He then transferred *Macadamia* into the *Helicia* genus in 1861, when he came upon the previously submitted 1843 specimen from Ludwig Leichhardt. This error was recognised by George Bentham, who reinstated it as *Macadamia* in 1870. The discovery and error will be covered in detail in Chapter 4.

The early botanists were largely influenced by leaf serrations and failed to recognise that the southern species, *M. tetraphylla*, had four leaves to each node[5] and that a clear difference between *M. integrifolia* and the true *M. ternifolia* was that the former has larger, edible fruit.

In 1897, Maiden and Betche described *M. integrifolia* as a new species, distinct from *M. ternifolia*.[6] F.M. Bailey in 1910 and 1911 expanded *M. ternifolia* into two species, *M. minor* and *M. lowii* and also *M. major*, but later taxonomists doubted this and considered them synonyms or variants of *M. ternifolia*.[7] In 1929, W.D. Francis commented that it was unclear how *M. minor* and *M. lowii* differed and he considered them small-fruited forms of *M. ternifolia* with unpalatable nuts.[8] Bailey's original description did not include the bitterness of the kernel.

In 1954, Dr L.A.S. Johnson described *M. tetraphylla* as a separate species but considered that *M. integrifolia* was a synonym of *M. ternifolia*.[9] He did not distinguish between the edible *M. integrifolia* and the intensely bitter, smaller tree and nut, *M. ternifolia*.

Finally, in 1956, Lindsay Stuart Smith, the Queensland Museum and Herbarium Botanist, collected sixty-seven macadamia specimens and ended the confusion by reinstating *M. integrifolia* and clearly describing the geographical distribution of the three then known species of the 'southern clade': *M. ternifolia* (the inedible Gympie or Maroochy nut), *M. integrifolia* (the edible, smooth-shelled Queensland or Bauple nut) and *M. tetraphylla* (the edible, rough-shelled Queensland or Bush nut). He defined the differences between the three species and noted that only *M. ternifolia* was cyanogenic.[10]

In this book, where the true *M. integrifolia* is clearly the subject, it will be called *M. integrifolia* even if the historical references record it as *M. ternifolia*.

With the addition of *M. jansenii*, described in 1991, these species and descriptions remain current. Even Smith's description can be criticised, as 'integrifolia' means 'entire leaves', whereas the leaf margins of this species are often serrated. Despite the species being now soundly described, the pharmaceutical and cosmetic industries persist in describing the commercial oil extracted from *M. integrifolia* as *M. ternifolia* oil.

Nuts of the four species. From left: *M. tetraphylla, integrifolia, ternifolia and jansenii.* Credit: I. McConachie.

M. integrifolia at Mary's Creek, Gympie, after rainforest cleared in the early twentieth century. Credit: I. McConachie.

The four macadamia species

Macadamia integrifolia

'Integrifolia' is derived from Latin. 'Integra' means 'entire', and 'folium' means 'leaf', referring to the edge of the adult leaves being smooth or 'entire', in contrast to the serrated edge of *M. tetraphylla* leaves. As mentioned above, 'entire' is somewhat misleading as most *M. integrifolia* have serrated leaves when young and many retain some serrations on their adult leaves. Its common names include Bauple nut, Queensland nut, Smooth-shelled nut and Australian nut. This is the main commercial species in Australia and globally, although hybrids with *M. tetraphylla* are being increasingly planted.

Trees generally grow up to 15 metres. Leaves have a petiole or stalk and young leaves are pale green. Each node or leaf whorl usually has three leaves. Flowers are creamy white and the nutshell is smooth. *M. integrifolia* grows in rainforest and on rainforest margins from the Gold Coast hinterland up to Mount Bauple in Queensland.

Painting of *M. integrifolia* by renowned botanical artist Dorothee Sampayo. Credit: D. Sampayo.

Macadamia tetraphylla

'Tetraphylla' is derived from Greek, with 'tetra' meaning 'four' and 'phyllon' meaning 'leaf'. It has four leaves at each node. It can also grow to 15 metres. Leaves have no petiole (stalk) and are strongly serrated, thus prickly. Usually, the leaf flush and flowers are pink to red. The nuts have bumpy shells. Common names are Rough-shelled nut or Bush nut, but it was also called Queensland or Australian nut.

It is generally accepted that *M. tetraphylla* kernel has a higher sugar content than *M. integrifolia*, which makes *M. tetraphylla* kernels darken more when roasted together, requiring the two species to be processed separately. In Kenya and California, *M. tetraphylla* became the preferred commercial species. Today few pure *M. tetraphylla* are cultivated in Australia.

This species is native to the New South Wales Big Scrub rainforest and the rainforest margins of the Richmond River district, Tweed River valleys and in southern coastal Queensland to Ormeau. On the eastern slopes of Mount Tamborine, it grows in conjunction with *M. integrifolia* and produces hybrids.

Painting of *M. tetraphylla* by renowned botanical artist Dorothee Sampayo. Credit: D. Sampayo.

M. tetraphylla at Crystal Creek, Northern Rivers, with Peter Norman. Credit: K. Dorey.

Macadamia ternifolia

'Ternifolia' is derived from the Latin 'ternus' meaning 'three' and 'folium' meaning 'leaf'. Its common names are Maroochy nut or Gympie nut. For many years, the name *Macadamia ternifolia* was used for any macadamia but this name now applies only to a non-commercial species, which has small nuts with intensely bitter kernels.

M. ternifolia is a smaller tree than the two commercial species and has smaller, serrated leaves. With its coppery leaf flush and delicate, light pink flowers, *M. ternifolia* makes an attractive garden plant. Fruit-spotting bug preferentially attacks this species, which may give it potential as an indicator tree for this major industry pest.

This species grows from the Pine River district north of Brisbane to elevated coastal habitats east of the Mary River at Gympie, with the westernmost

Painting of *M. ternifolia* by renowned botanical artist Dorothee Sampayo. Credit: D. Sampayo.

M. ternifolia showing leaves in flush and flowers at CREEC, Burpengary. Credit: I. McConachie.

Macadamia jansenii

This species was named for Ray Jansen, one of a group of field naturalists who discovered it in 1982, in a small rainforest gully in Bulburin National Park. Bulburin is the largest remaining area of rainforest in central Queensland, located between Gin Gin and Miriam Vale, north-west of Bundaberg. The site is 160 kilometres north of the other macadamia species habitat. The common name of *M. jansenii* is Bulburin nut. The tree is small and slow-growing, and the nuts are small and smooth, with a slight bitterness that makes them inedible. Like *M. ternifolia* and *M. integrifolia*, it has three leaves per whorl, but with consistently smooth leaf edges. New leaves can be green or pinkish and flowers are cream.

Initially only twenty-three mature trees were found, but surveys by dedicated naturalists and researchers, usually led by the most observant Keith Sarnadsky, have increased that number to above one hundred. Genetic research by Dr Alison Shapcott and Michael Powell of the University of the Sunshine Coast found the *M. jansenii* population to be relatively genetically diverse, prompting further searches and discovery of neighbouring trees.[11] The rare *M. jansenii* is listed by the Australian Government as endangered and by the Queensland Government as critically endangered.

population being in the Conondale Range. Most of the wild trees have been lost through clearing, but east of Gympie, there are at least two populations of more than thirty *M. ternifolia* in remnant rainforest. This rainforest covered possibly 800 hectares some 150 years ago and so may have contained a substantial population of this species. Recently recorded is a large population south of Kenilworth.

Because of its bitter kernels, *M. ternifolia* attracted no commercial interest, but an accidental cross between it and *M. integrifolia* in Hawaii, named HAES 791 or Fuji, has shown potential as a high producer in South Africa. When selectively crossed with *M. integrifolia,* the bitterness is recessive. Crosses involving *M. ternifolia* are being made by the Queensland Government Plant Breeding team, which may result in desirable hybrids.

To reduce risks to the trees and their habitat, the exact location is undisclosed and strategies are in place to protect the species. The main strategy is to protect the habitat from wildfire, invasive weeds and diseases, and Dr Shapcott has supervised reintroduction plantings to extend the current habitat.[12] Traditional Owners work with the Queensland Parks and Wildlife Service through the Gidarjil Land and Sea Rangers to protect *M. jansenii* habitat. The second strategy is to reproduce most of the population at alternative, managed sites and this has been undertaken at Tondoon Botanic Gardens at Gladstone. At the Tondoon Botanic Gardens nursery, six insurance populations have been propagated to plant at other secure locations to conserve the spe-

cies. These include the National Botanic Gardens at Black Mountain, Canberra. Severe bushfires in late 2019 burnt 25 per cent of *M. jansenii* habitat, bringing additional urgency to conservation efforts.

In 1991, Dr Cameron McConchie from the Commonwealth Scientific and Industrial Research Organisation (CSIRO) crossed *M. jansenii* pollen with *M. integrifolia*. The resultant trees and fruit, together with additional research, suggest future breeding potential with much higher kernel recovery. The smaller-size nut and kernel had enhanced fatty acids and a most acceptable appearance and flavour. Crosses with *M. jansenii* could result in macadamias being available as smaller kernels and able to grow in warmer temperatures with less winter chilling.[13]

Painting of *M. jansenii* by renowned botanical artist Dorothee Sampayo. Credit: D. Sampayo.

Headstone of Ray Jansen, showing his wish to be remembered for 'his' macadamia. Credit: I. McConachie.

M. jansenii with nuts. Credit: I. McConachie.

Chapter 3

ABORIGINAL PEOPLE: THE FIRST TO ENJOY MACADAMIAS

Aboriginal cultural and natural history of macadamias, names for the nut and engagement with settlers

… the cockatoo flew out and collected some nuts and scattered them around the mountain so Baphal could have food.

(Aunty Olga Miller, Senior Elder of the Butchulla People of Fraser Island, K'gari, January 1993)

For thousands of years, Aboriginal people would have known, studied, treasured, traded and eaten the macadamia. For fewer than 200 years, the macadamia has been shared with the rest of the world. For many years, I have searched historical documents and talked with Aboriginal people, asking about their knowledge of and interest in macadamias. In this chapter, I attempt to explain my understanding of Aboriginal cultural and natural history of macadamias, names for the nut and engagement with settlers. This is just a start and has limitations, but it is presented here in the hope that others will be inspired to continue this work and especially to support Aboriginal scholars in recording the relationship of Aboriginal peoples with macadamias.

However, with no written records, no known paintings or carvings, dispossession of their lands, population decline and the resulting loss of language and damage to culture, Traditional Owners have passed on – not surprisingly – very few reports about their knowledge of the macadamia. There is no doubt that the identity of many Aboriginal peo-

Left.: Lyndal Davis from the Gubbi Gubbi Dancers using traditional cracking stones. Credit: I. McConachie.

ple is shaped by a fundamental spiritual and cultural sense of belonging to the country that we know as Australia. Ongoing custodianship of its land, sea, flora and fauna is a foundation of their belief system. There are Memories sometimes called Dreamings, Legends or Songlines, which record spiritual beliefs and teachings.

Perhaps surprisingly, there are few early direct records from the colonists either. It was the 1840s when colonial settlers were first allowed into macadamia country after the convict era at Moreton Bay had ended and other settlers moved into today's northern New South Wales. Compared to the bunya nut, macadamias were a much less significant food source and limited by availability. Wild trees generally produced few nuts. Yet there is certainty that the macadamia was well known to Aboriginal people who had complex, sometimes close and trusting relationships with early settlers, botanists, officials and explorers. From the 1820s, some Aboriginal people would have shown the macadamia to Europeans, passed on oral knowledge and traded nuts. But most of what has been recorded comes from later letters and reports, and rarely were Aboriginal names recorded. No mentions of the nut in diaries, journals or letters from earlier than 1860 have been found. Yet it was in 1857 that the macadamia was named and described by the botanists, and when it was announced with surprise in 1867 that the macadamia was edible, several colonists were prompted to write to newspapers stating strongly that this was already well known. So at this current time, while there are at least 20,000 years of Aboriginal knowledge and history of macadamias, only some of it has been written down.

The first humans migrated to Australia at least 60,000 years ago and reached macadamia country much later. Over this time, Aboriginal people were largely quarantined from changes occurring in other parts of the world. As they migrated towards the east coast, they learnt of new foods by necessity. They would have observed that cockatoos and native rats ate macadamia nuts. Tasting the nuts, they discovered a delicacy that could be eaten raw or roasted in the shell in the dying embers of their cooking fires – for roasting the nut while still in its shell imparts an attractive, slightly different flavour to the kernel.

Early European settlers did not comprehend Australian Aboriginal culture or the richness and complexity of Aboriginal people's lives. They did not understand that most Aboriginal people had extensive knowledge of local fauna and flora and a more nutritious diet than most settlers. Aboriginal families ate foods of animal, marine and plant origin, which included flowers, leaves, fruits, tubers and nuts. After occupying the country for tens of thousands of years, their lives were dramatically disrupted by the arrival of Europeans.

Philip A. Clarke, in his book *Aboriginal Plant Collectors* (2008),[1] cited many broad examples of botanists, explorers and settlers using the skills of local Aboriginal people as guides and in collecting plants. He acknowledges that at least within local Aboriginal districts, Aboriginal people knew plants, particularly the edible ones. His book states that much of what explorers, botanists and early settlers learnt of Australian flora and fauna came from contact with Aboriginal people.

Aboriginal way of life

More Aboriginal people lived near the coast because of the greater range of food resources available. This particularly applied to the east coast of Australia, but even there the sustainable population was low. Each community was a group, often a kinship or clan of a number of families, who formed territorial bonds in a designated area. Predominantly they were agricultural hunter-gatherers,[2] and a typical Aboriginal group consisted of families who had the right to manage the food resources in the area where they lived. While the men mainly fished and hunted game, the women would take digging sticks and dilly bags to where they knew food was available.[3] Their extensive botanical knowledge meant that they knew of bearing macadamia trees, knew when the developing nuts would be mature, and possibly would have knocked them down with sticks to prevent their loss to native rats. It is likely that the husks were removed and the nuts dried in the sun for several days before they may have been roasted in the campfire, cracked and eaten, or maybe kept for trading.

Aboriginal people travelled long distances to keep up cultural connections with other groups and to trade. Macadamias, being rare, were most likely a minor part of their diet but it is likely that the nuts were used in trade and as gifts at gatherings and ceremonies. In some rainforests, for example the Amamoor Creek Valley south-west of Gympie, macadamias grew close to Bunya pine trees (*Araucaria bidwillii*). About every third year, when the majestic bunya trees produced large crops of nuts, Aboriginal people would travel long distances to feast on them.[4] Information from the Skyring family of 'Mumbeanna', a property at Mooloo near Gympie, describes 300 Aboriginal people in the early 1870s attending the 'great fighting and corroboree ground' of Chief Mumbea, which was 'thickly grown with the bunya pine and Queensland nut trees' and was 'a rich district for blacks; bunya and Queensland nuts were plentiful in the season'.[5]

Trade both within Aboriginal clans and with Europeans was common, with many foods involved. Early explorers, botanists and settlers communicated with local Aboriginal people, and macadamia nuts, when available, were one of the prized goods traded, often for axes (tommyhawks), rum and tobacco. The nuts were easily transported, stored well, and were valued for the nourishment, energy and wellbeing they provided.

Since the bunya season was just prior to macadamia maturity, macadamia nuts could have been collected and carried back to their home districts. Any that were dropped may have germinated and grown. Unconfirmed genetic studies indicate that sampled wild macadamia trees in a part of the Amamoor Valley had DNA markers characteristic of the wild populations south of Beenleigh.[6] This suggests that Aboriginal people may have carried nuts north as they came to the bunya feasts at Amamoor. Dr Ray Kerkhove from the University of Queensland believes that it was likely that the Bauple nut festivals were held at the mountain north of Gympie, but no primary records have been found. Based on genetic analysis, Dr Maurizio Rossetto, in collaboration with Traditional Owners and linguists, has demonstrated that the Black Bean (*Castanospernum australe*) was intentionally dispersed by Aboriginal people. Dr Rossetto considers it likely that this also applied to the macadamia.[7] Evidence of bunya nut shells aged between 1,320 and 1,400 years old have been found at Point Lookout on Stradbroke Island[8] and Dr Eleanor Crosby from the University of Queensland suggested that archaeological evidence of macadamias would probably be found in some middens if a flotation technique were to be used.[9]

Enjoying the macadamia

The outer husk split when dry or could be readily removed using rocks. The nut would have been cracked between rocks. One method was to place the nut in an indentation in a rock that served as an anvil, then to cover the nut with a flat stone, place it with the suture line of the shell upwards and strike it

An Aboriginal Elder, Lorelle Watcho, cracking macadamias in a traditional way using an anvil rock with flat stone and a hammer stone. Photo with permission of L. Watcho. Credit: I. McConachie.

with a hammer stone.[10] This is an effective technique that distributes the force evenly and minimises damage to the kernel. A number of cracking rocks have been found in macadamia country. In 1996, Senator Ron Boswell advised that he had seen cracking stones used by Aboriginal people under wild trees on the property of his father-in-law, Bill Beattie, at Calico Creek near Gympie.[11]

Sue Gallagher, from the Caboolture Regional Environment and Education Centre, in 1998 observed cracking stones on the ground beneath edible wild macadamia trees at Burpengary Creek near Caboolture.[12]

Although Aboriginal people were not sedentary farmers in the European sense, their skills to support plants that produced food led them to conserve macadamia trees, and there are reports that they-planted nuts along creek banks. In her book *White Beech,* Germaine Greer writes that Jenny Graham, a Kombumerri woman from Beechmont, told her grandchildren that as a young girl she had planted nuts as she walked along.[13] Michael Aird, an Aboriginal writer and leader who is a descendant of King Jackey (Bilin Bilin), advised that the grandmother of Elder Aunty Patricia O'Connor had planted macadamia seeds along the banks of the Logan River in the 1870s.[14] It is also likely that while carrying the nuts, some would have been dropped and occasionally they would have germinated and grown. Dr Rhys Jones, from the Australian National University's Department of Prehistory, stated in 1998 that he believed Aboriginal people practised widespread ringbarking to break up dense rainforest canopy.[15] Macadamias are mid-storey trees, so facilitating the penetration of sunlight through the canopy would increase their chance of flowering and fruiting. Fire was also used to prevent rainforest understorey from taking over wet sclerophyll forest that formed the boundary between rainforest and eucalypt forest.[16] Within the rainforest, low light and competition limited the ability to flower. Even where macadamia trees were more exposed, they did not always flower and fruit on an annual basis. Native insects attacked the flowers, nutlets and maturing nuts; rats, cockatoos, possibly possums, dingoes and other native fauna ate the nuts when they were mature, both on the ground and in the tree. In the Amamoor Valley near Gympie, it has been observed that flowering is induced only where sufficient rainfall is received. Over the last thirty years, there have been fewer than eight years when there was widespread cropping on trees in their rainforest habitat.

Dr Lennox Davidson, an agricultural scientist, in the 1970s studied Aboriginal medicines and advised the author that the oil was sometimes used as a carrier mixed with other plant extracts. The gum exuding from the bark of eucalypt trees was mixed with macadamia oil and used to treat chest complaints.[17] Dr Joanne Blanchfield, from the University of Queensland, advised that macadamias were used as a face cream, probably to allow clays and ochres to adhere, and as a carrier for ti-tree and eucalyptus oils.[18]

Lactating mothers would eat kernel that had commenced germinating, because, while bitter, it apparently contained a stimulant to aid breast milk production. Possibly the nuts were soaked in water to commence germination when mothers were feeding their babies.[19]

There are, as well, most unlikely reports that Aboriginal people may not have eaten the nuts at all. Examples are from King Bundy from Durundah near the Tweed, King Billy Engineer from the Brisbane Valley and Catchpenny from Enoggera, who were all quoted as saying that the nuts were poisonous.[20] However, it is probable they were referring to the true *M. ternifolia*, as there was confusion between this and the two edible species for a hundred years. It is not known if Aboriginal people consumed *M. ternifolia,* as the kernel of this species is both small and intensely bitter. While there are unconfirmed reports that people leached out the bitterness, this is considered most unlikely; rather, it is possible that people believed *M. ternifolia* trees had no nutritional significance.

There are many reports that Aboriginal people were well aware of the macadamia and advised the settlers of this.[21] Walter Petrie quoted his father Tom, who reported on the 'high food value set on it by the Blacks who made it a prominent food at their feasts'.[22] The *Sydney Morning Herald* in 1867 stated: 'The Queensland Nut – Macadamia – is very plentiful in NSW [New South Wales] especially in some locations at the Richmond and Tweed Rivers'.[23] In this case, they would have been referring to *M. tetraphylla,* as *M. integrifolia* does not grow south of the Queensland border.

Aboriginal names for the macadamia

At the time of European settlement, there were a number of distinct Indigenous language groups living in macadamia country. Tribes within these groups had their own dialect. From south to north, the main groups, while spelt variously, are: Bundjalung, Yuggera, Turrbal, Wakka Wakka, Gubbi Gubbi/ Kabi Kabi, Butchulla and Gooreng Gooreng. There were many tribes or clans within each language group.

Wayne Boldery, a most competent amateur researcher and historian, has compiled a list of names from the sources listed in Table 2 that were likely to have been used for the macadamia. Aboriginal languages were complex, translation is often imprecise, nuance is important and English-speaking people pronounced and spelt the names they heard in different ways. There were variants of the same name and broad generic names. As there is much current interest in Aboriginal names which refer to the macadamia, the table offers an extensive overview, although it is much simplified from Mr Boldery's detailed work. Words that are similar and for which early recorders used diacritical accent markers and apostrophes to show pronunciation have been summarised. Repetition which is common to signify the plural has been reduced to a single name, such as 'Kindal' for 'Kindal kindal'. There are some contradictions in locality which cast doubt over which species was intended, but as explained in other chapters, the application of scientific names has been prone to confusion ever since they were first allocated.

It seems highly likely that *M. integrifolia* and *M. ternifolia* were considered as different by Aboriginal people but that European botanists were confused over these species for a hundred years. It is less certain if Aboriginal people distinguished between *M. integrifolia* and *M. tetraphylla*, so the table should be considered as a guide and subject to revision and expansion. There are obvious overlaps in attempting to group names.

As an example of the difficulty arising between pronunciation and spelling, in 1980, Marjorie Oaks, an amateur anthropologist from the Richmond River Historical Society, reviewed the most commonly reported name 'Kindal' and its variants and concluded that the most likely pronunciation was 'gyndl'.

Table 1 lists the English common names. Table 2 summarises the Aboriginal names based on the work of Wayne Boldery and using his references (with the addition of reference 17 from the *Sydney Morning Herald*).

Table 1: English common names for the macadamia

Scientific name	English common names
Macadamia integrifolia	Macadamia nut, Queensland nut, Bauple [Bopple, Popple] nut, Smooth shelled nut and, in the 1930s, Australian nut
Macadamia tetraphylla	Macadamia nut, Queensland nut, Bush nut, Rough-shelled nut and, in the 1930s, Australian nut
Macadamia ternifolia	Maroochy nut, Gympie nut, Small-fruited Queensland nut
Macadamia jansenii	Bulburin /Bulberin nut. No records have been found of an Aboriginal name for this species.

Table 2: Aboriginal language names for the macadamia

Numbers in parentheses refer to the information sources listed directly below this table.

Aboriginal name: Baphal (4)		Species: *M. integrifolia*
Variants	**Language group**	**Comments**
Barpul (5, 6) Boppul (10)	Wakka or Dungaduau/ Dauwa Gubbi Gubbi	Baphal is used by the Butchulla for the Baphal nut and tree, Baphal Mountain and Baphal's lizard (4). Bauple is the anglicised version used as the name of the mountain and town, and time-honoured for macadamia nuts.
Aboriginal name: Barrum		**Species: mostly *M. integrifolia***
Variants	**Language group**	**Comments**
Barnum/ Barum/ Bar'rum (8, 9) Barroom	Gubbi Gubbi/ Kabi Kabi Barroom (7) Dauwa (tree) Bpa'rum/ Bar'rum (7) Dauwa (nut in husk)	
Aboriginal name: Kindal (17)		**Species: mostly *M. integrifolia***
Variants	**Language group**	**Comments**
Kindle (2) Kindel kindle (5, 8) Dal-kindal (14) Kgindhan (7)/ Kgindhul (7)	Butchulla Kabi Kabi Not known Dauwa (kernel)	Most widely reported name in nineteenth century Also *M. ternifolia* Dal-Kindal Possibly also used for *M. tetraphylla*
Aboriginal name: Gumburra		**Species: *M. integrifolia***
Variants	**Language group**	**Comments**
Gumbar/ Gumbur (1) Gumburra (12, 13) Kombum (1)	Mibinyah Yugambeh Yugambeh	

Table 2 continued

Aboriginal name: Jindilli (3,15)		Species: *M. integrifolia* and *M. ternifolia*
Variants	Language group	Comments
Yindilli (3, 13)	Dalla dialect of Wakka	Might also refer to *M. ternifolia*. May be a generic name and also refer to the Hairy Walnut *(Endriandra pubens)*. Jindilli and Kindle sound similar?

Aboriginal name: Worabill (3)		Species: *M. ternifolia*
Variants	Language group	Comments
Worabill	Yagara	May refer to Banksia

Aboriginal name: Babul (16)		Species: *M. integrifolia*
Variants	Language group	Comments
Babul	Dungidjawu	SE Queensland

Aboriginal name: Boombera		Species: *M. tetraphylla*
Variants	Language group	Comments
Boombera	Northern Rivers possibly Bundjalung	

Aboriginal name: Bumburra (12, 13)		Species: *M. tetraphylla*
Variants	Language group	Comments
Bumbera (11)	Yugambeh	

Aboriginal name: Dullubi (3)		Species: *M. ternifolia*
Variants	Language group	Comments
Dullubi	Dungidau	

Sources of information for the Table of Aboriginal language names:

1   Margaret Sharpe, *Gurgun Mibinyah – Yugambeh, Ngarahngwal, Ngahnduwal,* Canberra: Aboriginal Studies Press, 2020, p. 271. **Bambar, Gambar, gambur, gumbar, gumbur, kombum**.

2   F.J. Watson, *Vocabularies of Four Representative Tribes of South Eastern Queensland,* Brisbane: Royal Geographical Society of Australasia, Qld, 1944. **Barrum, kindle.**

3   T. Darragh and R. Fensham, *Leichhardt Diaries,* Memoirs of the Queensland Museum, Cultural Heritage Series, 7 (1) 2013, p. 370. **Dullabi, jindilli, yindilli, worabill.**

4   Olga Miller, 'Legend of the Macadamia Nut', 1993. **Baphal.**

5   Pat and Sim Symons, *Bush Heritage,* Queensland Complete Printing, 1994, pp. 73–4. **Bar'num, kindle, bar'pul.**

6   Mackenzie Willie; J.G. Steele, *Aboriginal Pathways in Southeast Queensland and Richmond River.* University of Queensland Press, 1984, p. 260. **Barpul.**

7   Granny Minnecon and Colleen Wall. **Bpa'rum, bar'rum, barroom, kgin'dhan, kgin'dhul.**

8   Shirley Foley, *Badtjala–English Word List,* Wondunna Aboriginal Corporation, 1996, p. 3. **Bar'rum, kindle.**

9   Jeannie Bell and Amanda Seed, March 1994, p. 13. **Ba'rum.**

10   Zachariah Skyring, *Aboriginal Dialect of the Gympie District* [Supplement to Royal Geographical Society of Australasia, Qld], 1870, p. 2. **Boppul.**

11   Francis Roberts (Surveyor to the Queensland–New South Wales Border, 1863), Diary No. 1, Entry for trees 430 and 440. **Bumbera.**

12   W.E. Hanlon, 'The Early History of the Logan and Albert Districts', *Journal of the Historical Society of Queensland* 2 (5), pp. 208–65. **Bumburra, gumburra.**

13   Yugambeh Museum Heritage Research Centre, Beenleigh. Patricia O'Connor. **Bumburra, gumburra.**

14   W.R. Guilfoyle, *Australian Plants,* Melbourne: Whitcombe and Tombs, 1911, p. 250. **Dal-kindal, kin-ternifolia** [unlikely].

15   Colin Roderick, *Ludwig Leichhardt: The Dauntless Explorer*, North Ryde, NSW: Angus & Robertson, 1988, p. 207. **Jindilli, yindilli.**

16   Suzanne Kite and Stephen A. Wurm, *Duungidjawu Language of Southeast Queensland*, Canberra: Australian National University, 2004. **Babul.**

17   *Sydney Morning Herald*, 'On the Land', 23 January 1936, p. 9. **Kindal Kindal.**

The number of names given to the macadamia suggest that it was widely known. The most common Aboriginal name for the nut was *kindal* (also *kindel* or *kindle*), *kindal kindal* or *kgindhul*.

'Eagle Eyes' Uncle Russell Bennet (Jirimir) finding nuts of *M. ternifolia* east of Gympie. Credit: I. McConachie.

Aboriginal people and Europeans

There are records that provide some evidence of the interrelationship between Aboriginal people and settlers. Some of these are:

- Redland City Council, in their publication *Southern Moreton Bay Islands Heritage Trail* (2008), reported that there were many middens on the islands and on the mainland but most of these were lost as they had been used in lime-burning to make mortar.[24] Some Aboriginal middens in Queensland contained large quantities of both marine and nut shells, and due to their locations, these may have been carried long distances. An unidentified local historian in a telephone conversation in the 1970s told how the Aboriginal communities living in the Redland Bay area of Moreton Bay left middens near the sea where there were large heaps of both seashells and macadamia shells. The nearest large populations of macadamia trees were at Mount Cotton, about 15 kilometres away, although one large macadamia at Redland Bay reported in the 1880s may have been of wild origin.

- Ian Fox, a Cultural Heritage consultant in New South Wales, advised that Aboriginal middens near Alstonville contained macadamia shells.[25]

- In 1870 and again in March 1877, the Queensland Lands Department enacted legislation to prevent both macadamia and bunya nut trees from being cut down, to preserve Aboriginal foods.[26]

- Penny Watsford quotes Walter Hill in telling of European and Aboriginal children in 1867 being employed regularly in collecting the nuts for food.[27]

- At Three Mile Scrub on Enoggera Creek in 1868, young boys were reported to have 'started for the creek to get nuts but were chased off by Aborigines'.[28] Macadamias grew wild near that area.

- Dr Ray Kerkhove referred to Aboriginal people collecting macadamia nuts in the Enoggera district. He believed that bunya nuts were treated as a significant food and that macadamias were considered more for trading and gifts.[29]

- Geographer Marcel Aurousseau in his chronology stated that Aboriginal people from Bribie Island visited the mainland to get nuts. Presumably these were both bunya and macadamia nuts.[30]

- In 1876, large quantities of '*Macadamia ternifolia*' [Edible Bauple nuts – *M. integrifolia*] were being brought down from the Boppel [Bauple] Ranges by Aboriginal people and offered for sale at Gympie and Tiaro, but only a comparatively small number found their way to Maryborough.[31]

Aboriginal people and their memories

In the 1860s, possibly the first marketer of macadamias to Europeans was Kawae Kawae or King Jackey Jackey of the Yuggera, Turrbal and Yugambeh tribal clans. He is mainly reported as Bilin Bilin (or Billemba) of the Yugambeh people, south of Beenleigh, a name that appears to have been used from the late nineteenth century and that will be used here. He was widely known as Jackey Jackey.[32] His wife Nellie was a respected Jagera Turrbal tribal clanswoman. Their great-great-grand-daughter Aunty Theresa Williams is recording their history.

An exuberant, enterprising man, Bilin Bilin became an Elder or leader of his people in about 1863. His tribal name may have meant 'king parrot' or possibly 'eagle'. German missionaries formed close relationships with Aboriginal people and, at the Bethesda Mission near Beenleigh, Pastor Johann Haussman paid the Yugambeh people led by Bilin Bilin £1 per acre to clear ten acres. Francis Roberts from 1863 surveyed the Queensland–New South Wales border, engaging Bilin Bilin and his men as guides and porters and, in recognition of his contribution, may have named the district where the Gold Coast airport now stands as Bilinga. This conflicts with the report that the name Bilin Bilin was given much later. A.E. Harch wrote in 1977 of King Jacky's group obtaining nuts from the scrub in the late 1870s.[33]

Pastor Haussman befriended Bilin Bilin, called him Jack, tried to convert him to Christianity and

Bilin Bilin (King Jackey Jackey or Kawae Kawae of the Albert and Logan rivers), about the 1870s. With permission of his descendants. Credit: John Oxley Library, State Library of Queensland, Brisbane, Neg. 63306.

they learnt from each other. Bilin Bilin promoted and co-ordinated the collection of nuts by Aboriginal communities, then bartered these with white settlers and traders usually for axes, rum and tobacco. With the axes, his people cut roofing shingles. In 1875, W.H. Walsh MLA presented Bilin Bilin with a brass 'King Plate' on behalf of the Queensland Government, and he became known as 'King Jacky of the Logan, Albert and Pimpama tribes'. He travelled widely and was reported from many different areas outside his district.[34]

Gurang Yugambeh Elder Uncle Wayne Fossey, in a letter to the author in September 2019, wrote of a Bora ring, Aboriginal graves and middens at Tallebudgera Creek, next to Burleigh Heads on the Gold Coast. This area had significance to his people as a site for energy, food and wellness. He is not aware if macadamia shells had been found in the

middens which are now a fraction of their original size. Near the coastal highway bridge that crosses Tallebudgera Creek is a large rock shelf which was sacred to his people. Macadamias from further up the creek and shellfish from near the creek mouth were eaten there in feasts. Uncle Wayne advises that some of the early cedar-cutters referred to the nut after being shown them by Aboriginal people. He believes that after firestick burning, any damaged macadamia trees were recovered by what he describes as 'approach grafting'.

From Gympie to Maryborough, macadamias have always been known as Bauple nuts, after the significant stands at Mount Bauple. This area was important both culturally and spiritually to the Butchulla and Dowarbara neighbours. It was steeped in legends and taboos.[35]

The wild trees assumed a significance beyond the amount of nuts available and the following Aboriginal legend is testament to this. This legend is also recorded in a painting by Nai Nai Bird.

The Legend of the Baphal

As re-told in January 1993 by Aunty Olga Miller, the senior Elder of the Butchulla People of Fraser Island.[36]

Way back in THE FIRST TIME [The Dreamtime] when Yindingie our Messenger God was leaving the Mountain, the Budjilla people had to decide who was to look after our Land.

There was someone to go to Burrum Heads to look after the north and someone to look after the south at Inskip Point. When it came to looking after the Mountain, nobody wanted to really leave and go to a far away place, so a man called Baphal said he would go.

So Baphal packed for his long journey and unbeknown to him his friend the jewel lizard stowed away in his pack. He had walked a long way, all the time he could see the Mountain in the distance. Finally he reached the Mountain and set up camp, when out jumped the little jewel lizard. Baphal said to him what are you doing here? The little lizard said I did not want to leave you so I hid in your pack and came with you.

One day when Baphal was walking along he fell and hurt his foot, he could not get to food and water. The little lizard could see that Baphal was hurt, so he went to the rock wallaby to ask him what to do. The rock wallaby said, we have to get him some water. So they got Baphal's eelamun and hurried to the water, but when they got there, the rock wallaby could not reach. So he took the eelamun to the kangaroo and the kangaroo filled up the eelamun with water and gave it back to the rock wallaby who with the lizard gave it to Baphal.

Then the lizard said we have to get him some food. The rock wallaby said we should ask the cockatoo. So the cockatoo flew out and collected some nuts and scattered them around the mountain so Baphal could have food.

Then the rock wallaby and the lizard decided that Baphal needed help from his people so they made a fire and asked the cockatoo to get some leaves. The cockatoo flew out once again and collected some green leaves from the nut tree, and this created smoke. Well, our people on the Island seen the smoke and they sent help to Baphal.

When our people saw what happened they called the Mountain, Baphal's Mountain. When our people seen the lizard they called him Baphal's lizard. When our people seen the nuts they call them Baphal's nuts.

Other family groups have recorded their Dreamings or Law Stories. Colleen Ma'run Wall, a Dauwa Kabi woman whose ancestral homelands are in the Munna Creek watershed of the Mary River, has documented family history from Mount Bauple. This was passed to her by Stirling Minniecon from the brother of her grandmother, Mary Beezley, who received it from their mother Katherine Minniecon (née Lingwoodock). She refers to the history as 'Story Lines or Story Strings that relate to what is sometimes called Dreamlike or The Dreaming', which she describes as 'Land, People or Sharing Laws'. Following is a summary of one of their Dauwa Kabi 'People Law' stories that teach children how to behave:

A very long time ago, before this land was settled by the English, the Bauple nut was called Kgin'del Kgin'del or the Moon Nut. At this time we had two moons and they were called Bpah'lun, a boy, and Bpar'uma, a girl, who were forever fighting. Their creator ancestor was so angry with them that he smashed them together and their mother cried continuously for them so he had to put them back together again but he could only gather together some of the bits so he only make one moon. Now this new moon had a dark side which was Bpar'uma, the sister, who wasn't allowed to look at her beautiful Mother Earth because she, as a female, had the responsibility of teaching young ones how to behave, and Bpah'lun, the boy, was the light side so his mother could keep

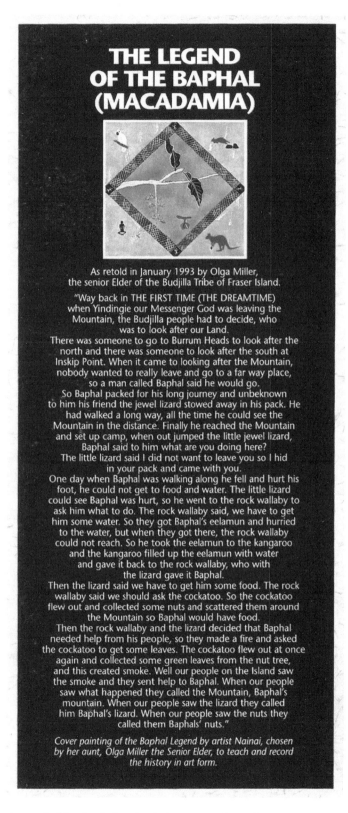

THE LEGEND OF THE BAPHAL (MACADAMIA)

As retold in January 1993 by Olga Miller, the senior Elder of the Budjilla Tribe of Fraser Island.

"Way back in THE FIRST TIME (THE DREAMTIME) when Yindingie our Messenger God was leaving the Mountain, the Budjilla people had to decide, who was to look after our Land. There was someone to go to Burrum Heads to look after the north and there was someone to look after the south at Inskip Point. When it came to looking after the Mountain, nobody wanted to really leave and go to a far way place, so a man called Baphal said he would go. So Baphal packed for his long journey and unbeknown to him his friend the jewel lizard stowed away in his pack. He had walked a long way, all the time he could see the Mountain in the distance. Finally he reached the Mountain and set up camp, when out jumped the little jewel lizard, Baphal said to him what are you doing here? The little lizard said I did not want to leave you so I hid in your pack and came with you. One day when Baphal was walking along he fell and hurt his foot, he could not get to food and water. The little lizard could see Baphal was hurt, so he went to the rock wallaby to ask him what to do. The rock wallaby said, we have to get him some water. So they got Baphal's eelamun and hurried to the water, but when they got there, the rock wallaby could not reach. So he took the eelamun to the kangaroo and the kangaroo filled up the eelamun with water and gave it back to the rock wallaby, who with the lizard gave it Baphal. Then the lizard said we have to get him some food. The rock wallaby said we should ask the cockatoo. So the cockatoo flew out and collected some nuts and scattered them around the Mountain so Baphal would have food. Then the rock wallaby and the lizard decided that Baphal needed help from his people, so they made a fire and asked the cockatoo to get some leaves. The cockatoo flew out at once again and collected some green leaves from the nut tree, and this created smoke. Well our people on the Island saw the smoke and they sent help to Baphal. When our people saw what happened they called the Mountain, Baphal's mountain. When our people saw the lizard they called him Baphal's lizard. When our people saw the nuts they called them Baphals' nuts."

Cover painting of the Baphal Legend by artist Nainai, chosen by her aunt, Olga Miller the Senior Elder, to teach and record the history in art form.

Legend of the Baphal and painting. Aunty Olga Miller and Nai Nai Bird. With permission from Uncle Glen Miller and Luke Barrowcliffe.

Ma'run. Written and painted by Colleen Ma'run Wall, a Dauwa Kabi woman. Permission by C. Wall.

an eye on him. The creator ancestor then made the macadamia tree with a kernel that had two halves so we could be reminded, each time we ate them, that brothers and sisters should look after and respect one another and always remember our sharing and caring law.[37]

Aboriginal people regularly managed large areas with fire deliberately using small, controlled burns to make wild game more visible and help to regenerate the land. This may have had a significant effect on changing the environment during Aboriginal occupancy.

Respecting Aboriginal knowledge and rights

Aboriginal groups are increasingly retrieving and preserving Traditional Knowledge of food and medicines. Some are rightfully seeking to have past injustices acknowledged, to have their knowledge and care for Country respected and to benefit from future development of their traditional resources.

In seeking Australian examples of traditional foods that have been developed and exploited by commercial interests, it is easy to single out the macadamia. There have been claims that macadamias as an Indigenous food have been 'stolen' and developers of the commercial industry could have a claim made against them. *The Bunya Mountains Aboriginal Aspirations and Caring for Country Plan* of 2010 includes a warning from Butchulla, Gubbi Gubbi and Wakka Wakka Elders to stop the white man getting control of the bunya nut, so that rights to harvest bunya nuts would not be lost overseas or to the horticulture industry, as happened with the Bauple nut.[38]

> ... We have got to stop the white man getting control of the Bunya nut, because they have already taken the Bauple nut, so we don't have any rights on the Bauple nuts anymore, that belongs to the overseas whoever ... Bunya nut harvesting we want rights to that ... But the Bunya nut its gonna, its getting too big, some horticulture is going to come in and just take that whole right off us unless we claim it, cause they done it with the Bauple, the Macadamias, they grafted them ... it's American now ... ! Aunty Nai Nai Bird (Butchulla Elder), Aunty Lerlene Henderson (Gubbi Gubbi Elder), Aunty Yvonne Chapman (Wakka Wakka Elder) and Selina Hill (Willi Willi/ Wakka Wakka Traditional Owner)

The Nagoya Protocol is a United Nations initiative, adopted by many countries, including Australia in 2012, that seeks to ensure that Indigenous peoples receive their rights and a benefit from native flora and fauna. How to implement commitments under the Protocol is a complex matter that is being considered by the Australian Government and for ethical and practical reasons should be taken seriously by all Australians.

All of us need to seek and embrace the long-term connection that many Aboriginal people had and still have with the macadamia. We need to better understand their history, custodianship and future aspirations, to work together to respect and advance their Law for the nut.

Walter Hill Macadamia Tree
Planted in 1858

*"This tree, a Macadamia integrifolia seedling, was planted by
Walter Hill, the Superintendent of the Brisbane Botanical Gardens
in 1858. It originated from a wild macadamia tree
in south-east Queensland rainforest
and is believed to be the first macadamia tree planted by Caucasians.
It represents the birth of the world macadamia industry".*

Chapter 4

EUROPEAN DISCOVERY: EXPLORERS AND BOTANISTS

How Australian explorers and botanists found, described and named the macadamias

It is by far the best nut that I've ever tasted; nothing comes near it, and I am most anxious that it should flourish in Europe, the Cape, etc.

(Dr Joseph Dalton Hooker, Director Royal Botanic Gardens, Kew Gardens, London, 1875)[1]

Early officials, explorers and botanists routinely kept records of all events. Apart from the occasional letter, the settlers who followed left few records of their lives, but some almost certainly tasted and appreciated macadamias before their edible qualities were officially recognised.

The Moreton Bay Penal Settlement, later to become Brisbane, existed from 1824 to 1842, after which free settlers arrived and ventured north and south into macadamia country. Many were timber-cutters, who lived in or near the scrubs, where Aboriginal people showed them plant and animal foods. Although these newcomers may have tasted the rare macadamias, they had little reason to regard them as a significant food.

When Ferdinand Mueller described and named the macadamia in 1857, he regarded it as a botanical curiosity, and it was another ten years before other botanists recognised its edible qualities. Three of Australia's well-known explorers, Allan Cunningham, Charles Fraser and Ludwig Leichhardt, may

have been part of the European discovery of macadamias; Ludwig Leichhardt certainly collected a specimen. All their reports pre-date Mueller's work.

By whom, when and where Europeans discovered the macadamia is still uncertain and the following is presented in some detail to explain this.

Allan Cunningham (1791–1839)

In the early days of the Moreton Bay Penal Settlement, the energetic King's Botanist, Allan Cunningham, explored the district and sent many plant specimens back to England.

Cunningham arrived in Sydney in 1816 and joined explorer John Oxley's expedition to the west of the Blue Mountains. From 1819 to 1822, he was involved in various explorations of the Australian coast, and in 1824 he accompanied Oxley's expedition to Moreton Bay to establish a penal settlement. In 1827, he journeyed overland to the Darling Downs, finding Spicers Gap in the Dividing Range. Returning to Moreton Bay by sea in 1828, he explored south to the Logan River, discovering Cunninghams Gap. In 1829, he explored the upper Brisbane River.

Allan Cunningham, King's Botanist, who may have found macadamias in 1828 at the Logan and Albert rivers. Credit: Mitchell Library, State Library of New South Wales, Sydney.

On his visit to Moreton Bay in 1828, Cunningham was accompanied by the Colonial Botanist, Charles Fraser (1788–1831), whose main role was to establish a garden for tropical plants. On 4 July 1828, Fraser, in company with Cunningham and the Commandant of the Moreton Bay Settlement (Brisbane), Captain Patrick Logan, were at Breakfast Creek about five miles east of the town. Fraser's diary records a new plant producing fruit larger than a Spanish chestnut:

> By the natives the fruit is eaten on all occasions: it has, when roasted, the flavour of a Spanish chestnut and I have been assured by Europeans who have subsisted on it exclusively for two days, that no other unpleasant effect was the result than a slight pain in the bowel and that only when it was eaten raw.

His diary and subsequent searches could suggest he was describing one of two separate plants, either the *Castanospernum australe* or Black Bean,

which had just been named by Allan Cunningham, or the macadamia. Fraser's diary refers to the abundance of 'Castanospermum' and its use by Aboriginal people as a food after roasting, and he calls its seed pods 'chestnuts'.[2]

Two weeks later, Cunningham, Fraser and Captain Logan were exploring and botanising south of Brisbane. On 30 July 1828, towards the heads of the Logan and Albert rivers, Cunningham wrote in his diary of the prolific *Castanospernum* or native Chestnut trees and noted fires kindled by the natives in the process of roasting nuts. In his report to Governor Darling, he stated:

> around were quantities of the large seed of that exceedingly ornamental tree of close woods called a Moreton Bay 'chestnut' for which, when roasted, it is by no means a bad substitute. Upon these nuts few natives who wander through these lonely regions chiefly subsist during some months of the year as like the English chestnut they contain evidently some saccharine and much farinaceous matter and by being well roasted are rendered easy of digestion.

His footnotes read:

> This tree, than which there is no plant, indigenous to the shore of Moreton Bay and adjacent country which the eye rests with greater pleasure, constitutes a genus perfectly distinct from any yet published; and, independent of its highly ornamental habit and refreshing shade afforded by its densely leafed branches, its nuts are produced in pods in such abundance as to be ere long worthy of the attention of the farmer, and its fruit could form nutritious food for pigs, etc. This tree effects a rich and moist soil.[3]

Captain Logan was killed, probably by Aboriginal people, near Logan Creek on 22 October 1830. Lieutenant G. Edwards, in his report on this death, stated: 'It had also appeared that he had roasted some chestnuts at his fire: the remains of the chestnuts lay about'.[4]

W.G. McMinn, in his biography *Allan Cunningham: Botanist and Explorer* (1970),[5] used the following words to claim that in July 1828, Cunningham:

> ... also examined the Queensland bush nut, *Macadamia ternifolia*, remarking that independent of its highly ornamental habit and refreshing shade afforded by its dense branches its nuts are produced in such abundance as to be ere long worthy the attention of the farmer.

McMinn stated that his opinion was confirmed by several experienced botanists in New South Wales, who believed that Cunningham had found the macadamia and was not describing the Moreton Bay chestnut (*Castanospernum australe*), also known as the Black Bean.[6]

Cunningham forwarded plant specimens to Sir Everard Home in England, and it is probable that they included these 'chestnuts'. He returned to England for five years in the early 1830s and spent much of this time working with fellow botanists to classify the plants he had collected. These specimens are so numerous that only a fraction were worked on in his lifetime, and it was reported in the 1860s that most of his dried specimens were still in their original parcels.[7]

In the 1830s, Kew Gardens existed as a large, expanding plant collection, repository of many of the botanical specimens collected throughout the world, but it was not declared a Botanic Garden until 1840. While undertaking research at Kew Gardens in 2012, Australian ecologist Dr Rob Lamond suggested that any specimen of macadamia that Cunningham may have collected was probably still in storage at a separate, not accessible site.[8]

In her book *White Beech*, Germaine Greer accepted that Cunningham found the macadamia, stating that 'the specimen of the Queensland nut Cunningham sent back to Kew in 1828 probably sank to the bottom of the vast mass of plant material arriving from all over the world, because it cannot now be traced'.[9]

So it remains uncertain whether Cunningham found and recorded the macadamia, or the true Moreton Bay chestnut, or perhaps both. Macadamias were a minor food and most but not all trees were small and insignificant in the rainforest. Some of Cunningham's words in his diary – namely 'called a Moreton Bay "chestnut"' and 'constitutes a genus perfectly distinct from any yet published' – invite examination. It is understandable that English-trained botanists might describe all nuts as 'chestnuts' and possible that Cunningham was distinguishing between *Castanospernum* and an unknown nut.

He had previously observed *Castanospernum* at Breakfast Creek and at the Albert and Logan rivers, and was possibly now describing an unknown nut. Cunningham and Fraser were competent botanists of that era and used the name *Castanospernum*, and both the widespread recordings of the trees and the shade they afforded suggest he meant the Black Bean. *Macadamia integrifolia* grew in the Logan and Albert River districts and also in Breakfast Creek, whose upper reach is known as Enoggera Creek, adding to the uncertainty.

Lastly in this discussion, many reports state that *Castanospernum australe* (Black Bean) was a common food for Aboriginal people.[10] But the nuts are poisonous when raw and preparing them so they are safely edible is a most involved process. The traditional process includes cooking the seeds for three hours or so, cutting them up finely, leaching them in a stream for between three and seven days, crushing the pulp into thick cakes and baking them.[11] An 'expert' comment on the 'Plants for a Future' website states that 'the mature seeds are deadly poisonous'. *Castanospernum australe* contains two very toxic constituents, a saponin and castanospermine, where one is soluble and removed by repeated washing and the other by roasting. Many native fruits and seeds are not edible and it is most unlikely that Europeans would have eaten the 'chestnut' unless they were advised by an Aboriginal person that it was safe to do so. In current times, the *Castanospernum* is not consumed at all, either by Aboriginal people or by Europeans.

Uncle Wayne Fossey, a Yugambeh Elder and President of Bunya Peoples Aboriginal Corporation, told this writer that Cunningham, Fraser and Logan could not have eaten *Castanospernum* without extensive preparation. He also believes that macadamias were likely to have been carried extensively by Aboriginal people as a food.[12]

So despite Europeans searching and botanising in macadamia country from the 1820s, there do not appear to be certain reports of the macadamia for almost forty years. As described later in this chapter, Ludwig Leichhardt collected a specimen of the true

Macadamia ternifolia in 1843, then Hill and Mueller found, named and described *M. ternifolia* in 1857. As described in the preceding chapter, it was 1867 before the first 'official' report of the nuts being edible. Yet there were many reports that Aboriginal people treasured macadamias as a food, and many settlers were aware of the nut's palatability. It could be expected that in the forty years after settlement and expansion north and south of Brisbane, with often a close relationship between settlers and the local Aboriginal people, the macadamia would have been known as a fine edible food. So, which European person or persons first discovered the macadamia and appreciated its qualities? Assessing the information available at this time does not provide an answer. If a specimen of macadamia were to be found in Cunningham's collection at Kew, we could be certain. Perhaps it was Fraser or Cunningham, perhaps an early unknown settler, convict or official who sometime after 1825 was shown the nut by an Aboriginal person and tasted it?

Ludwig Leichhardt (1813–1848)

In a tragedy that remains one of the mysteries of Australian history, German botanist and explorer Ludwig Leichhardt met an unknown end while attempting to cross Australia from east to west in 1848.

Leichhardt was born in Prussia and trained as a natural scientist. In a letter to his father when he was twenty-three years old, he indicated his determination to succeed, saying 'all of my striving is bent towards accomplishing something outstanding, to raise myself above the ordinary'.[13] He developed a fascination for the unknown, and after arriving in Sydney in 1842 began to explore the country to the north.

Arriving in Brisbane in June 1843, he was invited by the pioneer Archer brothers to make his base at their cattle station, 'Durundur', located about 100 kilometres north of Brisbane near present-day Kilcoy. For eight months, he explored the countryside in all directions, often accompanied only by Aboriginal guides.[14]

In September 1843, Leichhardt, David Archer, and three Aboriginal guides set off to explore the

LEICHHARDT
"Obstinate, valiant, strong-willed, fearless"
(Sydney Times, 13 October 1858)

Ludwig Leichhardt, who collected the first specimen of macadamia in 1843. Credit John Oxley Library, State Library of Queensland, Brisbane, Neg. 168799.

Blackall Range district, where he almost certainly found and collected the first recorded specimen of macadamia.[15]

Leichhardt's signature and the date '18 September 1843' are written on the paper accompanying a flowering specimen of the 'true' M. ternifolia, which is held in the Royal Botanic Gardens Herbarium in Melbourne. The species name '*ternifolia*' referred to *M. integrifolia, M. tetraphylla* and *M. ternifolia* until 1956. The main flowering period of this species is September, which fits the specimen date.

To Leichhardt, this plant was just one of many unnamed, new species he brought back from his time at Moreton Bay. The specimen was sent to the Herbarium in Melbourne, where it was not examined until 1860.

Leichhardt's specimen collected 1843 and held at Melbourne Herbarium, Royal Botanic Gardens, Melbourne.

Colin Roderick's detailed biography, *Leichhardt: The Dauntless Explorer*, includes a sketch taken from Leichhardt's diary of a nut which closely resembles the macadamia. Roderick's inscription reads:

> Leichhardt. Sketch and description of the Bauple – Queensland – Australian nut (macadamia) 18 September 1843 in its original restricted habitat near Tiaro, Queensland, thought to be the first scientific report on it and the first of the native name Jindilli [pr. Yindilli] for it.[16]

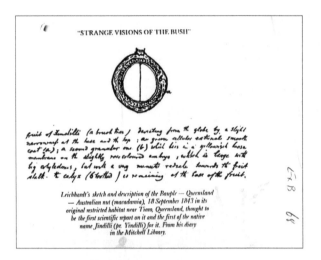

"STRANGE VISIONS OF THE BUSH"

Leichhardt's sketch and description of the Bauple — Queensland — Australian nut (macadamia), 18 September 1843 in its original restricted habitat near Tiaro, Queensland, thought to be the first scientific report on it and the first of the native name Jindilli (pr. Yindilli) for it. From his diary in the Mitchell Library.

Leichhardt's drawing and description of the macadamia, which has been controversial. The drawing is believed to be macadamia but the description relates to an unknown plant.

Leichhardt's diary shows that he was at Mount Bauple on 2 August 1843. It was six weeks later when he collected a branch of the species now named *Macadamia ternifolia* and dated it 18 September 1843. While the drawing of the nut does appear to show a macadamia, Leichhardt's accompanying description does not fit the macadamia.

> Fruit of jindilli (a brush tree). Deviating from the globe by a slight narrowing at the base and the top; a green cellular external smooth coat (a); a second granular one (b) which lies in a yellowish loose membrane on the slightly rose coloured embryo, which is large with big cotyledons, but with a very minute radicle towards the fruit stalk. The calyx (6 toothed) is remaining at the base of the fruit.

Dr Rod Fensham, of the Queensland Herbarium, has suggested that the sketch may be of *Endriana pubens*, known as the Hairy Walnut. He believes

that *jindilli* was a generic Aboriginal name for a plant with soft fleshy fruit, rather than a name for the macadamia. It could thus have applied also to *Endriana pubens*. The name *jindilli* is perhaps similar sounding to the most commonly reported Aboriginal name of *kindal*?

The *M. ternifolia* collected by Leichhardt does not exist at Mount Bauple, where there are only *M. integrifolia*, and two experienced botanists, Greg Smyrell and Ernie Rider, have not found *Endriana pubens* there either. Dr Fensham believes that Leichhardt's specimen was probably collected approximately 2 kilometres north-west of Commissioners Flat, between present-day Woodford and Beerwah, close to Durundur. Leichhardt called this area 'Bunya Bunya brush', where 'brush' was a term used for a type of rainforest. Although much of the rainforest has since been cleared, *M. ternifolia* still exists there in wild stands. The site of the annual Woodford Folk Festival is likely close to where the specimen was collected.

A study of Leichhardt's diary by botanists Dr Bill McDonald and Glenn Leiper sheds light on the confusion. They consider that the nut in Leichhardt's sketch was more likely to have been a macadamia than *Endriana pubens*. They also point out that the botanical description of the *jindilli* does not fit either plant.

Leichhardt's specimen at the Melbourne Herbarium is mounted on paper, where a number of notations have been made. A small label, in Leichhardt's handwriting, has the words 'Dullabi B Bunya Brush 18 September 43'. 'Dullabi' may have been a name provided by one of Leichhardt's Aboriginal guides.

On a separate, larger, undated label, headed 'Botanical Museum of Melbourne', the specimen is named 'Helicia ternifolia Dawson and Burnett Rivers' and was signed by Ferdinand Mueller. On this label, Mueller was twice in error, since in 1857 he had described the genus as 'Macadamia' and macadamias have never been recorded in the Burnett and Dawson catchments.[17]

Another handwritten note attached to the specimen, signed by Queensland Government Botanist Lindsay S. Smith and dated 4-11-1955, states:

Macadamia ternifolia, F. Mueller. As far as I am aware, the species does not occur on the Dawson and Burnett Rivers. 'Bunya Bunya brush' is not to be confused with the Bunya Mountains North of Dalby, in which tributaries of the Burnett River rise. Leichhardt's locality, from a consideration of dates and localities on his specimens, has been shown by Mr S.T. Blake ... to be in the Conondale Ranges area, the divide between the Stanley River, a tributary of the Brisbane River, and the upper Mary River.[18]

It appears that Mueller used the macadamia specimen provided by Walter Hill to name the species in 1857, but then used the Leichhardt specimen to name it *Helicia ternifolia*.

John Carne Bidwill (1815–1853)

John Carne Bidwill was an English botanist who documented plant life in Australia and New Zealand. He first visited the Moreton Bay District in 1841 where he botanised up to the Glass House Mountains.[19]

In 1848, Bidwill was appointed Commissioner for Crown Lands at Maryborough in the Wide Bay District. In 1849, at nearby Tinana, he established his own botanic gardens and imported and exported both seeds and plants. He made many collections, sending them to Kew Gardens, William Macarthur at Camden Park Estate near Sydney, and the Botanic Gardens at Brisbane, Sydney and Melbourne.

Bidwill's role as Commissioner allowed him to collect native plants while exploring the countryside. Given his botanising around the district and his association with local Aboriginal people, it would be surprising if he did not observe and collect the edible *M. integrifolia*, which grew prolifically in the forests and foothills of nearby Mount Bauple. If such a specimen was collected, it likely became one of the newly discovered plants that Bidwill sent away for other botanists to describe and name. Of the many plants he collected, it appears that he described and named only one, an orchid, *Dendrobium kingianum.*[20]

Bidwill's career was cut short with his death in 1853, partly as the result of deprivations suffered in 1851, after he became lost in the bush while seeking a direct route from Maryborough to Brisbane. His gardens were neglected, and in 1853, Charles Moore

from the Royal Botanic Gardens in Sydney came to Maryborough to assess Bidwill's plants.[21] Moore recommended that his plants not be forwarded to Sydney due to their size and the cost of moving them.

While we do not know whether Bidwill found the macadamia, which at that time had not been described or named, there is much to suggest that he did. In 1849–50, he sent away 120 'new' plants, which suggests that they were native species. Bidwill's diary went missing when he was lost in 1851, and he was then absent in Brisbane and Sydney seeking medical attention. His records are incomplete.

Bidwill had a mentor and friend in William Macarthur, later Sir William, son of John and Elizabeth Macarthur, with whom he regularly exchanged plants. Macarthur was actively involved with the Royal Botanic Garden in Sydney. From 1835, Macarthur established extensive botanic plantings at Camden Park, south-west of Sydney, which remain as an

Macadamia at Camden Park Gardens in the 1840 section that could have been provided by J.C. Bidwill to William Macarthur. Credit: Stuart Read.

excellent example of colonial gardens. An *M. integrifolia* whose planting records are not known stands in the section of these gardens with plants from the 1840s. The genetic provenance of this tree may be Mount Cotton, east of Brisbane, although there are no known early records of macadamias collected from that district. The genetic analysis also suggests an ancestor from the Gympie region. Bidwill visited Camden Park in 1841, so it is possible that he collected seed nuts and provided them to Camden Park. He could also have sent seed or plants from Wide Bay in 1849 or 1850.

Covered in more detail in Chapter 5 is an *M. integrifolia*, which may have come from Bidwill, that was recorded at the Sydney Royal Botanic Garden from the 1860s; it was removed in the 1980s to make way for new plantings. This tree had been labelled 'Camden Haven 1850–60 C. Moore',[22] which raises the question of its origin. It is believed that the tree was mislabelled and is likely to have come from Camden Park Gardens.

Walter Hill (1819–1904)

If there is one man above all who deserves to be recognised as the 'father' of the macadamia industry, it is Walter Hill. In the company of Ferdinand Mueller, in 1857, Hill found macadamias in the rainforest and then in 1858 planted the first domesticated tree, contributing to the description and naming of the species. He announced to the world that they were edible, grew them, promoted them and distributed them throughout much of Australia and the world. Walter Hill has received scant recognition, despite being an early champion of forest conservation and playing a major part in the discovery, description, growing and promotion of macadamias.

Born in Scotland, Hill trained in horticulture at Edinburgh and was a foreman at the Royal Botanic Gardens, Kew, before coming to Australia in 1852 to seek his fortune on the newly discovered goldfields. In a private botanical exploration in 1854, he was the only European survivor of an Aboriginal attack in north Queensland.

Walter Hill, Queensland Colonial Botanist, who probably deserves the title 'father' of macadamias. Credit: John Oxley Library, State Library of Queensland, Brisbane, Neg. 17483.

In 1855, he was appointed Superintendent of the Brisbane Botanic Gardens. Following the separation of Queensland from New South Wales in 1859, he was named Colonial Botanist, a position he held until his retirement in 1881.

Hill contributed to the development of Queensland by assessing commercial crops such as sugar, tobacco, grapes, wheat, fruits, tanning plants, tea and macadamias. With John Buhot, he demonstrated the commercial possibilities of sugar-cane when, in secrecy, late at night, he crystallised cane juice. When the Queensland Acclimatisation Society was formed in 1862, Hill worked with its members as the propagation and distribution co-ordinator for many potentially commercial plants. He continually assessed plants obtained from overseas and in return provided Australian plants of commercial and botanical interest.

In 1875, Hill recommended preservation of state

forest for reafforestation and scientific research, to be paid for by a timber export duty.[23] The government of the day was not impressed with his suggestion to curb the profits being made from timber extraction. A dour Scotsman, Hill had continual differences with those in authority, and his conflicts with the Queensland Government resulted in his forced retirement when he was just sixty-two. He continued to contribute to the Botanic Gardens without pay and developed a farm, 'Canonbie Lea', at Eight Mile Plains, today a suburb of Brisbane.

In the massive 1893 Brisbane flood, much of the Botanic Gardens was submerged, and records stored in an office were lost. This has resulted in limited knowledge of Hill's macadamia achievements.

In 1858, Hill planted what he recorded as a 'large-fruited macadamia' in the Brisbane Botanic Gardens. It is now a massive, healthy exemplar of the *M. integrifolia* species. Known as the Walter Hill Macadamia Tree, it is the most visited and photographed macadamia in the world. Over the years, various signs have been erected on or adjacent to the tree. An early sign stated that the tree had been relocated from the banks of the Pine River, but Hill's records show that in 1856 he was botanising south of Beenleigh, where *M. integrifolia grows*, and he also collected seed or plants from the Nundah Scrub during the 1860s.[24] Nundah Scrub was on the banks of Kedron Brook where *M. integrifolia* had been found, but the scrub was cleared for farming by the 1880s. Did he also collect a plant from John Bidwill's gardens in Maryborough? Ross McKinnon, the former Curator of the Brisbane Botanic Gardens, believed the tree was likely to have originated either in the Mary Valley or at Mount Bauple.[25]

But current DNA studies to determine the provenance of the Walter Hill Macadamia Tree confirm that it came from a population that has been almost completely lost. Dr Catherine Nock from the Southern Cross University at Lismore (NSW) investigated the geographic structure of genetic diversity in wild *M. integrifolia* using Short Sequence Repeat markers and found that the Walter Hill Tree shared the

Plaque on Walter Hill Tree in the 1970s. Credit: unknown.

greatest amount of Nuclear DNA with a few trees still growing at Whites Hill ('Sankeys Scrub') and at Mount Cotton in Brisbane.[26]

The first Governor of Queensland, Sir George Bowen, and his popular wife, Lady Diamantina Bowen, lived in Government House, adjacent to the Gardens. The now-named Lady Bowen Staircase, leading from the Gardens to the Brisbane River, passes close to the Walter Hill Tree. In 2008, a ceremony to celebrate the tree's sesqui-centenary was held. Participants had the opportunity of hugging the old tree.

It seems that Hill planted this tree believing the nuts to be poisonous, so how did he come to recognise the edible qualities of macadamia nuts? An often-repeated anecdote tells that Hill believed the macadamia would be difficult to germinate due to its thick, hard shell and asked a young assistant to use a vice to remove the macadamia kernels from their shells to improve germination, while instructing him not to eat the kernels as they were poisonous. He did not realise that this was a flawed way of proceeding, as germinating the nuts in the shell has a higher success rate because the inside of the shell contains a natural fungicide. Hill was upset and concerned when he came across the lad munching broken bits of kernel with great enjoyment. After anxiously observing the boy's health for several days, Hill tasted the nuts himself and from that day became an ardent enthusiast.[27]

On 5 March 1867, the *Brisbane Courier* published a letter from him that changed macadamia history.

> SIR, as you are doubtless aware, the varieties of Australian edible fruit are very few and far between, and when anything of the kind is discovered and can be made available I think it only right that the fact should be made public through the press. Some years ago I discovered in the coast scrubs a small but rather handsome tree, the botanic name of which was *Macadamia ternifolia*; but I was not aware until recently that it bore an edible fruit and singular to say the Aborigines appear to have been equally ignorant. The tree is now bearing abundantly in the scrubs both to the southward and northward and in several localities the children of the Europeans, as well as those of the black, are regularly employed in gathering the fruit for the purpose of food. The fruit of which I send you a sample for your own examination, is about the size of walnut, and contains within a thick pericarp a smooth brown coloured nut embracing a kernel of a remarkably rich and agreeable flavour, resembling in some respects that of a Filbert, but to my taste, much superior. I take this mode of addressing you with a view of affording information on a subject which is very little known indeed, in the hope also of drawing the attention of cultivators to the desirability of cultivating the tree, which I think, will greatly enhance its productivity, and the size and quality of its fruit.

> Yours, etc.

> Walter Hill, Botanic Gardens [28]

Responses to the letter revealed that the edible macadamia was already known and appreciated. A report in the *Queenslander* on 6 April 1867 pointed out that the nut was 'well known to timber getters, natives and others, and quantities are being daily gathered and eaten, thus proving its wholesome qualities'.

In a letter to the *Brisbane Courier* on 17 April 1867, Charles Prentice informed readers that the tree 'existed in considerable abundance, though circumscribed in locality, on high, exposed ground about 10 miles nearly due south of Brisbane; the ground is covered with the fallen nuts, most of which are perforated, and growing saplings are springing up under the old trees' (this site may have been Sankeys Mountain).

'Pomona' wrote to the *Queenslander* on 16 December, pointing out that 'the Queensland nut is already in our gardens, and bearing fruits under conditions favourable to its permanent improvement'.

Hill also sent his letter to newspapers in New South Wales, Victoria, South Australia and Western Australia, where its publication drew attention to the edible macadamia. From then on, macadamias were shown at colonial exhibitions and recommended for planting in private and botanic gardens. On 5 November 1867, the *Sydney Morning Herald* pointed

Left. Walter Hill Macadamia Tree, Brisbane Botanic Gardens. It is possibly the most photographed and best known macadamia tree in the world. Credit: J. McConachie.

out that nuts collected around Lismore (NSW) had been planted in the Sydney Botanic Garden three years earlier. A further report, on 3 February 1868, stated that 'within the year plants were available from commercial nurseries' (these would have been the edible *M. tetraphylla* from New South Wales).

Hill became inordinately proud of the macadamia, describing it as 'one of God's gifts to Queensland'.[29] There is an *M. tetraphylla* tree close to his 1858 tree, also believed to have been planted by Hill, and both are alive and healthy at the time of writing. He germinated edible macadamia nuts and grew them in metal cans. To ensure their distribution, he gave plants to friends and sent them to botanic gardens in other colonies. Probably he used the newly invented Wardian Cases – now called terrariums – to transport live plants. He rowed out to ships anchored in the Brisbane River, ascertained their destinations, and requested (and often paid) the captains to take plants to overseas countries.[30]

Reports had led botanists to believe the bitter *M. ternifolia* was poisonous. But it is probable that Aboriginal people told Europeans not to eat the kernel because it was bitter, which was misinterpreted by settlers to mean that it was poisonous. A CSIRO Masters project in 1995 assessed that a lethal amount of *M. ternifolia* kernel was approximately one kilogram.[31] It is impossible to eat even a minute amount, due to its bitterness.

Baron Ferdinand von Mueller

(1825–1896)

In 1848, a short, frail young man disembarked in Adelaide and began work in a chemist shop. In his subsequent career, he became recognised as a world-famous botanist and was commemorated by the issue of a special postage stamp. In 1871, the King of Württemberg created him a hereditary Baron and Queen Victoria honoured him as a Knight Commander in 1879. He received at least twenty knighthoods from other countries, becoming the most decorated citizen Australia had known. Mueller was a man of contradictions, but no one could

Baron Ferdinand von Mueller, Australia's greatest botanist. Credit: Mitchell Library, State Library of New South Wales, Sydney.

deny his skills, his prodigious work ethic or his devotion to the advancement of science in his adopted country.[32]

Ferdinand Jakob Heinrich Mueller was born in Prussia, Germany, and from an early age had a love of nature, particularly its plant life. After losing his parents and a sister with tuberculosis, he and his two surviving sisters sought better health in a warmer climate. In Australia, he embarked on a career that would have exceeded even his lofty ambitions.

After several years in South Australia, where he tried farming, he moved to Victoria. In 1853, Sir William Hooker from Kew Gardens recommended him as Victorian Government Botanist. He was based at the Melbourne Botanic Gardens, which had been established in 1846.

Mueller applied himself vigorously and undertook many explorations to collect botanical and animal specimens. In July 1855, he left Sydney by ship for Darwin as the botanist on A.C. Gregory's Northern Expedition. Stopping at Brisbane, he briefly explored the local area and Moreton Island but was unlikely to have collected the macadamia then.

In June 1856, after botanising around Darwin, he travelled overland on foot and horse, reaching Brisbane on 16 December. He had observed at least 2,000 plant species, of which almost 800 were new to science and about 500 were native only to Australia.

Until his departure from Brisbane on the steamer *Boomerang* on 29 December 1856, Mueller botanised around Brisbane with Walter Hill. On a trip to the Glass House Mountains, they observed macadamia trees on the Pine River. According to Mueller, Hill later supplied him with a specimen of *M. ternifolia*, which he then described and named.

On 5 August 1857, at a meeting of the Philosophical Institute of Victoria, Mueller presented thirty-two plant specimens, which had 'no other claim on your attention but their novelty'.[33] He said that some of the plants exhibited were 'selected from a Herbarium formed by Mr Hill, the Superintendent of the Brisbane Botanic Gardens, a gentleman of keen observation and great ardour for botanical research'. These plants included two of the Proteaceae family, which he named *Grevillea hilliana* (after Walter Hill) and *Macadamia ternifolia*. Stating that he found the latter in 'the forests of the Pine River of Moreton Bay', he dedicated it to 'John Macadam Esq. MD, the talented and deserving Secretary of our Institute'.[34] The following year, when the *Transactions* of the Institute were published, there were drawings of the two newly described Proteaceae. The first shows the flowers, leaves and fruit of *Grevillea hilliana*, incorrectly labelled as *Macadamia*

ternifolia. The second shows two macadamia leaves, one with serrated, the other with smooth leaf margins, also titled *Macadamia ternifolia*, although the smooth leaf may well have been collected from the true *Macadamia integrifolia*.

In 1977, Dr D.M. Churchill, the Victorian Government Botanist, wrote that Mueller's specimen was collected by Mueller and Hill in January 1857, was held at the Melbourne Herbarium, and consisted of a small flowering branch with flowers but no fruit.[35]

In 1860, Mueller had apparently forgotten that he had described and named the macadamia. Using the overlooked Leichhardt specimen, he named it as part of the South East Asian Proteaceae genus *Helicia*, species *ternifolia*. The English botanist George Bentham (1800–84), working on his seven-volume *Flora Australiensis* at Kew, saw that Mueller had been right the first time and reinstated the genus *Macadamia*. Mueller, although displeased not to have been given the job himself, provided all the specimens and descriptions that made the *Flora* possible, and it was a complete and credible record.[36]

Early Australian botanists must have felt overwhelmed at times by the rich variety of new plants, and numerous mistakes were made in describing and distinguishing their finds. In the years following Mueller's description, all southern species of macadamias were referred to as *Macadamia ternifolia*, and it took nearly a century to separate them correctly.

For the next twenty years, Mueller showed only a limited interest in the macadamia. He continued to research and study nature, seeking plants that would serve humankind. He backed the tragic Burke and Wills expedition, which was the first to cross the continent from south to north. His prodigious writings included 150,000 letters, 800 books and major articles on botany.[37] He named 2,000 species of plants and supported the acclimatisation of imported plants.[38]

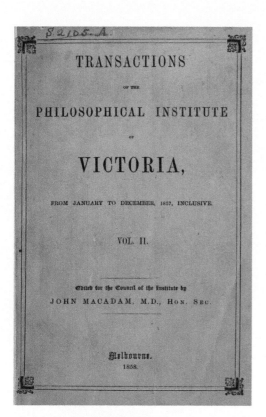

Cover, *Philosophical Institute of Victoria Transactions, 1857* (1858)

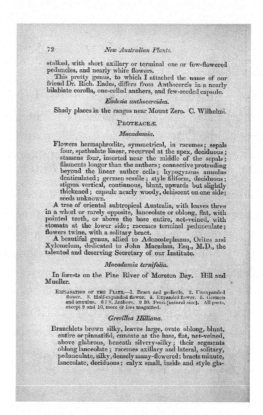

stalked, with short axillary or terminal one or few-flowered peduncles, and nearly white flowers.

This pretty genus, to which I attached the name of our friend Dr. Rich. Eades, differs from Anthocercis in a nearly bilabiate corolla, one-celled anthers, and few-seeded capsule.

Eadesia anthocercidea.

Shady places in the ranges near Mount Zero. C. Wilhelmi.

PROTEACEÆ.

Macadamia.

Flowers hermaphrodite, symmetrical, in racemes; sepals four, spathulate linear, recurved at the apex, deciduous; stamens four, inserted near the middle of the sepals; filaments longer than the anthers; connective protruding beyond the linear anther cells; hypogynous annulus denticulated; germen sessile; style filiform, deciduous; stigma vertical, continuous, blunt, upwards but slightly thickened; capsule nearly woody, dehiscent on one side; seeds unknown.

A tree of oriental subtropical Australia, with leaves three in a whorl or rarely opposite, lanceolate or oblong, flat, with pointed teeth, or above the base entire, net-veined, with stomata at the lower side; racemes terminal pedunculate; flowers twine, with a solitary bract.

A beautiful genus, allied to Adenostephanus, Orites and Xylomelum, dedicated to John Macadam, Esq., M.D., the talented and deserving Secretary of our Institute.

Macadamia ternifolia.

In forests on the Pine River of Moreton Bay. Hill and Mueller.

EXPLANATION OF THE PLATE.—1. Bract and pedicels. 2. Unexpanded flower. 3. Half-expanded flower. 4. Expanded flower. 5. Germen and annulus. 6 7 8, Anthers. 9 10. Fruit (natural size). All parts, except 9 and 10, more or less magnified.

Grevillea Hilliana.

Branchlets brown silky, leaves large, ovate oblong, blunt, entire or pinnatifid, cuneate at the base, flat, net-veined, above glabrous, beneath silvery-silky; their segments oblong lanceolate; racemes axillary and lateral, solitary, pedunculate, silky, densely many-flowered; bracts minute, lanceolate, deciduous; calyx small, inside and style gla-

Description and naming, *Philosophical Institute of Victoria Transactions, 1857* (1858), p. 72

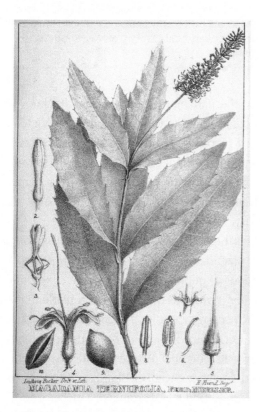

Mueller mistakenly confused this specimen, which is *Grevillea hilliana, Philosophical Institute of Victoria Transactions, 1857* (1858).

Two leaves collected, which are probably the species *M. ternifolia* and *M. integrifolia, Philosophical Institute of Victoria Transactions, 1857* (1858).

John Macadam (1827–1865)

Who was Australia's John Macadam? He was not, as is often stated, the famous Scottish engineer who developed the process of making stable roadways, but a Scottish doctor who came to Melbourne in 1855 and achieved much in his brief life.

Born near Glasgow in 1827, Macadam studied chemistry and then medicine. In Melbourne, he taught chemistry and natural science at Scotch College and was one of the first lecturers at Melbourne University's School of Medicine. In 1858, he was appointed Analytical Chemist to the recently formed Victorian Government. From 1857 to 1862, he served as Honorary Secretary of the Philosophical Institute of Victoria and in 1863 he was appointed Secretary of the committee that organised the ill-fated Burke and Wills expedition. A Member of the Legislative Assembly in the Victorian Parliament, he served as Postmaster-General in 1861. In 1865, while travelling to New Zealand to give evidence as a forensic scientist, he was injured and died of pneumonia.[39]

While a teacher at Scotch College, Macadam believed that physically invigorating sport would help to temper his students' 'energies'. Aware of a version of Scottish football that could be described as carrying the ball, he assisted Tom Wills to cobble together a mix of rules that developed the running style of football that today is known as Australian Rules (AFL). On 7 August 1858, the first game of Australian football was played between Scotch College and Melbourne Grammar near the present Melbourne Cricket Ground. The game, with forty players to a side, was played on a field 600 metres long, with eucalypt trees as goal posts. Macadam was an umpire in the competition, which took three weekends.[40] It has been suggested that the Gunditjmara Aboriginal people may have had a similar open game called Marngrook.

Although John Macadam saw the specimen of macadamia that Ferdinand Mueller named after him, it is unlikely that he would have seen a macadamia tree or had the pleasure of tasting its nuts. How-

Dr John Macadam. A great Australian in many ways, but one who never saw or tasted a macadamia. Credit: Mitchell Library, State Library of New South Wales, Sydney.

ever, the naming of the macadamia acknowledges his contribution to the development of Victoria.

Once recognised, the macadamia was increasingly planted in residences, botanic gardens and small orchards. During the twentieth century, it came from obscurity to develop as a commercial Australian industry.

Further research may provide information about the macadamia's 'discovery' by Europeans. While Aboriginal people were fully aware of the macadamia for thousands of years, it is probable that an unknown convict, settler, botanist or official was the first to be shown the nut, taste the kernel, and recognise its potential. At present, 'discovery' must be accorded to Ludwig Leichhardt in 1843, its naming and description to Ferdinand Mueller and Walter Hill in 1857, and its promotion and distribution to Walter Hill from 1867.

Part 2

DEVELOPING AN INDUSTRY

Chapter 5

INDUSTRY PIONEERS, 1860–1900

Realising the potential and planting the first orchards

But the most interesting of all is the macadamia ternifolia [sic], or Queensland nut, a beautiful tree unknown to the average Australian. Its nut is the sweetest and most nutritional in existence. When once tasted it will be preferred to all others, almonds, walnuts, Alberts, Brazilian nuts – all giving place to this, their superior. It is now being cultivated in California, and will probably be imported to this country for sale, without one in a thousand knowing it is Australian nut.

('R.B.' writing in *Sydney Morning Herald*, 30 April 1904)

By 1860, the macadamia had been discovered, described and named, but botanists had not realised that its nuts were edible. To most Australian colonists, the tree was either unknown or insignificant.

In 1860, Australia consisted of six separate colonies, all owing allegiance to the Mother Country, Great Britain. Sydney, Hobart, Melbourne, Adelaide, Brisbane and Perth were growing centres, surrounded by sparsely populated districts. The discovery of gold in Victoria and New South Wales in the early 1850s had created wealth and encouraged immigration.

Brisbane was opened for settlement from 1838, and from the 1840s, timber-getters and their families had been moving into macadamia country. In 1842, Brisbane was opened to free settlers, and European occupation steadily and irresistibly expanded. Settlers cleared the land of trees, often destroying the wild macadamias, which generally grew in more sought-after areas. Macadamias survived along creek banks and in steep and inaccessible areas unsuited to farming.

Cedar-cutters reached the Richmond River, in the Northern Rivers region, by 1842, and as early as 1844 ships from Sydney began to bring supplies and people into river ports, returning with timber. Until the 1860s, the Big Scrub, 74,000 hectares of almost continuous rainforest, remained largely undisturbed, and further north only small sections of forest had been cleared. In 1867, the *Sydney Morning Herald* reported that seed nuts collected from Lismore had been planted in the Sydney Botanic Garden by 1864.[1]

Squatters moved into the valley of the Mary River at Tiaro near Mount Bauple in 1843, and Maryborough was first settled in 1847. Macadamias, known locally as 'Bauple nuts', were then planted in Maryborough residences probably from the 1850s onwards. In the late nineteenth century, sugar-cane growing and a sugar mill were established at Mount

Bauple and many of the farmers were German. Often they and their families would collect nuts from the wild trees and even cut out competing trees to allow some of the trees to flourish. In some years, these settlers may have held a Macadamia Festival.[2] In 1977, Alice Graham wrote that her mother, born in Tiaro in 1874, remembered as a child coming down from Maryborough in a dray and filling it with nuts, dehusking them, drying them in the sun and selling them in Sydney.[3]

The creeks and rivers of the Gold Coast were searched by timber-getters in the 1860s. *M. tetraphylla* still grows prolifically along some stretches of Tallebudgera and Currumbin creeks, and it would be surprising if these macadamias had not been observed by the timber-getters. At least in the Logan and Albert rivers districts, the Aboriginal leader King Jackey Jackey (also known as Bilin Bilin and Kawae Kawae) traded macadamia nuts for European goods.

Early reports often commented on the macadamia's fine flavour, texture and potential. However, the nuts had thick, hard shells and were plagued by numerous insect pests. Their slow development as a commercial crop was due to lack of knowledge of their horticultural requirements and the inability to graft to replicate desired characteristics. The

Typical residential macadamia, planted well over a hundred years ago. Credit: I. McConachie.

macadamia industry would take another one hundred years to develop in Australia. Ironically it took another country, Hawaii, to see the nut's potential and take the sound business and horticultural steps needed to commercialise it.

Early reports

Walter Hill planted a macadamia in the Brisbane Botanic Gardens in 1858, and nine years later discovered to his surprise that its nuts were edible. Possibly William Macarthur had planted one even earlier at Camden Park Gardens, as discussed below. In 1861, Hill published a list of food plants growing in the Brisbane Botanic Gardens and did not include macadamias,[4] but from 1867 his reports created awareness of the nut's potential.[5]

A newspaper report in 1867 described him, with characteristic energy, germinating macadamias in a seed bed in a glasshouse with the intention of distributing plants.[6] In 1870, he sent macadamia seed nuts, along with other plants, to Hong Kong, Ceylon (Sri Lanka) and Bordeaux.[7]

In a report to the Queensland Government in 1875, Hill advised that a macadamia in the Botanic Gardens had fruited at five years of age.[8]

Macadamias were also planted in Toowoomba at the Queens Park and Botanic Gardens under the management of Mr R. Harding. A report in the *Brisbane Courier* of 15 May 1897 stated that they were 'fruiting plentifully' and further, 'The nuts are said to have won the appreciation of Lord Lamington when recently brought under his notice by Mr Harding – who, by the way, has great hopes for the future of this fruit'.[9]

The Queensland Acclimatisation Society, later subjected to much criticism, played a role in recognising and distributing the macadamia. ('Acclimatisation' refers to the adaptation of plants and animals to different environments.) After creating environmental disasters by introducing sparrows to Brisbane and rabbits to the Moreton Bay Islands, their zeal was tempered and they achieved success by cultivating crops, including macadamias. They were granted 30 acres near the Brisbane Exhibition Grounds, a small part of which still exists as Bowen Park, named after their first patron, Governor Sir George Bowen. Moving to Lawnton in 1905, they acted as an experimental and wholesale nursery, which included encouraging the growing of macadamias. The Society took on the role now played by the Queensland Government's Department of Agriculture, and after this body was established, the Society slowly declined.[10]

As described in Chapter 4, a specimen of *M. integrifolia*, labelled 'Camden Haven 1850–1860 C. Moore', was grown at the Sydney Botanic Garden. Camden Haven, near Port Macquarie, is 300 kilometres south of the known range of *M. tetraphylla*, and 500 kilometres south of the range of *M. integrifolia*. In the 1850s, Camden Haven consisted of an abandoned military hut and a base for timber-cutters. Could the tree in the Royal Botanic Garden have been mislabelled 'Camden Haven', when actually it came from Camden Park?

Removed over one hundred years later, the tree had likely been mislabelled and come from private botanic gardens at Camden, south-west of Sydney, owned by Sir William Macarthur at the time when Charles Moore was Curator at Sydney. These early botanic gardens were established from 1835 by Sir William, the son of John and Elizabeth Macarthur.

If the Royal Botanic Garden tree was sourced from Camden Gardens, it could be the first recorded macadamia and was probably planted earlier than the Walter Hill Macadamia. Given the dates, it could have come from John Carne Bidwill's collection at Maryborough. Supporting this, there is an *M. integrifolia* in the 1840s section of Camden Park Gardens for which there are no planting records. Genetic testing suggests that this tree is related to trees that grew in the Gympie region as well as trees around Brisbane.

It is known that Bidwill (who was Commissioner of Lands at Maryborough from 1848 and also a botanist) collected plants or seed and sent them to Sir William, but there is no record that they included macadamia. If the date of planting for the macadamia at Camden House could be confirmed, it might

again be the first domesticated macadamia.

As for the Sydney Botanic Garden, Charles Moore kept limited records and after the regrettable loss of many files and artefacts as well as plants in the 1882 fire that destroyed the immense Garden Palace (built for the Sydney International Exhibition of 1879), the Garden deteriorated.

At the Alstonville Sub-Tropical Fruit Research Station in New South Wales, a possibly wild, historic *M. tetraphylla* could be traced back to the 1860s. Tim Trochoulias, a horticulturist from the Station, reported that an old letter had stated that nuts were collected from this tree in 1872. Consultant historian Arthur Lowndes reported that Drew Leigh, the Director of the Station was told of an early farmer who had chipped weeds from around this tree in the 1870s.[11] Sadly, in recent times, an ill-informed bureaucrat from Sydney instructed that this tree be cut out to provide space for other trials.

Tim Trochoulias with *M. tetraphylla*, probably a wild tree at the New South Wales Sub-Tropical Fruit Research Station at Alstonville, dating back to at least the 1860s. Photo taken in 1978. Credit: I. McConachie.

In 1869 and 1870, accounts of the edible macadamia were reported in an English newspaper.[12] These accounts, illustrated with drawings of the nut and kernel, revealed that the macadamia was introduced to English gardens by W. Bull and was exhibited by E.G. Henderson and Son. Specimens were shown to the Royal Society by Dr Joseph Hooker, who described the nuts and their flavour as being 'fully equal to the Kentish filbert' (known today as the hazelnut). Since macadamias can only tolerate light frosts, their survival in England would have depended on being grown in a hothouse.

In 1938, Gympie pioneer Zachariah Skyring Jnr recalled that, in the 1870s, 'Mumbeanna', his family's property at Mooloo, was 'a rich district for blacks; bunya and Queensland nuts were plentiful in the season'.[13] Humanitarian concerns about the survival of Aboriginal people, because their native foods were being lost, led to the Queensland Government's gazettal in 1870 of Timber Regulations which prohibited the cutting-down of bunya nut and Queensland nut trees.[14] These regulations were re-gazetted in further Acts in 1877 and 1884.[15]

The *Capricornian* newspaper on 25 March 1876 reported: 'Large quantities of the Queensland Nut were being brought down from the Boppel Range by the Aborigines and offered for sale at Gympie and Tiaro but only a comparatively small number have as yet found their way to town'.[16] The article (reproduced in a number of papers on the east coast) expressed concern that the trees were rapidly becoming extinct and urged anyone desirous of perpetuating the tree that produces such a delicious nut to plant a tree. Walter Hill expressed concern that South Sea Islanders brought to Queensland as indentured labourers were cutting down Bauple nut trees to get nuts.[17]

In 1888, on a trip to Sankeys Mountain, south of Brisbane (now in Whites Hill Reserve), field naturalists described young boys collecting 'Queensland nuts'. Much of this mountain is now a quarry and only a tiny area of scrub is left. In the early 2000s, there were possibly thirty healthy wild macadamias there but by 2023, only two remained. This was the best known site of wild macadamias close to Brisbane in the nineteenth century.[18]

From the 1870s, the New South Wales Government produced basic advice sheets about the macadamia. In 1893, the Department of Agriculture produced a booklet, *The Cultivation of the 'Australian Nut'* by Fred Turner,[19] which detailed all aspects of the nut and its culture, and stimulated interest in the planting of macadamias.

Turner's description was quoted later that year in an article, 'The Queensland Nut', in the *Brisbane Courier*: 'each nut produces one or two kernels of excellent flavour, resembling, but superior to the filbert'. He commented on the desirability of bringing it under cultivation and planting more in gardens, and described how to germinate, grow and plant seedlings. He noted the advantage of its long shelf-life and the disadvantage of its hard shell.[20] This is early evidence of the losing battle to have macadamias known as the 'Australian nut' rather than the more parochial 'Queensland nut'.

Turner was enthusiastic about the macadamia's potential as an economic plant. In 1911, in the *Sydney Morning Herald* he described the macadamia as a 'beautiful, evergreen, nut producing tree, which produces abundant crops, which contain a remarkably rich edible kernel of excellent flavour resembling but superior to the filbert'. He said he had planted a macadamia in Hyde Park, Sydney, several years earlier and gave the opinion that 'it was difficult to understand why the tree had not been more extensively planted'.[21]

Frederick Manson Bailey (1827–1915) had a commercial seed business in Brisbane in 1861 and became responsible for the botanical section of the Queensland Museum, then Colonial Botanist in 1881, a position he held until just before his death. His explorations added to plant knowledge, and he published a large number of botanical lithographs, including some featuring macadamias, which depicted plant features more clearly than photographs of the time. These were printed and made available.

Joseph Maiden (1859–1925) and Ernst Betche (1851–1913) were systematic botanists in New South Wales. Between 1899 and 1905, and as a separate part of their extensive study, *The Queensland Flora*, they described a new species of macadamia, *M. integrifolia* – as distinct from *M. ternifolia* – and as a type specimen. They considered this species as the basis of the genus. In 1889, Maiden stated that the macadamia was 'well worth extensive cultivation as the nuts are eagerly bought'. [22]

Early growers

The Petrie family

Andrew Petrie (1798–1872) was the patriarch of a pioneering Queensland family, who later contributed greatly to the macadamia industry. He arrived at the Moreton Bay Penal Settlement in 1837 as Foreman of Government Works. A stonemason by trade, he constructed many of Brisbane's early public and commercial buildings.

In May 1842, Petrie, accompanied by an exploring party, left Brisbane in a whaleboat to travel north up the Queensland coast. They ventured up the Wide Bay (now Mary) River, reaching present-day Tiaro.

Jan Petrie Hall, a Petrie descendant, believes, based on family stories told when she was young, that Petrie also found the macadamia.[23] This was never recorded, possibly because the plant's significance was not recognised.

Tom Petrie (1831–1910), a son of Andrew Petrie, planted an *M. integrifolia* at his family home in 1866 at Murrumba, which is now in the suburb named after the family. This tree has been photographed many times and remains both massive and healthy.[24] Tom

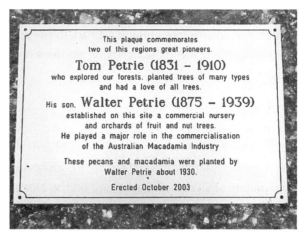

Plaque at the former 'Yebri' nursery and orchard commemorating the Petrie family. Credit: I. McConachie.

is also reported to have planted macadamias on his separate farm at Murrumba in 1866, from seed nuts from trees in Buderim given to him by Aboriginal people. Only the 'true' *M. ternifolia* is endemic to Buderim, but there is no evidence that he planted this small, inedible macadamia. Possibly Aboriginal people from the Buderim district collected *M. integrifolia* from the Blackall Range, a little to the west. Tom's son and grandson continued the family passion for the macadamia.

Dr Joseph Bancroft (1836–1894)

Dr Bancroft arrived in Brisbane in 1864 and built his home, 'Kelvin Grove', near Enoggera Creek, where his gardens are now preserved as Bancroft Park. Apart from his medical practice, he was an avid agricultural experimentalist and amateur botanist, who contributed to the commercialisation of many crops. Regarding macadamias, he stated in 1899, in a letter to a journalist at the *Australian Town and Country Journal*:

> Some years ago I planted two Australian nut trees at Kelvin Grove. Both give immense crops of nuts every year and some hang on for months, and give nuts the greater part of the year. If struck carefully with a knife and mallet on the small spot, and along the furrow, the two halves separate as when the seed bursts them. If warmed a little on the fire-range the kernels shrink slightly and then fall out readily. They should be brought to table so prepared. When one finds such splendid yields from trees of ten to twenty years old, one regrets not to have planted more.[25]

HERE IS A QUEENSLAND NUT TREE 65 YEARS OLD.

Tom Petrie Tree at Murrumba, 1931. Credit: unknown.

In the early 1900s, the local council cut down a large macadamia tree in Bancroft Park to prevent children climbing it and risking injury. No trees now remain.

'Restdown', Sunshine Coast

Do the words 'a grove of macadamias' make 'Restdown' the world's first macadamia orchard? In 1872, brothers James and Edward Smith, successful gold miners at Gympie, bought land on the Noosa River between Lake Cooroibah and Lake Cootharaba. Wishing to create an apiary, they planted a wide range of fruit trees at 'Restdown' and called their business an 'orangery'. A detailed report in 1895 describes their 'grove of macadamias'. Most trees were young, but one mature tree was stated to regularly produce over 10 kilograms of nut-in-shell annually, so some trees may have been planted in the 1870s.[26]

In 1892, 'Restdown' was sold to John Johns and has remained in the Johns family for several generations. In 2012, at least ten very large trees remained, slowly being overwhelmed by creepers. In a raised seed bed, which Alan Johns said existed when his grandfather bought 'Restdown', grew macadamia trees which have survived for at least 120 years. They grew about 25 centimetres apart and all had diameters of about 25 centimetres. The Johns family also planted between fifty and one hundred trees on an adjoining property known as 'Johns Landing'. About fifteen of these trees survive.

An historic macadamia in the Gold Coast hinterland is the Jordan Tree, known for decades as the 'Mother of all Macadamias'. In 1892, the Lahey family, who had settled at 'Sunnyside', Hotham Creek, near Pimpama in 1872, were visited by Robert Jordan and his niece Queenie.[27] Impressed by the wild nuts growing on the property and in the district, Jordan took half a sugarbag back to his brother Edward in Hawaii.[28] It was thought that these trees may have been the source of the cultivars later developed in Hawaii, but Dr Craig Hardner and Dr Cathy Nock's studies on the 'Domestication of the Macadamia in Hawaii', did not find any DNA linkage between Hawaiian cultivars and the Jordan Tree or other wild trees in the Hotham Creek district.[29] Sadly the tree was lost to fire around 2020, although there are shoots growing from the stump.

From the early 1880s, James Collins planted citrus and macadamia trees at 'Mt Carmel Orchard'

One of the remaining trees at 'Restdown', planted after 1870s. Credit: I. McConachie.

The Jordan Tree at Willowvale, Hotham Creek, recently lost to fire and believed to be the source of nuts sent to Hawaii in 1892. Credit: I. McConachie.

around what is now the oldest pioneer house in Redland Bay, south-east of Brisbane. It is unlikely that he planted enough trees to be considered a commercial orchard, but Tom Petrie and his son Walter considered Collins' nuts to be of high quality and sought them for their nursery.[30] In 1954, botanists from Hawaii, led by Dr John Beaumont, took cuttings of some of Collins' trees. A tree at Redland Bay, described as being large enough for children to play under in 1890, was either a wild tree or one that had been planted by early settlers.[31]

New South Wales

There are reports from the Richmond River Historical Society that in 1870 Jim Armstrong from Dorroughby, near Lismore, planted a small number of *M. tetraphylla* trees. He served in the Sudan War and on his return was known as 'Sudan Jim'. There is some uncertainty about the date, but he was certainly one of the first settlers to plant macadamias.[32]

What is considered to be the world's first commercial orchard was planted at Rous Mill, near Lismore. Charlie Staff selected the property during the 1880s and planted an orchard of 'Bush Nuts' (*M. tetraphylla*), initially consisting of 213 seedling trees on two acres. Later he adapted an old corn-sheller to remove the husks and spread the nuts on sheets of iron in the sun to dry.

'Rous Mill Orchard' was sold to Mr Austin, then to Mr Collard and subsequently to Mr Fredrickson, who doubled its size. It remained in the Fredrickson family until the 1990s, becoming famous for the suitability of its nuts for germination as nursery rootstock. Many of the industry's trees from the 1960s to the 1980s are grafted on to seedlings from 'Rous Mill'. Keith Fredrickson owned the orchard until the 1990s, but after it was sold, most of the trees were removed. Only four of these historic trees now remain.

Other growers were Richard Barlow at Teven Creek, near the Richmond River, in 1888 and W. Hansford, who had twelve trees growing at Dunoon.

Thomas George ('T.G.') Hewitt (1841–1915) and his son Norman Cowan Hewitt (1879–1956) were passionate about the potential of macadamias, and actively promoted their planting. Businessman T.G. purchased the Lismore newspaper *Northern Star*, which enabled him and Norman, a journalist, to expound on macadamias.

In 1890 they established an orchard of different types of fruit trees, including macadamias, and by the turn of the century this was one of only five macadamia orchards in New South Wales. T.G. Hewitt became actively involved in farming and reportedly sent seed nuts to the West Texas Nursery in the USA in 1892.[33]

In an article in *Walkabout* magazine in 1939, Norman Hewitt described twenty-six trees that, due to neglect, averaged only 62 pounds (weight) of nut-in-shell. Since 200 pounds (about 90 kilograms) per tree was expected, he considered this disappointing.

From 1890 to 1940, the Hewitts collected everything they could read about macadamias, and the Hewitt Collection, donated by Florence Valentine, a granddaughter of T.G. Hewitt, is an invaluable source for the early history of macadamias. The Hewitt family have contributed to the macadamia industry for several generations and, in 2016, Lindsay Hewitt and other family members were still macadamia growers.

'Rous Mill Orchard' in 1978. Generally considered the first certain commercial macadamia orchard in the world. Credit: I. McConachie.

Chapter 6

EARLY COMMERCIAL YEARS TO 1940

Enthusiasm, struggles and a cottage industry

The Queensland Nut is one of the finest edible nuts, at the same time the hardest to crack.The spherical shell is iron-like, thick and smooth, and can be broken only by an irresistible hammer on an immovable surface. It is a nut particularly suitable for high-class confectionery, but the hardness of the shell is an effective obstacle to its commercialization.

(E.H.F. Swain, for Queensland Forest Service, 1928)[1]

In 1901, the six Australian colonies combined to become one nation, but recovery from the floods and economic depression of the 1890s and from the great drought of 1899–1902 took time. Many farmers and householders in Queensland planted both a mango and a macadamia tree in their backyards but only a few macadamia orchards were established in the early part of the century. Then concerns about marketing any increased production, together with the distractions of World War I (1914–18) and the Spanish Flu epidemic, resulted in a lack of confidence in the industry's future.

Farming in this period was basic, labour intensive, and performed by hand or with horse-drawn implements. There were no 'modern' fertilisers, no effective treatments for pests and diseases, and no herbicides to control grass and weeds. Destructive insects attacked the flowers and bored through the shell of developing nuts, even when they were fully

formed. All planted trees were seedlings, and thus had variable characteristics.

Nuts were dried in simple racks, on concrete floors, on wooden verandas or in the sun, but the need to fully dry them to obtain crispness and flavour was generally not recognised. The nuts usually had thick shells that were hard to open and had a low percentage of kernel. There was no commercial processing into kernel and consumers cracked them in a vice or by placing them in a small hole in a concrete floor and hitting them with a hammer. The kernels were either eaten raw or roasted in a frying pan with butter and usually liberally salted. Nuts of the *M. tetraphylla* and *M. integrifolia* species were sometimes mixed together, causing uneven roasting.

Wherever macadamias were planted, the nuts were appreciated and their reputation began to spread, even as far as England. After receiving a box of 'Queensland nuts' from Charles Staff of 'Rous

Typical orchard and standard during the 1930s. Credit: *Walkabout*, date unknown.

'Johns Landing' on the Noosa River, planted in the early 1900s. Photo taken in 2011. Credit: I. McConachie.

Mill', King Edward VII, who had succeeded Queen Victoria in 1901, obtained information from Kew Gardens to learn more about them.[2]

In 1932, John Waldron wrote in the *Tweed Daily* that macadamias which had survived the severe winter of 1890–91 were growing in gardens in the south of France, as well as the Botanical Gardens in Hobart.[3] In northern New South Wales, the small, established orchards of seedling *M. tetraphylla* trees were starting to produce. These included 'Rous Mill', Sudan Jim's, T.G. Hewitt's, Barlow's at Teven and Hansford's twelve trees at Dunoon. In 1906, with an acre of bearing trees, G. Austin sold nuts locally for 6–7 pence per pound, and sent some to Sydney, where they retailed for 1 shilling per pound.[4] In Queensland, Johns' 'Restdown' apiary was marketing small quantities of nuts.[5]

New growers in New South Wales in the early 1900s were the Gaggin brothers at The Settlement, Mullumbimby, and Kellie McCallum at Coorabell, who developed a nutcracker and a kernel oil roasting process.[6] In Queensland, Fred Adams planted between ten to twenty macadamias in his orchard on the Ebersleigh Estate at Kowbi, near Childers, and these trees were still growing well in the 1940s.[7] Between 1904 and 1910, Ernest Fisher planted thirty to forty *M. tetraphylla* trees in a mixed orchard near Coolum, which cannot be traced.

Other growers in this period were the Benfer family at Bahrs Scrub, near Beenleigh, E. O'Mara at Bauple, and Wragge and Staid at Maryborough. At the delightfully named Foreign Legion district at Rocky Hole Creek, Imbil, the new Queensland Forestry Department planted twelve seedling macadamias around 1920 to assess their potential for timber and veneer. These trees were still healthy in the 1980s.[8]

The small crop marketed from these orchards was supplemented by nuts collected from wild and backyard trees. Most nuts were only partly dried and were of variable quality. They were mainly sold as nut-in-shell (NIS) through produce merchants in Sydney and Brisbane, and in smaller towns near where they were collected.

Many older residents recall that, as children, they picked up nuts and sold them. As a young girl at Maleny in the early 1910s, Marjorie Puregger collected 'Queensland nuts' and sold them for pocket money.[9] Miss S.G. Fitzgerald recalled a bearing 'Queensland nut' tree on a property bought by her father in Clayfield, Brisbane, where, during the 1920s, his children collected and sold the nuts. In her history of the Tweed Valley, *Forest Bountiful,* Penny Watsford recorded that children at Uki and Limpinwood knew where 'Bush nuts' were plentiful, which was often on dairy farms.[10] When Hazel Long celebrated her 100th birthday in 2021, she reminisced that, as a young girl living in Dagun, she had collected macadamia nuts in season and sold them through a convenience store in Channon Street, Gympie.[11]

In 1907, a party of citizens and Members of Parliament, who were visiting the Amamoor Creek district in connection with a proposed railway in the Mary Valley, were presented with 'Queensland nuts', 'for a consideration', collected by local Aboriginal people. The journalist who reported this incident commented that 'it was somewhat remarkable that many of the visiting members of Parliament had not previously tasted this fine, edible nut'.[12]

From the early 1920s, most macadamias in New South Wales were planted on steep slopes as small commercial ventures, often in conjunction with bananas. Accepting that bananas had a life of about seven years, some farmers interplanted macadamias in the belief that they would be bearing and providing income when the bananas died. However, due to insect pests, bushfires and low yields, the macadamias were often abandoned. Farmers who interplanted *M. tetraphylla* seedlings collected from local trees included E. Blanch, T. Leadbetter, J. Wilson and P. and F. Crawford.[13]

MACADAMIA NUT TREES INTERPLANTED WITH PAPAWS.

Macadamias and papaws growing together in Queensland. Credit: *Queensland Agricultural Journal,* June 1939.

Early growers, New South Wales

Jack Middleton (1919–2004) grew up at The Channon near Lismore. He remembered a neighbour, Duncan Curry, who had planted two *M. tetraphylla* trees around 1920, telling him that these nuts would become well known and widely planted one day. During the Great Depression of the 1930s, Jack collected nuts from backyard and wild trees and made more money selling them than by receiving government benefits. In the late 1930s, he grew a few macadamias, but his main interest commenced in 1945 at 'Dorroughby', a property formerly owned by 'Sudan Jim' Armstrong, who had planted a small number of trees in the 1870s. In 2010, at least one of Sudan Jim's trees remained.[14]

Later, Norm Greber, whose story is told below, gave Jack Middleton experimental varieties to plant, and inspired him to develop his own grafting technique and plant a wide selection of cultivars to assess. Jack spread the nuts on the veranda to dry, and

John Waldron in 'Stoke Orchard' with Mount Warning in background, in the Tweed Valley in the early 1930s. Credit: E. Cheel and F.R. Morrison, *The Cultivation and Exploitation of the Australian Nut,* 1935.

his children were expected to walk over them (in shoes) to break up the husks.

Keith Parry, who grew up at Stony Creek near Nimbin in the 1930s, remembered that his father planted over 300 macadamias of both species, and sold his nuts to John Waldron. In 1933, Mr Pfluckis grew bananas at Tomewin in the Tweed Valley. He planted *M. tetraphyllas* separate from his bananas, which was not the usual practice. In 1940, this orchard, containing 475 trees, was bought by Gordon Rodesky. In 2005, about 200 trees remained, making this one of the few Tweed orchards to survive. Rodesky built a novel dehusker, which is displayed at the Tweed Heads Historical Society.[15]

In the 1930s, George Gaggin interplanted macadamias with bananas on 40 acres at the Upper Main Arm of the Tweed Valley. He claimed that the macadamias were profitable. The New South Wales Department of Agriculture undertook trials to control insect pests using nicotine sulphate and arsenate of lead, and when oil was added to assist coverage, sound control was achieved.

Seedlings of *Macadamia ternifolia* planted between bananas, Brunswick River, New South Wales.

Macadamias grown in conjunction with bananas in Tweed River Valley. Bananas usually had a life of seven years and the macadamias were then cropping. Credit: E. Cheel and F.R. Morrison, *The Cultivation and Exploitation of the Australian Nut*, 1935.

Early growers, Queensland

From 1917, Alex Probert planted twenty acres of macadamias at Mapleton, on the Blackall Range. In the 1920s, he distributed seed nuts from promising trees to a number of growers. Several of his trees were chosen for assessment by the Department of Agriculture and Stock, and one selection, named Probert, was planted by some growers. Other Queensland growers included Fred Peachey, who planted over 150 macadamias at Cooloothin near Noosa, and Werner Laube, who planted trees at Upper Kandanga in the Mary Valley.[16]

In 1925, Jack Alcorn planted 400 macadamias of both *M. integrifolia* and *M. tetraphylla* at Witta, on the Blackall Range near Maleny. The *M. tetraphylla* seed came from Alcorn's relatives in the Lismore district. Eric Howard, who lived next door, managed the trees and had some of his own, of which L1, an *M. tetraphylla* selection, was assessed by the Department of Primary Industries. Jack Hurwood, who lived on the other side of the Alcorn Estate, also planted macadamias. In 2022, some of these trees remain. Other local growers were Andy Jacobson and Mr Collard.[17] Many large, old trees can still be seen in properties and along the roadsides on the Blackall Range.

In 1926, the Queensland Government germinated several hundred macadamia nuts for assessment at their Bribie Island nursery. John Waldron, from northern New South Wales, had sold them 200 seed nuts, some of which were claimed to have very thin shells.[18] This project, which was part of the winding-up of the Acclimatisation Society, appears to have been abandoned.

Commencing in the 1920s, Cecil and Morley Rickard planted 2.5 hectares at the Three Mile at Tinana, Maryborough, in conjunction with papaws, pineapples and tomatoes. They irrigated the trees by building a dam and using a Fordson tractor and overhead sprinklers. In the late 1940s, they used DDT to control insect pests. They introduced floating off the immature nuts in water. This orchard was eventually abandoned because the brothers believed they needed to invest more capital to make the venture viable.[19] There is no doubt they were ahead of their time.

A well-known identity from Mount Tamborine in the 1920s and 1930s was George Rankin Snr, who was active in selecting and growing macadamias. He sold seed nuts to South Africa, Fiji and America.[20]

Godfrey McCullough, a grower at Montville in the 1980s, reported that his grandfather, Charles Williams, obtained seedling trees from the Queensland Agriculture Department and planted five acres on the eastern side of Russell Island in Moreton Bay, in the 1930s. He claimed that his grandfather was one of the earliest Australians to process the nuts into kernels.[21] During the 1970s, Bella Slack had a grafted macadamia orchard on nearby Karragarra Island.

A member of the Queensland Parliament, William Drayton Armstrong, planted macadamias at his property 'Adare', near Gatton. These were bearing in the 1930s. At Woongoolba near Beenleigh, the Mark brothers had one acre of trees.

At Gilldora in the Mary Valley near Gympie, Vincent Fitzpatrick planted several hundred macadamias in the 1930s. In 1952, teenager Len Weller worked on the farm for pocket money and cracked nuts with a hammer and a modified corn-sheller. The kernels were given to ladies at the local Catholic Church, who used them in toffee to raise funds. This orchard no longer exists.

In 1935, David Tulloch, advised by Jack McGregor Wills from the Southport branch of the Department of Agriculture and Stock, developed an orchard at Mudgeeraba in the Gold Coast hinterland from seed nuts collected in the scrub, resulting in 3,000 seedling trees. He also supplied seed nuts to Fitzroy Nurseries in Rockhampton. This orchard was abandoned ten years later.[22]

Investment management schemes for macadamias commenced in 1933, when Mutual Estates Ltd in Brisbane promoted the growing of macadamias and offered to sell trees, supply the land, plant and care for the trees, pay 5 per cent interest, harvest and only retain 25 per cent of the crop proceeds. During the depression years, this attracted little interest but it was revived after World War II.[23] This proposal was supported by the Australian Nut Association, whose President, J.C. Sibbald, suggested that £500 ($1,000) from each Golden Casket lottery should be allocated to unemployed families to build a home on the land and plant half of it to macadamias.

Other sources of information about the cultivation of macadamias were a brochure by M. Barnes, Director of Fruit Culture for the Queensland Government in 1933, and a section on 'The Queensland Nut' by J.M. Wills, published in the *Queensland Agricultural Journal* in 1939. This was a detailed report on all known aspects of growing macadamias commercially.

In the 1930s, there were an estimated 20,000 residential macadamia trees in Brisbane and 30,000 between Grafton and Maryborough.[24] New growers were entering the industry, but losses through insect pests, bushfires or poor management resulted in the abandonment of many orchards.

Visionaries

The 1920s saw the emergence of visionary leaders, whose enthusiasm for macadamias never waned: they were John Waldron, Herbert Rumsey, Foreman Crawford, Norman Greber and Walter Petrie.

John Bucknell Waldron (1870–1960)

John Waldron was born and grew up in Brisbane and worked as a printer at the Government Printing Office in George Street. A competitive boxer who became Bantamweight Champion of Brisbane, he trained in the nearby Botanic Gardens. One day he sat under the Walter Hill Macadamia Tree and cracked a few nuts. From then on, his life was dominated by macadamias, and he became known as 'Nutty' Waldron.

With his fight earnings and savings, Waldron and his wife Elizabeth bought a 200-acre property at Eungella, west of Murwillumbah, which became 'Stoke Orchard'. The property had wild *M. tetraphylla* growing on it, and Waldron searched the local scrub for thin-shelled nuts, which he planted, believing they would produce true to type over several generations. By 1932, he had 500 thin-shelled, bearing trees and 1,500 other macadamias. He claimed success for producing thin-shelled nuts, sending samples to the USA, and receiving reports that these nuts were the finest they had ever tasted.[25]

Waldron collected his own nuts and bought others. He dried the nuts in the sun in his backyard on sloping, corrugated, galvanised-iron roof sheets. In fine weather, he took dried nuts from the bottom of the roof sheets and placed undried nuts at the top. He cracked the nuts with a hammer in his wife's laundry.

By 1934, he was drying nuts and kernels on racks in a metal drum, using a Primus kerosene stove. At a conservative estimate, he cracked six million individual nuts with a hammer. He improved on hand cracking by using a foot-operated pedal to drop nuts down a chute, one at a time, and opening them with a chisel activated by a lever. Late in life, he used a small mechanical cracker.

Waldron roasted and salted the kernels, packing them in 2-ounce (57-gram) glassine brown paper bags, which he delivered in a suitcase from his horse and dray. They sold for one shilling a bag from almost every store in Murwillumbah and Tweed Heads. He built a press to produce oil and patented Macadamia Nut Coffee, made from coffee beans and broken kernels roasted in macadamia oil. He roasted kernels with honey, and also made nut butter and relish. In 1937, he made a brooch for his wife out of macadamia shells. This was the start of a continuing hobby, carving the shells to make costume jewellery, buttons, Scout badges, chess sets and dress rings set with pieces of opal and sapphire.[26]

During the 1920s and 1930s, Waldron's nuts were much in demand. He was the industry's most tireless promoter, writing widely and lobbying politicians. In the 1940s, he even sent nuts to Australia's Governor-General, the Duke of Gloucester, and his wife. An eccentric, legendary figure, Waldron retired to Kingscliff, where his passion remained undiminished until his death at the age of ninety.[27]

John Bucknell Waldron cracking macadamia nuts at Kingscliff, New South Wales, 1957. Carter, Jeff, 1928–2010. Permission granted by copyright holder and NLA, Ref. NLARef185697 and HPE R24/22644. http://nla.gov.au/nla.obj-148668675.

Herbert John Rumsey (1866–1956)

Herbert Rumsey was a macadamia enthusiast who advocated a commercial industry. In 1895 he established Rumsey Seeds, a plant nursery near Sydney, where he imported, exported, and grew plants from seeds.

Rumsey believed that macadamia growing would prosper in many countries. In 1910, he exported 10,000 macadamia seed nuts to Hawaii, Texas, California and Florida, where the resultant trees may have been used to select many of the Hawaiian cultivars. In his 1927 book, *Australian Nuts and Nut Growing in Australia*, which assisted the commercial industry, Rumsey enthusiastically recommended macadamias and summarised contemporary knowledge. Acknowledging the nut's weakness to be its thick shell, Rumsey stated that growers such as John Waldron of Eungella, Mr Banner of Mullumbimby and Max Reynolds of Murwillumbah had produced thin-shelled nuts.

Rumsey was a leading promoter of the Australian Nut Association in 1932. He acknowledged the 'consensus of opinion' that the nut was one of the finest the world produced and prophesied that it would make 'a most valuable food product'.[28] After his death in 1956, his business was acquired by the well-known seed merchants, Arthur Yates and Co.

Foreman (A.F.) Crawford (1891–1993)

Foreman Crawford lived a very long life, providing confidence and support to other macadamia growers. In his book *Duck Creek Mountain now Alstonville* (1983), he praised the industry's other pioneers – Charles Staff, John Waldron, Norman Greber and Steve Angus – for their vision, faith and initiative. Maintaining a life-long interest in macadamias with small, ongoing plantings, he was rewarded by seeing an industry growing where he lived.

Foreman's family farmed in northern New South Wales at Duck Creek Mountain (Alstonville dis-

trict), moving from property to property to get established. His brother, Pearson Crawford, owned 19 acres near Alstonville, and by 1921 had planted 3 acres of *M. tetraphylla* among his bananas. Foreman, who had commenced planting macadamias on his own farm in 1920–21, bought Pearson's property in 1925 and planted more *M. tetraphylla*. In 1931, he planted some seed nuts he had been given. To his surprise, the trees were smooth-leafed and the nuts smooth-shelled – the first *M. integrifolia* nuts he had seen. In his book, he wrote of macadamia contractors in the 1920s and 1930s, who offered a service in hand harvesting, dehusking and basic drying.[29]

About 1970, Foreman sold a farm at Uralba and retained one hectare, where he planted grafted *M. integrifolias* at 15-foot by 15-foot spacings. This, one of the first grafted orchards, was sold ten years later due to Foreman's declining health. In 1983, he replanted a macadamia in a 'Big Scrub' restoration project above the Lismore Show Grounds. The family home in Alstonville is now the Crawford House Museum.

Foreman Crawford at ninety-two years of age, planting a commemorative macadamia. Credit: *Northern Star* (Lismore).

Norman Rae Greber (1902–1993)

Norman Greber was dedicated to macadamias for over seventy years. One of fifteen children, he grew up near the 'Rous Mill Orchard' in the Northern Rivers district, sampling nuts from local orchards and wild trees from an early age. At seventeen, he travelled north, eventually settling in the newly opened Amamoor district near Gympie, where he bought a farm in virgin rainforest. While working with timber-cutters in the scrub, he discovered wild macadamias.

Finding thin-shelled nuts, he planted them, and by 1930 had 1,000 trees. By observation and experiment, he became the first Australian to master grafting. Observing that grafting success was much higher in spring, Norm realised that cincturing or girdling (removing two centimetres of bark) on a limb resulted in a build-up of carbohydrate reserves that fed the graft until it had combined with the rootstock. When Queensland Department of Primary Industries horticulturists visited his orchard and told him macadamias could not be grafted, he replied, 'No one told me that' and showed them his successful technique.[30]

Norm's further contributions to developing the macadamia industry are described in the next chapter.

Andrew, Tom, Walter and Rollo Petrie (1875–1939)

From 1838 to the 1940s, over four generations, Andrew, Tom, Walter and Rollo Petrie contributed to the development and promotion of macadamias.

Andrew Petrie's 'discovery' of macadamias, when shown them by Aborigines in the early 1840s, may have pre-dated Ludwig Leichhardt, but this claim is hearsay. However, after taking up 'Murrumba' at Petrie in 1859, his son Tom (1831–1910) planted macadamias, several of which survived until the early 2000s. One massive tree remains. In 1931, its height was over 12 metres and its diameter 17 metres.[31] By 2022, it had grown only a little larger.

Tom's son Walter Petrie was an outstanding industry pioneer, whose energy, experiments and advice made major contributions to the macadamia industry during the 1930s. At 'Yebri', his home, farm and nursery at Petrie, Walter planted a wide range of fruit trees, including an orchard of pecans and macadamias. As pecans were wind hardy and deciduous, he reasoned that interplanted macadamias would be sheltered during the summer winds, yet receive winter and spring sunshine.

Walter put all his funds into 'Yebri', and his dairy cows provided the income to allow him to experiment with plants. After obtaining his first thin-shelled nuts from the Blackall Range in 1901, which may have been from domesticated trees, he planted macadamias and selected those with thin shells and large crops. Between 1916 and 1919, he supplied selected seed nuts to Maryborough and other places.

In 1935, trading as Yebri Nursery, he was offering eight seedlings or their nuts – named Eggshell, Smooth Queen, Venus, Pearl, Comet, Rough King, Planet and Large Everbearer – all of which he believed would produce thin-shelled nuts and which he called 'varieties'. Eggshell could be cracked by the teeth, but it failed because the nut germinated on the tree.

Walter exhibited and promoted his 'Queensland nuts' at the annual Brisbane Agricultural Show. He provided advice and supplied large quantities of seed nuts within Australia and to several overseas countries, including Hawaii and South Africa.

In the 1950s, nurseryman Norm Greber would select trees from Petrie's seed nuts that went on to produce some of the most successful Australian varieties.

At the time of his unexpected death in 1939, Walter Petrie was confident that he was on the threshold of a commercial breakthrough. His son Rollo (1910–96) continued the nursery for a time, but after World War II, 'Yebri' was sold off in parts.[32]

Between 1951 and 2013, Australian Paper Mills (APM) operated a large factory on the site. Many of the macadamias were lost, but in 2003 APM erected a plaque in front of the remaining orchard of pecans and one macadamia tree to commemorate Tom and Walter Petrie. In 2019, the University of the

Sunshine Coast established a precinct on the site of the APM paper mill, but Walter's original macadamias have been lost.

Parochial Australian reaction to Hawaii planting our nuts?

While Australian growers struggled through the first half of the twentieth century, Hawaiian businessmen were commercialising the macadamia with significant plantings. Their efforts led to Hawaiian dominance of the global industry until the 1990s. The American plant propagator Luther Burbank described the macadamia as 'without doubt ... the finest of all nuts in quality ... Here they are esteemed as the finest dessert nuts available.'[33]

Australians may have enthused about 'our nut' and accused the Hawaiians of 'stealing' it, but neither local farmers and businessmen nor state governments were prepared to lead a commercial industry. Parochial resentment was reflected in newspaper articles such as 'Another Industry for the Foreigner', which lamented that Hawaiians were about to unfairly exploit the macadamia nut.[34]

Another article, headed 'Queensland Nut – Neglected in Australia – an American Delicacy' stated:

> A prophet has no honour in his own country, neither has a tree, apparently, if that tree happens to be Australian. Thus it is that a visitor to the north coast of NSW or south Queensland may happen across the Queensland Nut, find it good, enquire as to its origin, and meet with the reply, 'Oh, it grows on a tree in the bush.[35]

Journalist Norman Hewitt, the son of T.G. Hewitt, wrote a scathing report, criticising the New South Wales Government for doing so little to develop the crop:

> ... the Queensland Nut will pass into history as an Australian achievement, but this delectable nut is likely to yield its fortune, not to the original experimenters but to the American exploiters.[36]

Experiments to find other uses for macadamia nuts were undertaken. For a short period, there was interest in using macadamia husks for tanning purposes, as the smell was not offensive. This process

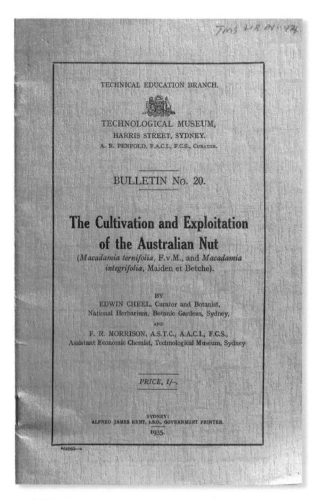

The Cultivation and Exploitation of the Australian Nut, 1935, was widely recognised as the authoritative handbook on cultivation and processing.

proving ineffective, so its use was discontinued. In Queensland, macadamias were sometimes interplanted with tung oil trees, the seeds from which produced a drying oil used in paint finishes. Growing tung oil trees proved unviable, so these mixed orchards disappeared.

In 1935, Edwin Cheel, a botanist and curator at the National Herbarium, and the Government Analyst, F.R. Morrison, published *The Cultivation and Exploitation of the Australian Nut*, a book that was widely sold. Cheel and Morrison reported that, although propagation by grafting and striking cuttings had been tried, these were not viable commercial tools. Like Walter Petrie and John Waldron, they incorrectly believed that there were some trees where the nuts produced true to type after several generations. Edwin Cheel had obtained thin-shelled seed nuts from John Waldron in 1925 and grew some trees at his home in Ashfield, Sydney.[37] He was frustrated in obtaining data from them because young boys delighted in stealing the nuts, even before they were mature.

Professor Henry de Montmorency

While there have been many colourful industry characters, most too recent to mention, one of the most interesting from the past was H.D. Montmorency. Henry was one, if not the first, in Australia and maybe in the world to produce and market macadamia confectionery. If the name he used was indeed the one he was given at birth, Henry may well have been a descendant of Henri I de Montmorency (1534–1614), Marshal of France who committed the unpardonable sin of allowing French Roman Catholic items to fall into the hands of Protestants. Henri's son, Duke Henri II de Montmorency, was beheaded for treason in 1632.

In 1906, we have a record of our Henry being charged in Ballarat with being a rogue and a vagabond by imposing on Sergeant Edward Francis Britt with fraudulent representations for the purpose of obtaining money. He sold medicine which was supposed to cure unmentionable diseases. The case against him was dismissed.

He then turned up in north Queensland in 1907, while travelling around Australia as Professor Henry de Montmorency, Phrenologist, Palmist, Astrologist and Crystal Reader. Later that year he had reached Lismore, where he established the Federal Herbal Institute. In 1910, he was charged with unlawfully using an instrument on a young woman in order to bring about a certain event. There's a full account of the committal, but the case did not proceed. In this early period, de Montmorency published numerous advertisements with dozens of testimonials about the efficacy of his herbal treatments for many diseases.

The Federal Herbal Institute became the North Coast Confectionery Works, and de Montmorency was on his way to becoming an important citizen of Lismore. He was reported just as 'H.D. Montmorency' and was a prominent member of the Rowing Club, the Chamber of Commerce and the Show Society. Although considered a 'foreigner', he had strong opinions and stood up for his rights.

Taking advantage of the local sugar industry, he built up an expanding confectionery business. There are accounts of his factory in 1926, and in 1931 he decided to form a public company and build a new factory. Probably this was bad timing with the Great Depression soon dominating the world. It appears that his macadamia and confectionery business failed, as he was, in 1931, sued for £252, being arrears in rent.

He was one of the founders of the Australian Nut Association in 1932 and active for several years in the industry and in promoting macadamias. He described macadamias as being very tasty. Most of his products were chocolate coated and one was named 'Nutty Cubes'.

He later lived at Pretty Gully near Tabulam, where he grew crops and searched for gold. He passed away in Brisbane in 1965.[38]

Organisations

During the 1930s, people interested in growing, marketing and promoting macadamias came together to form associations. David Harrison, for example, formed the Lismore Australian Nut Growers Association, but this small organisation ceased during World War II. Harrison maintained his interest, and in the 1950s worked for the New South Wales Department of Agriculture, promoting macadamias.[39]

In 1932, to stabilise a situation where prices for nuts were declining because supply was greater than demand, fifty delegates, led by Herbert Rumsey and Norman Hewitt, met in Murwillumbah to form the Australian Nut Association. The meeting was facilitated by the Banana Growers Federation, and chaired by its President, H.L. (Larry) Anthony, a leading banana grower but one who did not plant macadamias. Many years later, Larry's son, Doug Anthony, as Deputy Prime Minister, studied and supported the industry's development. Doug met Steve Angus in Murwillumbah, Dr Beaumont from Hawaii, and visited Hawaii to study the macadamia industry.

Delegates included representatives of the Queensland and New South Wales governments and local councils, produce buyers, and value-adding businesses such as Sanitarium Foods. Growers included Walter Petrie from Pine River, Fred Whittle from Murwillumbah, H.J.R. Gaggin from Mullumbimby, John Waldron from Eungella and Sid Greer from Dungay. Henry de Montmorency, the Lismore confectioner who used macadamias, also attended.

Herbert Rumsey was elected President and Miss A. Stevens, a delegate from Brisbane, became Secretary/ Treasurer. Vice Presidents were Larry Anthony, Walter Petrie, Foreman Crawford, John Waldron and Jens Fredericksen.

The meeting affirmed that the macadamia nut should 'henceforth be known as the Australian Nut, owing to its patriotic sentiment and the suitability of the name for advertising purposes'. Members were told not to use the name 'Bush nut', which was considered 'common'. If they did, they would first be warned and then expelled!

Frank Elliot, an inventor from Woodburn, stated that he had designed a machine that would crack 11 tons of NIS per hour with no damage. It was never adopted. He urged farmers to 'plant macadamias as soon as possible'.[40]

The following year, Sir Leslie Wilson, the Governor of Queensland, was appointed Patron of the Association, John Cecil Sibbald was elected President, and Norman Hewitt, Sid Greer and George Gaggin were elected to the committee.

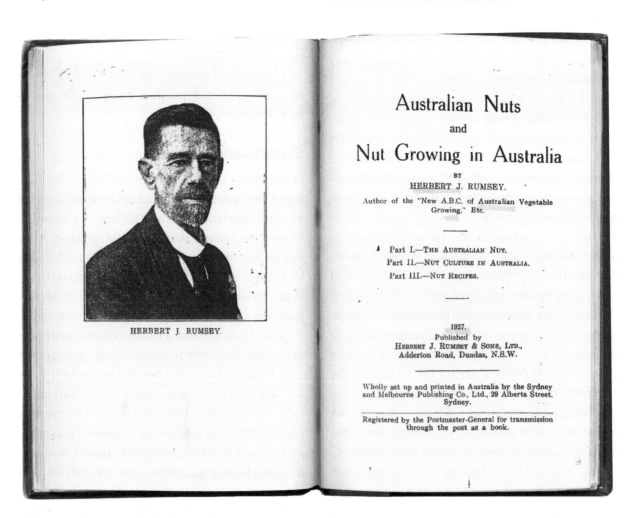

HERBERT J. RUMSEY.

Australian Nuts
and
Nut Growing in Australia
BY
HERBERT J. RUMSEY.
Author of the "New A.B.C. of Australian Vegetable Growing," Etc.

Part I.—THE AUSTRALIAN NUT.
Part II.—NUT CULTURE IN AUSTRALIA.
Part III.—NUT RECIPES.

1927.
Published by
HERBERT J. RUMSEY & SONS, LTD.,
Adderton Road, Dundas, N.S.W.

Wholly set up and printed in Australia by the Sydney and Melbourne Publishing Co., Ltd., 29 Alberta Street, Sydney.

Registered by the Postmaster-General for transmission through the post as a book.

Herbert Rumsey, President of the Australian Nut Association and author of an informative book, 1927.

The Australian Nut Association prepared quality standards that required 90 per cent of NIS to be sound. It also undertook the exporting of NIS, but in 1932–33, the United States rejected a consignment on the grounds of low quality. Dealing with this issue dragged on for years, putting pressure on the Association.[41]

Realising that the ability to market kernel as well as NIS was essential, the Association promoted the locally made Red Devil macadamia cracker. Dr E. Wall from Hawaii, who assessed this cracker, was not impressed with the commercial machine, but recommended the small version. There is no record of its adoption.[42]

Later in the 1930s, the Australian Nut Association expanded to form the Nut Association of Australia, which also promoted pecan, tung oil and candle nuts. Sir Leslie Wilson remained Patron, with Walter Petrie and Herbert Rumsey as Vice Patrons. John Sibbald was President, and Larry Anthony, Sid Greer, George Gaggin, Foreman Crawford, John Waldron, Jens Fredericksen and Norman Hewitt were Vice Presidents. Initial enthusiasm for this or-

ganisation declined. During World War II, it went into recess and never recommenced. There was no formal organisation until 1974. It took a new decade and recovery from the aftermath of war for a resurgence of interest and growth.

John Waldron, depicted in this newspaper cutting, became a Vice President of the Australian Nut Association. Credit: unknown newspaper.

Chapter 7

THE AFTERMATH OF WORLD WAR II TO 1970

Lessons from Hawaii, government support, and the first commercial orchards

In the late 1950's Australia had one orchard of 20 acres with the rest six largest less than three acres. All were overgrown with weeds, no fertiliser was used and pests were usually unchecked.

(Arthur Lowndes, *California Macadamia Society Yearbook*, 1964)

War dominated in most of the world from 1939 to 1945. Due to the war effort, many industries and activities in Australia were neglected. The available workforce was either serving in the war or focused on growing crops essential for survival. The emerging macadamia industry had no priority, and most orchards received little care during this period. The price to growers was a low 7 pence per pound and, even then, demand was slight.

Most macadamias were sold in shell, often in packets of mixed nuts, which were a traditional Christmas treat. Cottage-scale retail roasted and salted kernel packs were produced in small amounts and sold locally. Even then, macadamias were much harder to crack than other tree nuts. Attempts to process the nuts into kernel were at cottage scale, and the need to fully dry the nuts was not understood.

Despite predictions of the trees' cropping ability and the anticipated global demand, macadamia growers found realities discouraging. Tree yields and kernel recovery were low. There was an incorrect belief that seedlings could produce true to type, when

Steve and Mrs Angus collecting nuts from a wild *M. tetraphylla* at Billinudgel in the Northern Rivers, reputed to produce 170 kilograms of NIS annually. Car enthusiasts will recognise the iconic FJ Holden. Credit: unknown.

in reality there was much variability in the characteristics of trees and nuts. In Hawaii during World War II, sales to military personnel had triggered interest in macadamias as premium snack foods or gifts. Unlike Hawaii, where there was early scientific research, macadamia growing in Australia was in the hands of enthusiastic amateurs, whose belief in the nut's potential did not relate to the realities of the time.

However, Hawaii's success had shown how a commercial industry might be developed in Australia, and after the war, the Queensland Department of Agriculture and Stock gave renewed attention to macadamias. Orchards were inspected, productive trees sampled and grafted, and trials were undertaken at the Maroochy Horticultural Research Station near Nambour. This resulted in the naming of the first Australian cultivars in 1960. The Department of Agriculture in New South Wales also provided extension services to growers and undertook limited trials.

In 1948, the Queensland Department of Agriculture and Stock commenced plant breeding with a survey that showed the state had approximately 150 acres of bearing macadamias plus a further 150 acres not yet bearing.

This led to a plant selection field trial at the Glass House Mountains in 1963. The researchers selected sixty-four seedling trees that were reported to have superior qualities, from which they selected twenty. They sought thin-shelled nuts for home or table cracking, medium to thick-shelled selections for processing to kernel, and large nuts with moderately thick shells for which there was apparently a demand. Many selections were named and several were grafted in modest quantities. These were made

"A RICHMOND FARM, CHEESE, NUTS ONIONS AND HYDROELECTRICITY — £168 ESTIMATED FROM 240 TREES"

'This year no fewer than 3 ton were harvested from 240 macadamia nut trees: In other words, 28 lb nuts per tree. If we put down the net profit at 6d. lb, we find that these 240 trees returned the handsome sum of £168 in this, their 50th year. There is money in Australian Nut growing to be sure!'

Orchard for sale. Probably late 1940s. Credit: *Northern Star* newspaper (Lismore).

available to growers over the next fifteen years, but in time all were discarded.

The image and availability of Hawaiian cultivars progressively from the late 1950s resulted in a false belief that these must be superior. Of the Australian varieties, only H2, Own Choice and HY (Rankine) were widely planted until the 1990s. Beaumont was collected by an Australian, selected by a Californian, assessed in Hawaii and widely planted internationally, but not in Australia. H2 became the preferred choice for nurseries as a rootstock.

Investment schemes

From the 1930s to the 1950s, several optimistic ventures in planting and managing small orchards were unsuccessful. In 1933, Mutual Estates Ltd offered to supply investors with land, provide and maintain trees, and harvest the macadamia crop, but this attracted little interest.

In 1944, probably as a follow-on to the unsuccessful 1933 Mutual Estates offer, William Taylor Anderson of the Macadamia Production Co. offered a similar scheme, which lasted until 1949, when the company became Macadamia Production Pty Ltd and was operated by commission agent Raymond Spinks. In 1950, this company issued a detailed prospectus, advising that they held 3,199 acres of suitable land at Archerfield, Wacol and Kingston, and that 500,000 seedling trees were available. Land would be provided free in four-acre lots, and 550 lots had been allocated. The main cost to investors would be the purchase of trees.[1]

The promoter's goal was to establish a commercial macadamia industry, but there were legal disputes over title deeds and the suitability of the land, and the company did not survive. A list of macadamia growers, prepared by the Queensland Department of Agriculture and Stock in 1955, showed only one orchard that could have originated as a result of the Macadamia Production Pty Ltd prospectus.[2]

Example of a high-standard orchard of the 1940s and used to promote investment. Credit: Macadamia Production Pty Ltd.

Developments

In 1957, a Queensland Nut Growers Conference was held in Brisbane under the auspices of the Department of Agriculture and Stock and the Committee of Direction of Fruit Marketing (COD). The sixteen growers who attended sought a government-supported marketing scheme and a minimum price for macadamias. They also discussed changing the product name from Australian nut, Queensland nut, Bauple nut or Bush nut to Macadamia, a move that was gradually adopted,[3] though probably to the detriment of the Australian industry.

Norm Greber

Norm experimented in many ways. As described previously, through observation and trials he mastered the skills to propagate. When told that macadamias could not be grafted, an astonished Norm replied that no one had told him that.

At 'Nutty Glen' near Amamoor, Norm Greber continued to select and graft promising trees. He farmed using a horse, and from the top of the steep

'Nutty Glen' farmhouse and orchard in 1952.
Credit: Graham Smith.

ridges DDT would be drifted down to control insect pests. In 1951, the hills at Amamoor seemed to be getting steeper, so he retired to a flat farmlet at Beerwah, where he experimented with breeding and selecting superior varieties. When Norm sold 'Nutty Glen' to Jim Smith, an English migrant, it contained 1,000 mainly seedling macadamias, 300 mangoes and 300 pecans. The Smiths replaced the horse with a tractor and continued the use of DDT without protective clothing.

Throughout the 1950s, Norm selected trees from Petrie's seed nuts. One, named both D4 and later Renown, was the female parent tree for many of the 'A' series varieties later developed by the inventor and breeder, Henry Bell. His A4 and A16 became the first macadamias to be protected by Plant Breeder Rights.

Norm Greber's hand-crafted entrance sign.
Credit: I. McConachie

CSR Ltd entered the industry in 1963 and realised the need to reliably propagate macadamias, so they engaged Norm as an advisor. This allowed him to be part of a successful commercial industry, which he fiercely defended against any unfair criticism. He contributed to discussion and provided written information on all aspects of the macadamia industry.

As a guiding force and mentor, Norm Greber was a founding father of the macadamia industry. In 1980, he was made the first Patron of the Australian Macadamia Society and he was later given Honorary Life Membership. In 1993, he was awarded an Order of Australia. After his death, the industry created the Norman R. Greber Award for outstanding contributions to the industry he loved.

Almost blind in his later years, he would often be found in his farm shed, surrounded by his macadamia memorabilia, receiving friends and people interested in talking with him about macadamias.

Norm Greber was allocated two Italian World War II internees. This photo shows the hut they lived in for two years. Credit: Graham Smith.

Norm Greber inspecting one of his experimental varieties in the 1960s. Credit: CSR Ltd.

Other growers and developments

At the Glass House Mountains in Queensland, grower George Schulz was one of the early nurserymen to produce grafted trees. He grew and grafted the trees in seedbeds before cutting the taproots and stabilising them in plastic bags for delivery. Due to no fault of his own, he sold the variety HAES 660 as being HAES 333 and vice versa, which caused confusion and angst.

In 1941, farmer Bernie Mason purchased a property at Lagoon Pocket, near Gympie. Influenced by Norm Greber, whose orchard was visible in the distance, he decided to plant macadamias. Although the Department of Agriculture and Stock advised him against macadamias, he was determined to proceed.

When he was assigned two Italian prisoners of war, he was able to search the nearby scrub for wild trees, collect the nuts, and assess the thickness of their shells. While it is reported that he germinated selected nuts and planted them out,[4] eventually having 1,000 widely spaced, healthy trees, his family were certain that in about 1944, Bernie with his son Barry also collected thin-shelled nuts from Montville and grew these.[5] By the 1950s, the trees were cropping well. Under his house, he built a small drier and cracked the nuts, which he marketed in small cellophane packs.

The Italian workers were fine men, but false rumours spread that they had a hidden camera and were Fascist supporters. One of them, Guido Vaccarini, became a hero when he saved the lives of two of Bernie's daughters who were overcome by fumes in a truck powered by a charcoal burner.

At one time, Bernie's orchard, named 'Integrifolia', was probably the largest producer of nuts in the country. In the 1970s, it was bought by Geoff Garratt, who owned it for over forty years. Close by, at Calico Creek, Bill Beattie, also using Italian prisoners of war, selected nuts from local wild trees and established a small orchard.

In the 1950s, growing macadamias was an industry-in-waiting, a cottage industry. Grafted trees were becoming available, and the first commercial processor, Steve Angus, was operating his factory in Murwillumbah.

In the Northern Rivers of New South Wales, dairying and butter manufacturing, the mainstays of the local economy, were declining even before the United Kingdom entered the European Common Market in 1973. Dairying largely relied on the valleys, and there was little demand for the hills and tablelands, where small orchards of macadamias were being planted. Over the next thirty years, larger orchards changed the landscape. A progressive grower, Surrey Bogg, believes that from the 1960s macadamias played a major role in improving the economy of the Northern Rivers.

Tony Conte, who came from northern Italy in the 1920s and settled in Nimbin, was a quiet pioneer. Noticing wild macadamias, he was told they were an Aboriginal food. In 1948, he bought land at Binna Burra, near Bangalow, selected *M. tetraphylla* nuts and planted about 100 trees, which produced well and added to family income. Tony's daughter married Aldo Viola, and their sons, Ash and Bill, and daughter Sylvia, expanded the orchard to 1,100 trees. The family established a nursery, secured scion wood of the Hawaiian cultivars, and supplied grafted trees to Macadamia Plantations of Australia at Dunoon. As the industry standardised on the *M. integrifolia* species, they removed 1,000 trees and replaced them with modern varieties, retaining the original 100 in memory of their grandfather. The nursery continued until the early 2000s.

Bruce Chick from Murwillumbah grew a few macadamias, but in the 1940s and 1950s greatly supplemented these by collecting nuts from wild trees. In 1950, Sid Greer from Upper Dungay near Murwillumbah was promoting macadamias and offering four different selected seedlings at 8 pence per tree.

Meanwhile, the New South Wales Department of Agriculture was providing a range of extension services at their Duranbah Experimental Station near Murwillumbah. Drew Leigh started at Duranbah about 1950 and in 1960 was transferred to the newly opened Alstonville Sub-Tropical Fruit Research Station, where he experimented with propagation and advised macadamia growers into the 1970s.

In 1952, Alan Ross, from the Department of Agriculture and Stock in Queensland, prepared a detailed booklet 'The Macadamia Nut', which covered the cultivation of macadamias and became an industry reference.[6]

In 1954, the Maroochy Horticultural Research Station obtained promising seedling trees from growers in New South Wales and Queensland and assessed their productivity and suitability of their nuts for processing. By the 1970s, most of the original seedling trees had been replaced, due to the availability of grafted trees, and some of the original plantings were lost due to urban expansion.

A 1954 survey listed sixty Queensland growers who had more than one acre of trees, including Taylor at Currumbin, Sewell at Mount Tamborine, Hampsen at Eight Mile Plains, Ardrey at Flaxton, and the Rickard brothers at Maryborough. It also mentioned James Thompson and Les Moore on the Hendersons' property at Victoria Point, Eric Howard and John Hurwood at Maleny, and Norm Greber's orchard at Gympie, then owned by James Smith. At Gilston, where the Hinde H2 variety was selected, lived the Hill family and another grower, Alfred Powell.[7]

A popular variety widely planted from the mid-1960s to the late 1980s was the Hinde tree or H2. Until the 2020s, H2 was the preferred source of seed nuts for nursery propagation in Australia. Its story spans almost a hundred years. It is still sought after for nursery rootstocks but is being gradually replaced by Beaumont. Michael Hinde selected what became 'Colliston' near Gilston in the Nerang Valley and is reported to have collected nuts from wild trees while walking to meet his sweetheart. He planted a range of fruit trees about 1880, and when his citrus were declining, replaced them with macadamias. In the 1950s, his thirty trees were inspected by the local Department of Agriculture and Stock and selected for assessment. In 2022, the original H2 trees still remain at 'Colliston', where they are being well looked after.[8]

As the industry became commercialised in the 1960s, macadamias received more attention. This included visits from researchers in Hawaii and California, where macadamias had a high image and were in strong demand. These visits were assisted by government officers and created interest at grower and government level.

An ongoing problem in Australia, which both attracted growers into the industry and sometimes caused their downfall, was Hawaii's apparently high-yield figures. Hawaii's more equable climate, better tree horticulture and few insect pests resulted in consistent cropping, and Australia's average yields proved to be lower than expected.

As part of their planning for an expanding industry, Hawaiian researchers sought an enlarged germplasm base from wild trees in Australia. In 1953, Dr John Beaumont, Senior Horticulturist at the Hawaii Agricultural Experiment Station, had received a Fulbright research grant to study wild macadamias in Australia and seek selections for assessment in Hawaii.

In 1959, Dr W.B. (Bill) Storey, from the University of California, reported that macadamias were being considered as a replacement for avocados suffering from large losses due to *Phytophthora* root rot as the macadamia was much more resistant.[9] In 1960, Bill and Colonel Wells Miller, President of the California Macadamia Society, selected cuttings from promising wild, backyard and orchard trees in Australia and studied the climate and soils of wild macadamia habitats. In his report, Bill Storey noted that, in Australia, macadamias were not grown as a crop in their own right.

In 1961, the Queensland Department of Agriculture and Stock published a list and description of sixteen varietal selections. Despite poor grafting results from the Redland Research Station in 1964, a 1.5-acre trial was planted at the Maroochy Horticultural Research Station. It consisted of thirteen cultivars, including the varieties HAES 246 and 508, which Hawaii had provided to Australia and other countries. By 1968, the Department was recommending HAES 246 and 508, and five of the Australian selections. Of this five, only Own Choice and H2 were planted in any numbers.

In New South Wales, about 1960, W.C. Gray bought 80 acres at Dunoon, and established a commercial orchard known as 'The Alamo'. By 1966, he had planted 600 trees and had plans for another 1,800. The first section he planted was known as the 'Heinz block' after the American Heinz brand of fifty-seven varieties. Gray's son Kevin later produced macadamia chocolates. 'The Alamo' remains one of the oldest orchards in Australia.

From 1960 to 1964, Dan Mills, representing G.J. Egin, Produce Merchants of Murwillumbah, was buying 'Bush nuts' for between 1 shilling and 6 pence and 2 shillings per pound.

In 1961, the Queensland Department of Agriculture and Stock reported that there were approximately 300 acres of seedling macadamias in Queensland.

At Goomboorian, east of Gympie, Griffith and Jean Grace had a large orchard of 1,500 trees, which was later abandoned. When assessed by Dr Bill Storey in 1960, some very high-recovery nuts were found in the drier but the parent could not be located.

At 'Sahara Farms' in the Glass House Mountains, Jack and Beau Gowen grew a range of crops, including tobacco and pineapples. The properties were named by their father in 1904, when he cleared the land and observed that it looked just like a desert. In 1964, the brothers planted about 100 macadamias, expanding to 80,000 trees over the next twenty years, mainly by purchasing and converting pineapple farms, becoming major industry producers. The brothers led the industry with their willingness to purchase and experiment with mechanisation. Jack Gowen was a founder of the Australian Macadamia Society, President from 1980 for ten years and then Patron. His leadership, generosity and enthusiasm, coupled with intolerance for any criticism of the industry, were parts of his legacy.

More than fifty years later, Beau's sons, Max and Drewe, continue to farm these picturesque, efficient orchards, which are among the most photographed in Australia.

'Sahara Farms' when first planted in 1967. Credit: Heather Gowen.

An unacknowledged pioneer was John Bell, who purchased 480 acres between Shute Harbour and Airlie Beach in north Queensland in 1959. A young solicitor with farming in his blood, he had just hung his shingle in Proserpine. After researching potential horticultural crops, he selected macadamias, taught himself to graft, and planted 4,400 trees over the next seven years. When success seemed in the offing, Cyclone Ada, with wind gusts over 314 kilometres per hour, hit the coast on 18 January 1970. Nature tore nature apart, destroying the Whitsunday Island resorts and the town of Airlie Beach. The cyclone took fourteen lives and all but two of Bell's trees. His former orchard site is now the Whitsunday Airport.

In the late 1960s, pioneer nurserymen Horst Dargel and Bill Van Allmen, from the Pink Lily Nursery near Rockhampton, were producing 4,000 grafted macadamia trees per year and exporting to Rhodesia (now Zimbabwe), New Zealand and Fiji.

Processing

From the 1950s to the 1990s, research into the quality of Queensland's horticultural crops was led by Dr S.A. Trout's team at the Sandy Trout Food Preservation Research Laboratory at Hamilton, in Brisbane. The laboratory's macadamia section was under the control of food technologist Rowland Leverington, who had studied macadamias in Hawaii and was able to provide support in varietal assessment and sound processing procedures. Until his early death in 1990, he continued his applied research into varietal assessments, crop-handling practices, quality, shelf-life, and processing and packaging.

Richard Mason continued Rowland Leverington's applied research both at the Queensland Department of Primary Industries (QDPI) food laboratories and at the University of Queensland. In 1962, Dr Trout and Rowland Leverington published a detailed report, *The Processing of Macadamias in Australia,* which served as a guideline for many years.

Sir Frank Nicklin, the Member for Landsborough on the Sunshine Coast, was Premier of Queensland from 1957 to 1968. A farmer himself, he saw the potential of macadamias and local interest in planting this crop. He encouraged the Queensland Department of Agriculture and Stock to prioritise the cultivation of the tree, and, through Dr Trout's laboratory, to develop quality and processing technology.

Apart from small quantities marketed by John Waldron and Fred Whittle in Murwillumbah, most macadamias were sold to the consumer as nut-in-shell (NIS) until the early 1950s, when Steve Angus commenced processing and produced a range of retail packs.

Steve Angus was the pioneer of commercial processing and value-adding of macadamias in Australia. With his brothers Nick and George, and bearing the Greek name Angouras, he tried to sound more Australian by changing the family name to Angus. The brothers had fruit shops, first in Lismore and then in Murwillumbah. Steve's interest in macadamias started in the late 1930s, largely by knowing John Waldron and local banana and macadamia farmers. Initially, he bought NIS from John Waldron, which he cracked with a hammer and sold in his shop. He then bought raw kernel and Waldron's roasted and salted packs, before sourcing NIS from other suppliers.

Until Steve commenced processing, sales of kernel were limited, due to the difficulty of cracking the nuts. Steve experimented with mechanical crackers, and in 1939 ordered a cracker from Hawaii, which arrived, in parts and with no instructions. When the cracker was assembled, it did a poor job. In 1954, Steve bought a more effective Wiley Cracker from the United States. The size-graded nuts slid down a solid, fixed steel plate. This acted as an anvil where they were held, and an adjustable steel plate moved up and down and cracked the shells.

Steve had observed that nuts dried in the sun cracked better, and Rowland Leverington advised him to fully dehydrate the nuts before cracking. In 1956, Rowland provided Steve with a sack of fully dried nuts to crack. Dehydration caused the kernel to shrink and reduced adherence to the inside of the shell. Steve was convinced, and this was the breakthrough he needed to become the first commercial Australian processor. His factory at Murwillumbah employed twelve people and cracked 20 tons of nuts a year. Most were oil roasted and salted, then packed in 1¾-ounce (50-gram) cellophane packs and in 6-ounce and 1-pound tins. Like Aboriginal people had done, Steve also began to roast them in shell. His brother Nick's speciality was confectionery, and his macadamia chocolate blocks became well known.

Attracted by a Queensland Government incentive, Steve moved Macadamia Nuts Pty Ltd to Slacks Creek, south of Brisbane. In his first full year there, he processed almost 45 tons of nuts (100,000 pounds). His business expanded, and he purchased nuts from farmers from Lismore to Maryborough. By the late 1950s, he was processing more than half the Australian crop. In both 1957 and 1958, Steve paid the average price of 1 shilling and 8 pence per pound. His growers were held to stringent standards, but poor horticultural practices and mixing the two edible species resulted in variable quality. He allocated his better-quality kernel to roasted and salted packs and the lesser-quality kernel to confectionery.

In 1965, as the industry expanded, Steve sold an interest in his business to CSR, whose orchards were about to produce. After he suffered a stroke in 1969, he and his wife relinquished the whole business to CSR. Honoured with Life Membership of the Australian Macadamia Society, Steve retained his interest in macadamias until his death in 1985.

Fred Whittle and his son Noel from Murwillumbah bought a property at Dungay, and in 1930 planted 40 acres of *M. tetraphylla* with seed from 'Rous Mill'. With a view to processing and marketing macadamias, Fred built his own dehusker, and in 1936 he asked a local engineer to design and build a nutcracker of the flywheel-and-blade type. This design, later reinvented by Paul Shaw and others, became the world's most widely used machine. Fred only partly dried the nuts before cracking, which may have limited the nutcracker's efficiency.

After many of their orchard trees were lost in bushfires, the Whittles accepted that macadamias would not be viable and replaced the trees with bananas. Their old machinery was stored in a shed at the rear of the Whittles' pharmacy in the main street of Murwillumbah until the 1980s, when apparently it was dumped.

Geoff Shrapnel, an entrepreneur who founded Buderim Ginger, incorporated macadamia processing and roasting at his ginger factory, and led an informal Queensland Macadamia Nut Association during the 1950s.

In 1960, he chaired a meeting of thirty-four growers to form the Australian Macadamia Nut Society, which did not survive. The goals of this organisation were to research grafting and to import Hawaiian cultivars, which they optimistically believed would average 77 kilograms per tree and 36 per cent kernel recovery, compared to seedling yields in Australia of 18 kilograms at 24 per cent kernel recovery. Geoff Shrapnel announced that Buderim Ginger would buy NIS and process it, but would not compete with the existing processor, Steve Angus.

From 1963 to 1967, the New South Wales Banana Growers Federation, recognising that many banana growers had interplanted macadamias, undertook to buy and process them in a factory at Chippendale, in Sydney. Their facilities were of a high standard, with kernel separated from shell by hand. The kernel was marketed in 4-gallon, solder-sealed metal tins of the type used to pack kerosene. The cracker they purchased, built by Mr Chipenoi of New York, had come to Australia earlier and

THE "B.G.D." MACADAMIA NUT PLANT

NORTH COAST Macadamias are inspected by Mr. W. E. Clements, General Manager of B.G.D., on arrival at the processing plant. Each bag returns growers about £10.

DEHYDRATION (left) is carried out in a gas-fired kiln at controlled temperatures. Later (right) the nuts are cracked in a machine especially imported from New York.

SORTING. The nuts pass along a moving belt where skilled staff expertly pick out the quality kernels. Finally (right) the B.G.D. label goes on the tins for despatch to selected retail outlets.

Banana Growers Federation processing macadamias in Chippendale, Sydney, in 1964. Credit: unknown

may have been the first unit bought by Steve Angus.

The Mendels family came to Australia from Austria before World War II and in 1939 established The Nut Shop in Sydney's Strand Arcade. Known initially for its Viennese almonds and still today for the aroma of roasted nuts, this prestigious business was expanded by Karl and his son Paul to at one stage eleven retail outlets.

It was well after the war when the Mendels became aware of macadamias in a government scheme to encourage exports by issuing generous licences, which granted subsidies of up to 20 per cent of goods sold.

In the early 1950s, The Nut Shop was offered twenty kerosene tins containing macadamia NIS, which had been collected from small orchards and backyard trees in northern New South Wales. The intended sale of these nuts to India had fallen through, and Paul Mendels acquired them reluctantly, not knowing what they were. Tasting the kernels convinced him to make macadamias a major part of his business. He found it difficult to obtain bulk NIS and kernel from Steve Angus, but eventually received both from the Banana Growers Federation and imported kernel from Hawaii. Soon the demand for his chocolate-coated macadamias exceeded supply.

In 1974, Paul purchased an orchard in the Northern Rivers, enticed by the claim that its planted trees were three years old. He soon realised that the trees had just been planted after three years in the nursery. This orchard is still part of the Mendels' business, now managed by third-generation son Dan and his family. Paul Mendels remains an enthusiastic industry pioneer.

Christopher Joyce is a long-time marketer of tree nuts and Australia's largest grower of pistachios. His father Ken Joyce owned the Summerland Nut Company near Lismore, built a macadamia processing plant in 1967 and relied on young lads to harvest nuts from backyard and wild trees. However, the supply was insufficient to sustain the business.

From the 1940s to 1980, the bohemian entrepreneur Max Harris promoted Australian art, established the Mary Martin bookshops, and was known for his endorsement of the controversial Ern Malley affair, a poetry hoax intended to ridicule the literary elite. Keen to promote all things Australian, Harris marketed roasted and salted macadamias by mail order from his Mary Martin bookstore in South Australia. These were supplied by Steve Angus, who could not cope with the demand.

In 1969, the Queensland Government estimated that there were approximately 300 acres planted in New South Wales, and 500 acres in Queensland plus 320 acres owned by CSR. NIS production was about 50 tonnes.

Part 3

GROWING AN AUSTRALIAN INDUSTRY

Chapter 8

CSR ENTERS AND LEAVES, 1960–1986

Leading by investment and example to commercialise
the industry, but no fairy-tale ending

Might CSR's claimed best rum in the world, 'General Managers Reserve'
later to be marketed as 'Inner Circle', have been responsible for
founding the commercial Australian macadamia industry?

(A whispered CSR staff anecdote, late 1950s)

The power of investment

The single factor that resulted in the commercialisation of the Australian macadamia industry was the entry of the Colonial Sugar Refining Co. (later CSR Limited). Founded in 1855, CSR had become the largest grower, miller and refiner of sugar in Australia, and had also diversified into other industries. Many of their farms and much of their sugar infrastructure was in macadamia country. CSR's senior managers regularly attended meetings with Hawaiian sugar manufacturers, who boasted of their successful macadamia industry.[1]

John Smith, Manager of the Condong Sugar Mill near Murwillumbah, visited Hawaii for a sugar conference and noted that some of the sugar mill companies there were growing macadamias. An anecdote told by a CSR staff member recalls an occasion when Ian Dixon, a Deputy General Manager from Sydney, was invited to open the Condong Bowls

Club. At least two CSR directors, Mark Hertzberg and Barry Aldrich, were present. John Smith's friend Steve Angus provided his best quality macadamias, which, together with CSR's own brand of 'Inner Circle' rum, were the highlight of (and quite possibly influenced) the evening.[2] Sensing an opportunity, the directors declared a strong intention to enter the macadamia industry, partly as they were wrongly informed that macadamias could be processed in sugar-cane mills during the off-season.[3] Over the next twenty-five years, other CSR directors were to question the wisdom of that memorable night.

In the late 1950s, an agricultural consultant, Dr Lennox Davidson, had seen the potential of macadamias. Taking options on land near Port Macquarie, he registered a company, Macadam Pty Ltd, and unsuccessfully offered his services to CSR.[4]

Then in 1960, Ian Dixon engaged Arthur Lowndes to assess the potential of macadamias in Australia. With a background in geography and agricultural economics, Arthur had established an agricultural consulting business. His diverse career included serving as a Commissioner of the then Australian Broadcasting Commission and for eighteen years its Deputy Chair. He edited a book on the history of CSR[5] and his geography textbook was used for many years in New South Wales schools.[6]

Lowndes initially studied the Australian, Californian and Hawaiian macadamia industries. He reported that Australia's existing orchards were small and of a low standard. When overseas, he was embarrassed that Australia was struggling to produce 30 tons of nut-in-shell (NIS) per year. He believed that the enterprise, research and investment of large companies was needed to make the industry viable.[7]

Through chance, Arthur Lowndes met Norm Greber and they became life-long friends. Arthur again visited Hawaii, where the larger growers were reluctant to assist him, but smaller growers and the Hawaii Agricultural Experiment Station (HAES) provided valuable information.[8] Then in 1962, CSR engaged Arthur to lead their investment in macadamias. This involved a major research programme and the purchase of suitable tracts of land. Norm Greber was employed to establish a large nursery on his property at Beerwah, which CSR later bought and named 'Greberfield'.

Arthur was directed to seek land where its use would not compete with sugar-cane, and due to price, he bought land only in Queensland. In 1963, at Maleny on the Blackall Range, CSR purchased five dairy farms totalling 770 acres for £90,000. The resultant combined orchard was named 'Dixonfield', after Ian Dixon. Six years later, CSR announced that some of their Maleny land would be sold and no more planting of trees would occur, due to damaging frost.

Next, CSR purchased three properties at Baffle Creek, north of Bundaberg, and named them 'Fernfield', 'Kenfield' and 'Fingerfield'. Each property was quickly planted with about 15,000 trees. 'Mt Bauple Orchard', a property at Bauple, was purchased but not planted until 1969.[9]

Arthur Lowndes, Norm and Kathleen Greber celebrating in 1973. Credit: unknown, possibly CSR.

At Yaamba, north of Rockhampton, 1,200 acres were purchased in 1971 and named 'Macfields'. One hundred trees were planted initially, half on good soil and half on marginal soil, followed by larger plantings, but doubt about soil suitability meant that numbers did not significantly increase. In 1986, CSR sold the property, and from 1989, the new owners gradually increased plantings to 40,000 trees.

In 1979, CSR attempted unsuccessfully to purchase the orchards of Macadamia Plantations of Australia in northern New South Wales. They then bought part of the Gowen family's established orchard at Peachester and named it 'Mt Beerwah'.

Norm Greber and Stan Henry, Nursery Manager, at Beerwah original nursery, c. 1975. Credit: CSR.

'Dixonfield', Maleny, when first purchased, 1964. Credit: CSR.

Baffle Creek orchard in 1975, with Manager Ken Kleinschmidt. Credit: CSR.

'Dixonfield', Maleny. First plantings, 1966. Credit: CSR.

Baffle Creek land being cleared for macadamias, 1964. Credit: CSR.

'Dixonfield', Maleny, 1966. Preparation for planting and frost protection. Credit: CSR.

Lincoln Doggrell, Ken Kleinschmidt and Arthur Lowndes at Baffle Creek, c. 1969. Credit: CSR

Arthur Lowndes summarised the issues that faced the developing industry:

- Australia's insect pests were a massive problem.

- Improved cultivars from Hawaii were not initially available, and varietal requirements for Australian conditions were not understood. There was ignorance of the grafting techniques needed to produce planting stocks.

- There was no reliable information on potential yields under Australian conditions.[10]

Arthur lived in Sydney at Strathfield, entertained frequently, and had a special macadamia tree with limbs grafted to different varieties. A ritual was that, late in the evening, his male guests were required to pay homage and fertilise the macadamia by urinating on it. In retirement, Arthur bought five acres at Dural, north of Sydney, and named it 'Nutty Glen' after Norm Greber's orchard at Amamoor. Here he planted about eighty grafted macadamias, and he and his wife developed a small retail business selling nuts-in-shell.

After Arthur Lowndes retired in 1976, Ted Davenport, who had transferred from the Fiji Sugar Division, became a successful Manager of CSR's Macadamia Division and later played a leading role in the Australian Macadamia Society (AMS). A colourful character with a wide knowledge of macadamias, believed to have been born wearing a bow tie, Ted actively travelled the world representing CSR, retired in 1985 and later became the AMS Executive Officer. After retiring from all roles in the AMS, he maintained an interest in the industry and its people until his passing in 2021.

Keith Ainsbury, Lincoln Doggrell, Simon Newitt and John Stock were part of CSR's orchard management team for many years. Keith Henderson managed the nursery, where Stan Henry pioneered bud punching to propagate macadamias. John Simpson was manager at the Slacks Creek and 'Sunshine Plantation' factories, and also served in Hawaii. Financial and company manager Jim Twentyman continued in the industry for another twenty years. Long-serving staff who continued after CSR was sold included Bri-

E.R. (Ted) Davenport, industry pioneer, AMS leader and raconteur, c. 2008. Credit: Kerry Davenport.

an Loader, whose career exceeded forty years, and Rick Vidgen, who became a global visionary and leader in Hawaii. At one time, they had five entomologists researching insect problems.

As a blue-ribbon Australian company, CSR's macadamias received ongoing favourable publicity, which influenced others to consider growing macadamias. By promoting the industry and subsidising government departments to undertake research and provide extension services, CSR for several years spent almost as much in applied research as in direct orchard development. In the early 1970s, their orchards were beginning to produce, and they believed that NIS production would double each year from 1971 to 1976. In 1974, their crop reached 82 tonnes of NIS.[11] Although production steadily increased, reaching over 600 tonnes in 1979, it fell to 200 tonnes in 1982 due to seasonal factors. In 1979, as part of its global vision, CSR bought MacFarms of Hawaii. Their factory Mac Nuts Co. at Slacks Creek,

Brisbane, was wholly owned in 1970–71 and they were marketing retail packs in a range of sizes and packaging.[12]

CSR was anxious to protect its heavy investment and retain its controlling position. But its dominance was threatened by the increasing crop produced by new growers, some of whom had founded the Australian Macadamia Society in 1974. In 1980, CSR approached the AMS, seeking to provide a single-desk marketing body, which would handle all exports. Growers would be encouraged to supply NIS directly to a market pool. However, marketers and large growers wanted to control their own destiny, and although the scheme operated for several years, it was abandoned when CSR sold its Macadamia Division.

As a necessary step towards processing and marketing its crops, CSR acquired an interest in Steve Angus's Macadamia Nuts Pty Ltd in 1968. When Steve suffered a stroke in 1970, CSR bought the company outright and operated the Slacks Creek factory in Brisbane until 1978.

CSR then entered into a leasing agreement with Bill and Lyn Taylor to construct a modern, tourist-oriented, macadamia processing factory at their 'Sunshine Plantation' on the Bruce Highway, south of Nambour. With its giant, fibreglass Big Pineapple, the 'Sunshine Plantation' was an agri-tourist attraction, established in 1971. Based on the Sunshine Coast's agricultural and horticultural heritage, it promoted subtropical fruits, especially the macadamia.

Trading as MacFarms of Australia, CSR opened the new factory on 19 July 1979. Designed as a colonial-style plantation homestead, the factory incorporated a state-of-the-art processing plant, an information centre, a walk-through gallery and a retail shop. During the 1980s, almost a million tourists a year visited the 'Sunshine Plantation' and saw the macadamia displays.[13]

Travelling between the main tourist complex and the macadamia factory was the Nutmobile, a modified tractor drawing fibreglass carriages shaped as half-macadamia nuts and carrying six people each.

MacFarms of Australia's new factory after completion, 1979. Credit: CSR.

On the walkway to the factory, The Magic Macadamia structure sold refreshments and enticed visitors into the world of macadamias. In April 1983, Prince Charles and Princess Diana visited the 'Sunshine Plantation', and learnt about the macadamia industry.

Most of CSR's kernel was marketed in 50-gram flexible packs and cans of 150 or 500 grams. In 1980, its retail division, Macadamia Nuts Pty Ltd, received the International Food/ Europe Award for their outstanding range of retail packs. In the early 1980s, CSR entered a joint venture with the Reckitt and Colman Planters Group to market an expanded range.[14]

Caption: Macadamia Week at 'Sunshine Plantation', 1980. Credit: *Sunshine Coast Chronicle.*

Challenges of being a pioneer

During the early 1980s, CSR continued as a forceful leader of the Australian industry, but as the industry expanded, their influence declined. After twenty years of involvement, the challenges of pioneering the commercial industry had not resulted in meeting the financial goals of the company. In 1981, CSR reported that Australia produced 460 tonnes of kernel; Hawaii, 3,200 tonnes; Guatemala, 100 tonnes; South Africa, 50 tonnes; Kenya, 150 tonnes; Brazil, Venezuela, Costa Rica and Malawi, 140 tonnes in total. They advised that the average sound kernel recovery in Australia was 28 per cent. Realising that with a new, long-term crop they were facing unknown and emerging challenges, the Board reluctantly concluded that they should sell the Macadamia Division and focus on their core businesses. For several years, however, it was imprudent to announce this, due to concern that it would damage shareholder confidence.

MacFarms box pack, 1970s. Credit: CSR

Stepping aside

Finally, in September 1986, a leveraged management team, led by Lance O'Connor in Australia and Tom Modisette from the RAM Group in the USA, purchased Macadamia Nuts Pty Ltd, the holding company trading as MacFarms of Australia and MacFarms of Hawaii. In Hawaii, there was a close association with the California Almond Growers Exchange (CAGE), who undertook bulk marketing.

Arnott's Limited, the Australian biscuit and food manufacturer, had become a minority shareholder in the new company. Differences between the management of Macadamia Nuts Pty Ltd and Arnott's resulted in Arnott's acquiring 51 per cent of shares and taking control of the Australian and Hawaiian businesses. In a severe global and macadamia market downturn, resulting in revaluation and restructuring, Arnott's reported macadamia losses of $3,808,000 in 1990 and $4,632,000 in 1991. However, these were not trading losses but largely the writing-down of assets.

Campbell Soup Company International acquired Arnott's in 1997 and quickly sold the Macadamia Division to The Peninsular Group, an Australian-Hong Kong company led by Brian Findlay and Winson Woo. This company was undercapitalised and went into liquidation in 1998. Andy Burnside, the General Manager of MacFarms and then The Peninsular Group, had the difficult job of ensuring all suppliers were treated fairly and in fairly representing all parties of a struggling business.

During 1996, Graham Hayes, trading as Rough-end Pineapples Pty Ltd, purchased the 'Sunshine Plantation'. In 1998, when The Peninsular Group sold its macadamia orchards, he acquired the MacFarms factory and processed and marketed macadamias as 'The King of Nuts'. In a competitive market, his lack of experience and the business's turbulent image resulted in the factory's closure in 2000.

A characteristic of CSR's twenty-five years in the macadamia industry was the enthusiasm, not only of key staff but of all who worked on this challenging project. Employees had a sense of purpose and pride in creating an industry for a native Australian product that had an unknown future.

Without doubt, the commencement and development of the commercial Australian industry was in large part due to the vision and resources of CSR. Without CSR's involvement, the Australian industry would have developed more slowly and might never have reached the scale of today.

What went wrong with CSR's achievements, plans and vision? As a public company, they eventually accepted that they would struggle to obtain a satisfactory return on their investment. They underestimated the problems and overestimated the returns in a volatile, undeveloped market. Their goal of being a major Australian and global grower and marketer became unlikely as competition from four large rival processors and new entrepreneurs in Australia developed. CSR suffered the fate of many pioneers, but its contribution greatly supported the emerging macadamia industry.

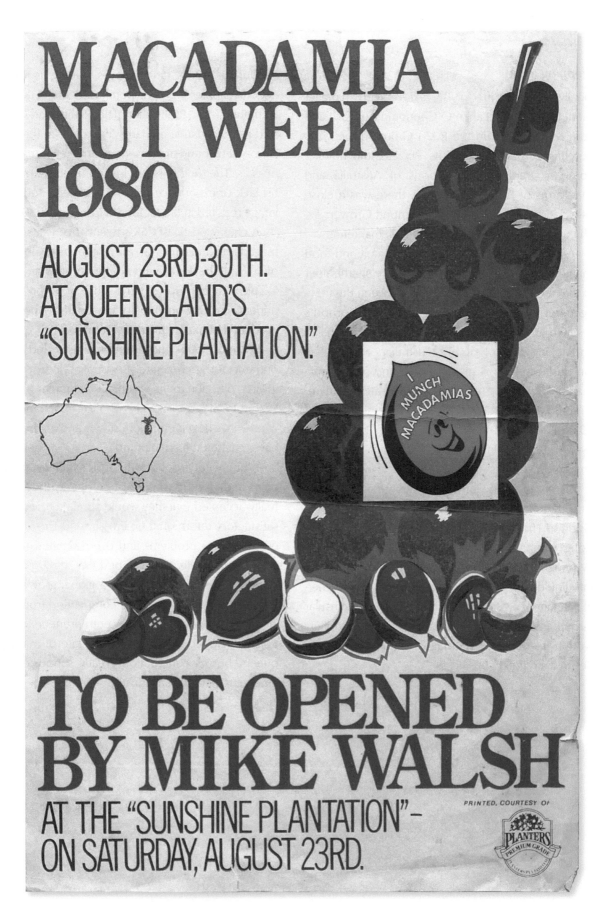

Poster for Macadamia Nut Week at 'Sunshine Plantation', 1980. Credit: I. McConachie.

Chapter 9

THE AUSTRALIAN MACADAMIA SOCIETY

Leading, uniting and stabilising the industry through promotion and research

To promote all aspects of the industry, support a free interchange of ideas and information and encourage good will.

(Objects of the AMS)

Every successful industry needs a representative body to lead and co-ordinate it. The Australian Nut Association, formed in 1932, was the first industry association for macadamias. Other short-lived or regional groups that did not gain the support of the whole industry came and went. In 1958, the Macadamia Industry Society was formed under the direction of the Queensland Department of Primary Industries, but this body did not last.

In 1974, with the certainty that a commercial industry was developing and because there was no industry spokesperson, a group of growers formed the Australian Macadamia Society (AMS). These growers were, in part, concerned that CSR, which was starting to produce large crops, would come to dominate and control the industry.

Formation

Invitations to discuss an industry representative body were widely circulated.[1] The AMS was formed on 21 March 1974 at a meeting of twenty growers at Beerwah, in Queensland. Col Heselwood was elect-

ed President, Eric Cottam Vice President, and Norm Richards Secretary-Treasurer. There was support at the meeting for a growers-only society, but common sense prevailed and the AMS became open to all.[2]

Subsequently, thirteen people from New South Wales, mainly growers, applied to join, and by October 1975 the membership, excluding partners, was 130. This rose to 300 in 1980 and 500 in 1990. By 1995, there were 600 members, including forty from overseas.

In 1976, a New South Wales group joined the AMS and was recognised as a southern sub-branch.[3] Later a Nambucca group joined, but over several years these branches merged into the single organisation.

Role

In accordance with its official Objects (as cited in the epigram above), the AMS brought together growers, processors and researchers, defined priorities, and, as time passed, addressed them.[4]

From the beginning, the AMS Committee was ac-

Steve Angus, Col Heselwood and Norm Greber at AMS Annual General Meeting, c. 1984. Credit: unknown, but a local photographer engaged by the AMS

Keith Ainsbury, Steve Angus, Norm Greber, Col Heselwood and Jack Gowen. AMS Annual General Meeting, c. 1984. Credit: unknown, but a local photographer engaged by the AMS

tive, publishing newsletters and organising field days and well-attended Annual General Meetings and dinners. The first *News Bulletin*, distributed in July 1974, recommended the development of a mechanical harvester and the introduction of uniform nut-in-shell (NIS) purchasing and quality standards.[5] This publication continues as the main means of communication and information, not only in Australia but throughout the world.

By 1976, a Macadamia Research Committee had been formed, chaired by Henry Bell. This committee's priorities were to research maturity testing, leaf deficiencies, varietal breeding, chemical and mechanical harvesting; to produce an insect recognition booklet; and to establish a reference library.

To make information about macadamias available to the public (before the internet), the AMS provided 800 items, including fifty-four books or booklets, to the Caloundra Library, where Librarian Dawn Madden maintained the collection. A similar library was later established at the New South Wales Research Station at Wollongbar. Due to its limited use, the Caloundra collection was passed back to the AMS in the late 1980s.[6]

Technical, Finance, Marketing and Promotion committees had been formed and two success-

ful Research Workshops were held, in 1983 and in 1987.[7]

Since Australian production was certain to increase, it was essential to expand both domestic and international markets and to promote the quality and image of Australian macadamias. An ongoing concern was that as production increased, price to the grower could fall.

The AMS began its promotions at successful information centres at the 1975 and 1976 Royal Queensland Shows (Brisbane Ekkas). From 1975 to 1986, the 'Sunshine Plantation' at Woombye supported the AMS by holding an annual Macadamia Nut Week, which introduced tourists to macadamias as a premium Australian food.[8] During the first half of the 1990s, macadamias were promoted to the public at annual Agricultural Shows. Volunteers Ross and Robbie Litchfield travelled far and wide, setting up displays at major shows and enthusiastically telling the macadamia story.

In 1978, the AMS wrote to Australia Post, the Federal Government body responsible for postal services, requesting the issue of an Australian stamp featuring the macadamia. Surprisingly, this request was denied on the grounds that Australia Post's policy was to emphasise 'subjects dealing with Australian flora and fauna'. After thirty-six years of repeated requests, an Australian stamp featuring macadamias was finally issued in 2011.

An early AMS initiative was support for a Tree Insurance cover, which was a priority with many growers. However, growers discovered that each tree had to be totally lost and a high excess payment applied to claims. These problems, together with large initial claims, resulted in the withdrawal of this insurance. Subsequently, various companies offered Tree Insurance and considered crop loss, but grower uptake was not strong enough to make it viable.

Australia Post's 60-cent stamp, 2011. Credit: I. McConachie.

Industry levy

A priority in the early 1980s was to seek funds for research into orchard management, raising productivity, and controlling pests and diseases. The AMS approached the Queensland and New South Wales governments, requesting research and offering to contribute funding, an unusual offer that was welcomed by government departments. In the words of a senior public servant, 'The Government will help those who help themselves'.[9] CSR Limited had set an example by contributing to government research.

In 1983, to finance research, the AMS introduced what was initially a voluntary $5 research levy, to be paid by members. Amended to 5 cents per tree per year, this levy was initially strongly supported by growers. However, as most orchards were not yet income producing, many growers were not prepared to support an additional cost, even as an investment in research for the future. By 1985, only 45 per cent of growers were paying the research levy.

At the same time, the AMS proposed a voluntary promotion levy of 1.5 cents per kilogram of NIS collected by the processor and this was debated for several years. This levy was finally introduced in 1985. In 1987 and 1988, there was a voluntary levy of 1 cent per kilogram of NIS paid directly to the AMS for research, and a 'compulsory' levy of 0.5 cents per kilogram collected by the processor for promotion. For the 1988 season, $6,000 was received for research and $51,000 for promotion. The industry had to have certainty in securing research and promotional funding and to ensure there was equity in its collection.

In 1987–88, the Australian Government established the Horticultural Research and Development Corporation (HRDC) to provide levy-based, matching funding for approved research, generally on a dollar-for-dollar basis, and the Australian Horticultural Corporation (AHC) to manage promotion but not provide matching funds. These were statutory bodies, so, if adopted, the industry's levies were enforceable by Commonwealth legislation.

It became obvious that levies needed to be compulsory, so eventually a Statutory Levy proposal was developed, to be collected at a central point, usually by the processors. In retrospect, all funding initiatives were heading towards this being accepted by the industry.

Doubts about putting industry research and promotion into the hands of a quasi-government organisation created robust debate, and the vigorous discussion on the funding of levies revealed differing viewpoints. Some AMS members believed that the industry had established sound programmes and should maintain full control.

In 1991, a motion to adopt a Statutory Levy was put to AMS members and carried by 93 per cent of those attending and by 84 per cent of proxy votes. From 1992, a Statutory Levy was in place, to be paid by growers but collected at the point of NIS sale and remitted to the Australian Government. The levy was set at 3 cents per kilogram of NIS, with 2 cents to the AHC for promotion and 1 cent to the HRDC for research and development (R&D). The HRDC and AHC deducted management costs, and levies for R&D were matched by the Australian Government.

The Statutory Levy has proved to be a major advance for both the government and the industry, and a model for Australian horticulture. The AMS was acknowledged as a progressive, professional and successful leadership body. Since then, the total levy and its distribution has varied, depending on current priorities. For example, due to low kernel stocks and strong demand, in 1995 the promotions levy dropped to 1.5 cents per kilogram, R&D increased to 3.3 cents and the residue testing programme remained at 0.2 cents.

Australian macadamias are claimed to be the only nut crop in the world – possibly the only horticulture crop in Australia – that has had twenty-four years of 100 per cent compliance with a government-run residue testing programme. This is an extraordinary achievement that has added cents per kilogram to Australian export prices, at least in Japan. There were many research priorities, but the limited funds available from the industry restricted applications to the HRDC. The AMS Research Sub-Committee determined priorities and invited universities, state governments, the Commonwealth Scientific and Industrial Research Organisation (CSIRO) and other bodies to submit projects. In 1990, there were twenty-seven requests for funding and $130,000 was allocated. In 1991, research projects totalling $225,000 were approved, with the industry levy contributing half and the HRDC matching this. The resulting research led to improvements in quality, yield, varieties, plant nutrition, health benefits and mechanisation. Emphasis was on higher-density plantings, the treatment of pests and diseases, and the conservation of wild genetics.

With the support of the AMS, CSR led the industry with discussions on future processing and marketing structures. Competition among existing kernel marketers limited the development of new markets. The AMS investigated single-desk marketing and engaged CSR to establish a co-ordinated trial. In 1984, thirty-nine growers supplied a marketing pool. To permit competitors to work together, the Australian Government granted a trial exemption from the *Trade Practices Act*.[10]

During the next five years, the co-ordinated marketing pool was refined and expanded, but the trial failed to demonstrate that growers received higher returns, and there was a gradual loss of support. In addition, marketers wished to control their own destinies.

Incorporation

From the establishment of the AMS, its Constitution, with amendments, had served the industry well. In the 1980s, however, voluntary bodies became aware of the risk that their members could be held personally responsible for debts or claims against their industry body. The AMS formed a sub-committee, held discussions with members and obtained legal advice. In 1987, a completely new Constitution, under which the AMS became an incorporated body with liability limited by guarantee, was accepted by members. This meant that members became shareholders and would, in the event of any claim or dissolution, have a maximum liability of $10. Because at that time the voluntary payment of levies had not obtained full support, the new Constitution included the ability to impose a compulsory levy, enforceable on members – a change that paved the way for statutory levies.

Representing the industry for fifty years

The AMS played a major role in advancing global recognition of macadamias by its membership of the International Nut and Dried Fruit Council (INC). This body ably represented the major tree nut industries, mainly in Europe and the USA, so the inclusion of macadamias assisted both macadamias and Australia to be globally represented.[11]

The AMS promoted the health benefits of macadamias through specific projects that it initiated and, from 1988, through the Nuts for Life Program of the Australian Nut Industry Council (ANIC), which represents the seven nuts grown commercially in Australia. Promotion to create demand and enhance the image of macadamias continues through the media, cooking shows and recipes, social media, regional coverage and school programmes.

During the 1990s, the industry relied heavily on AMS leadership, beginning with a plan to overcome the loss of confidence that resulted from a market downturn in the late 1980s. This involved encouraging market development and maintaining orchard standards and was successfully implemented over several years. Although all sectors of the industry supported working together, buyers of NIS were competing against each other for supply.

A second downturn occurred in the late 1990s and the AMS did everything in its power to assist the industry's recovery. When The Peninsular Group, a major growing, processing and marketing company, collapsed in 1998, the AMS organised a co-ordinated approach for growers who sought to recover crop payments.

In 2003, the AMS updated its Strategic Plan to determine the industry's future directions. These included monitoring farm performance and processing, utilising competitive advantages, and expanding domestic and export marketing. Priorities were implemented, subject to funding. The AMS continued to hold conferences, where presentations and updated information were provided and published. This resulted in networking for members and technology transfer to all sectors of the industry.

Planning for the future included recognising the importance of conserving remaining wild macadamias as a gene bank for the industry, so in 2004, the AMS formed the Macadamia Conservation Committee with the goal of conserving the four wild species of macadamias. In 2006, this committee became the Macadamia Conservation Trust (MCT), with the AMS as its trustee.

In September 2012, AMS hosted an International Symposium in Brisbane, attended by 300 delegates including 100 from overseas. A wide range of topics was covered and global co-operation was promoted.

In 2021, AMS CEO Jolyon Burnett announced that the AMS's future goals were advocacy, social responsibilities, training for growers, non-levy-funded R&D, regional support, and international leadership and collaboration.

Women in AMS

The early days of the AMS were male dominated, but from the 1980s women began to be elected to the Board. This was seen most positively, as wives, partners, employees and researchers were a major part and asset of the industry and often presented a new perspective.

The first woman to be elected to the Board was Bev Atkinson, who farmed at Nambucca with her husband Don. For many years, Bev edited the *News Bulletin*, assisted in conference planning and management, and strongly supported the direction of the AMS. Over time, Fay Bogg, Morag Page, Yolande Bromet, Kay Spencer, Pam Woods, Sandra Lindstrom and Aimee Thomas all served with distinction as well.

The AMS management team has included many competent and dedicated women. Lynne Ziehlke served as Industry Marketing Manager from 2010 to 2020, when Jacqui Price took up the role. Nyree Epplett is Communication Manager, Susan Vallis was Business Manager until 2022 and Leoni Kojetin is Development Manager, who has led and supported growers in applied research and become the recognised industry advisor in all horticultural matters.

In 2021, the AMS formed a 'Women in Macadamias' group to encourage women to have active roles in the AMS and the industry. Staff members Leoni Kojetin and Nyree Epplett had the vision and drive to form the group, strongly supported by Aimee Thomas, the first female Deputy Chair of the Board.

Australian Macadamia Handlers Association

In 2008, processors and purchasers of NIS formed the Australian Macadamia Handlers Association (AMHA). The AMHA is independent of the AMS but works in liaison with it, maintaining a database of production and marketing statistics so that processors can share industry knowledge in confidence.

World Macadamia Organisation

In 2018, the AMS was instrumental in forming the Global Macadamia Council whose establishment goals were to develop international collaboration. Larry McHugh was the inaugural Chair. In 2021 this became the World Macadamia Organisation (WMO) – with the slogan 'Here for the love of macadamias' – which in 2023 is supported by eight macadamia-producing countries. To raise funds, the WMO initiated an International Voluntary Marketing Fund in 2020, to be raised by a grower levy. Growers were asked to contribute 3 cents per kilogram of NIS to address new product development, to develop global markets particularly in Asia, and to promote the health benefits of macadamia nuts. Jillian Laing is the founding CEO of the WMO. The organisation and its goals are described in more detail in Chapter 14, and its funding, following the market downturn from 2022, has been increasingly sourced from marketers and industry monies.

Some of the many people who have contributed to the success of AMS

Over just on fifty years, there have been many who have served on the Board, its committees and as employees. Overwhelmingly, they have been dedicated and competent, and contributed to the success of the organisation; regrettably, space constraints prevent proper acknowledgement of everyone in-

Table 3: Presidents and Chairs of the AMS from its inauguration to 2023

Name	Year	President/Chair	Additional
Colin Heselwood	1974	President	Continued as *News Bulletin* Editor
Eric Cottam	1977	President	
Jack Gowen	1980	President	Medal of the Order of Australia (OAM) and Patron
Keith Ainsbury	1989	President	Vice President 10 years
Andrew Stapleton	1991	President	
Geoff Garratt	1994	President	
Andrew Stapleton	1995	President	
Rod Fayle	1997	President	OAM
Cliff James	2000	President	
Graeme Hargreaves	2003	President	
Andrew Pearce	2008	Chair	
Kim Wilson	2008	Chair	
Andrew Starkey	2012	Chair	
Richard Doggett	2014	Chair	
Craig Mills	2020	Chair	
Mark Napper	2022	Chair	

volved. What follows is a sample of the many fine people who have been part of the AMS.

At its formation, all AMS Executive positions were unpaid. The first Treasurer of the AMS was Frank Rivers, a grower from the Glass House Mountains. After Frank's sudden death in 1988, E.R. (Ted) Davenport and his wife Anne undertook the role with style and skill. Ted's active leadership as Secretary-Treasurer/Executive Officer ended with his resignation in 1995, but he remained on the AMS Board. He received the Graeme Gregory Award from the HRDC for excellence in horticultural research and development.

A permanent location was established in Lismore, first in a dingy back office in Woodlark Street but soon after in Dawson Street, where it remained until 2022. The first paid employee was Ted Davenport, then Jonathan Edwards who served as Administration Officer from 1995 to 1998. Andrew Heap was appointed General Manager in 1999 and left the AMS after nine years of dedicated, passionate and skilled service to the industry.

He was succeeded in 2008 by Jolyon Burnett as Chief Executive Officer. Jolyon's intellect, professionalism and leadership, in Australia and internationally, added to the reputation of the AMS as a respected and successful organisation and he led Australia and guided the global industry through major changes. Jolyon retired to a well-earned new life in 2022, to be succeeded by Clare Hamilton-Bate who has a successful background in horticulture and strong leadership skills.

Col Heselwood, the first President of the AMS, was a man of vision. After suffering declining health for a number of years, he passed away in 1991. The Colin Heselwood Marketing Award commemorated his passion and focus. In 1977, Col was succeeded by his respected Vice President, Eric Cottam. Jack Gowen followed Eric in 1980 and gave the AMS strong, impartial leadership until 1989, when, due to health concerns, he retired as President but continued in the role of Vice President. Jack was succeeded by Keith Ainsbury, whose oratory skills made him a natural speaker.

AMS leaders

Keith Ainsbury was a larger-than-life visionary, whose leadership skills shone in everything he did. He commenced his macadamia career with CSR in 1973, then became Manager of Braham and Hoult's Macadamia Plantations of Australia in 1978. He was a founder and CEO of the processor and marketer Agrimac International. His early death in Malawi in the late 1990s was a loss to both his family and the global industry. Concerned that his employment by a processor might be perceived as a conflict of interest, Keith stood down in 1991 and was replaced by Andy Stapleton.

Andy had established his Dorroughby orchard in 1975, and in 1988 had been seconded to the AMS Board to replace a retiring Director. A strong leader, he contributed discipline, the clear identification of goals, and by writing software for the AMS and the industry, he brought many members into the digital era. As a retired Qantas Senior Check Captain, he believed that the concept and application of quality assurance was the reason that flying had become so safe, and he introduced this philosophy to the macadamia industry.

In 1994, Andy stood down as President and Geoff Garratt accepted the role. Geoff owned the small orchard at Gympie planted at the end of World War II by Bernie Mason. He was skilled in uniting the different sectors of the industry to focus on common goals and became known as the 'People's President'. After a year in office, Geoff was prepared to allow Andy Stapleton to take over again, but in May 1996 the stress of the role led Andy to resign, and a few weeks later he suffered a major heart attack. Geoff Garratt then resumed the Presidency until he retired in 1997. Fortunately, after Andy's recovery, a quieter lifestyle allowed him to live another twenty years.

Over these years, Board members have included: Geoff Garratt, John Jodvalkis, Ross Litchfield, Berry Spooner, Andy Stapleton, Ian McConachie, Bev Atkinson and many others. Ian Mulligan was elected to the Board in 1994 and, with his direct and engaging style, became the Treasurer.

Rod Fayle, a former Submarine Captain and a leader of the Royal Australian Navy, served as President from 1997 to 2000, and continued to address industry needs and support growth. He was an advocate of single-desk export marketing, a concept that had been debated for several years. He promoted the importance of guaranteed crop payment (particularly after the collapse of The Peninsular Group) and a Strategic Plan that would stabilise the industry. His proposal to guarantee payment to growers by guaranteeing the assets of processors could not, however, be realised. Rod did not seek re-nomination at the end of 2000 and was succeeded by Cliff James, an articulate grower from the Northern Rivers, whose direct style and open addressing of issues resulted in many progressive outcomes. When Cliff retired in 2003, Graeme Hargreaves, who had contributed to marketing and promotion at Macadamia Processing Australia, became Chair, with Kim Wilson as Vice Chair and Andrew Burnside, John Gillett, David Hughes, Phil Montgomery and Andrew Pearce as Board members.

Graeme, whose strength was in marketing, retired in 2007 and was succeeded by Andrew Pearce, a young visionary and an able leader from Bundaberg.

In late 2008, Kim Wilson was elected Chair, bringing to the AMS his thirty years' experience of growing and consulting in macadamias. Andrew Starkey was Vice Chair, and Board members who have played an ongoing role include Richard Doggett, Pam Woods, Doug Rowley, Greg Woods, Larry McHugh, Andrew Leslie, Aimee Thomas, Sandra Lindstrom, Kay Spencer, Morag Page, Brice Kaddatz, Matthew Durack, Michael Waring, Yolande Bromet, Jim Twentyman and Trevor Steinhardt. In 2013, Andrew Starkey was elected Chair and led with dedication and strategic thinking. He resigned in 2014 and was followed by Richard Doggett, who led the Board competently until his tragic loss in a farm accident in 2020. Following Richard's death, Craig Mills capably led the AMS until 2022 when he was succeeded by Mark Napper, ex-Deputy Chair of Horticulture Australia Limited with a long history of board and chair roles and a proven record in guiding and transforming organisations. Mark is the first AMS Chair who is not a macadamia farmer and thus the first non-AMS member to hold the position.

Over the years, there were many who contributed so much to the industry and AMS. First was long-term grower, nurseryman and researcher Norm Greber, whose vision and writings had a major influence on the AMS, became the Society's first Patron

Participants at the inaugural Women in Macadamias event, Glass House Mountains, 2022. Credit: AMS Annual Report 2021/22.

in 1980 and later an Honorary Life Member.

Matthew Durack, a Veterinary Scientist, has worked in the industry for more than thirty years. He managed Stahmann Farms' pecan orchard and became General Manager of the company. He continued on the Board of Stahmann Websters, chaired the AMS R&D Committee, and in 2021 is a pioneer, growing macadamias on a large scale on the coastal flats south of Ballina.

Michael Waring, long-term Vice Chair of the AMS, retired from the Board in 2019, and was elected Chair of the International Nut and Dried Fruit Council (INC), where he displays strong leadership. Michael's father, John Waring, founded Michael Waring Trading (MWT) with its associated companies, and received the Golden Tree Nut Award from the INC. Michael and his brothers Andrew and Christopher continue to lead MWT as marketers, growers and processors.

Kim Jones, a Research Horticulturalist, was appointed Macadamia Industry Development Officer in 1999, due to the availability of levy funding. As the interface between the AMS and growers, Kim led many cultural and regulatory advances. After leaving the AMS, he purchased the company Cropwatch from Mary Burton and provides services to assess and optimise the quality of NIS.

In 2011, Robbie Commens became Industry Productivity Officer, a role where he applied his skills and enthusiasm in an entertaining, challenging style. In 2019 he resigned to manage a large orchard on the coastal flats south of Ballina.

Brice Kaddatz ably served as Treasurer and Secretary of the AMS Board and as Chair and Treasurer of the Macadamia Conservation Trust (MCT). Brice continues to serve the industry in a number of roles within AMS, as well as being a respected horticultural consultant and managing the introduction of new cultivar MCT1 on behalf of the MCT.

The AMS expanded to provide more services from humble beginnings with only voluntary staff, to having part-time administrators, then a permanent office, and eventually seven full-time staff plus three part-time employees. The devastating floods in early 2022 in Lismore severely damaged the office in Dawson Street, requiring temporary facilities and some staff working from home. In late 2023, the temporary office was closed and a 'modern virtual' office created.

Without the influence and action of the AMS, the industry would not have developed the structure, cohesion and outcomes that have made it the global leader. Now with the growth in South Africa, China and other countries, it is time to share the load.

Chapter 10

AUSTRALIA, 1970–1980

Rapid growth from 50 to 1,400 tonnes per year

The rightful place for such a gem [as the macadamia] is up top
amongst our other rare and beautiful native species such as the
Emu, the Kangaroo, Koala, Echidna, Platypus, etc..

(Norm Greber, letter to AMS, June 1976)

Industry overview

Largely because of CSR's initiative, the coming decade was obviously going to be one of growth and development of commercial farms rather than a continuation of a cottage industry. Australia's macadamia industry began to visualise a future as successful as that of Hawaii. The University of Hawaii was sharing its commercial varieties and its cultural and marketing research. Study tours by researchers and interest from growers and entrepreneurs were demonstrating the success, initiatives and potential of the Hawaiian industry and indicating that Australia had the opportunity to follow suit.

In Australia, the substantial orchards planted by CSR gave the local industry confidence and direction, although CSR's research was not always shared with other growers. Governments were providing applied research and extension services, and nurseries were propagating both Hawaiian and Australian cultivars. Growers were accepting the need for higher standards of tree culture and considering mechanical harvesting.

Trevor Grant's integrated small orchard from the late 1970s. Credit: I. McConachie.

In 1972, the New South Wales Department of Agriculture published *Investment in a Macadamia Nut Orchard*,[1] estimating a likely price of 38 cents per kilogram and a yield (rather inflated) of 50 kilograms of NIS per tree at Year 20. As other processors entered the industry, price to the grower increased significantly, so that by 1978 the price per kilogram had increased to $1.40. The labour cost of engaging hand harvesters varied from 25 to 40 cents per kilogram of NIS.

Farming: a growing industry

There was no formal reporting of orchard statistics at this time, so published figures from different sources tended to vary widely. In 1973, the Queensland Government published *The Australian Macadamia Nut Industry Review,* which provided a detailed statistical and written analysis, including projections. This report stated that in 1969–70, Australia had 62,500 non-bearing trees and 15,600 bearing trees, which produced 54 tonnes of NIS. No orchard produced more than 10 tonnes. By 1975–76, production was 400 tonnes, and by 1979–80 it was 1,740 tonnes. A detailed survey by the Queensland Department of Primary Industries (QDPI) for 1977–78 listed 164 growers in that state, who had 54,000 non-bearing and 98,000 bearing trees, producing 714 tonnes of NIS and averaging 7.35 kilograms of NIS per bearing tree.

The Australian Macadamia Society, established in 1974, took the lead as macadamia growing developed from a cottage-style to a commercial industry. By 1980, four commercial factories and several small shed-scale businesses were processing NIS into kernel, marketing locally and internationally, and producing a range of retail products.

Increasing crop production from CSR, Macadamia Plantations of Australia and private growers attracted entrepreneurs who could see the potential of macadamias. The early 1970s saw the start of promo-

Macadamia orchard at Glass House Mountains, typical of increasing standards. Credit: Facebook.

tion of macadamias to investors with high incomes, who were likely to be paying up to 50 per cent of their incomes in tax. Investing in macadamia farming was a legitimate tax shelter, but a perception of some of the public that it constituted tax evasion meant that the industry received some negative publicity.

For the next twenty years, however, the development of investor-owned orchards drove growth in the industry. Real money was invested, most operating expenditure was an allowable tax deduction, and tax was payable when the orchards became profitable. Jobs were created and long-term assets were established that produced export earnings for the country.

The first person to promote investment in macadamias and offer management services was Peter Cloutier, of Cloutier and Brosgarth, Lismore.[2] In 1968, Peter, an Agricultural Scientist based in Lismore, recognised the potential of macadamias while assessing 'The Alamo' plantation for a possible buyer. In 1971, he launched a prospectus inviting investment in macadamias and appeared on the television programme *This Day Tonight.* Peter's prospectus created interest and was copied by several competitors, who were more successful in attracting investors. They even copied some of the mistakes in the prospectus. In the late 1970s, Peter moved overseas and his business was acquired by John Wilkie Snr, who developed many large orchards and provided consultancies in Australia and overseas, particularly in China.

During the 1970s, several agricultural management companies offered services to investor owners. From the mid-1970s, Berry Spooner developed many orchards in New South Wales and from 1978, Mac Food Consultants, later Australian Macadamia Management, developed orchards in Queensland.

The most prominent, colourful and successful entrepreneurs were two young New Zealanders, Mel Braham and Tom Hoult, who came to Australia in the mid-1960s with $6,000 to invest. In Queen Street, Brisbane, in 1965, they were tantalised by a delightful aroma from Allan and Stark's emporium.[3] Tom bought four kernels for 20 cents and was entranced. He was told they were the rare, almost unprocurable, macadamia nuts. Tom realised that 'there had to be money in growing something that tasted so good and cost so much'.

Tom and Mel headed for the nearby QDPI office and were told that macadamias were difficult to grow and they should speak to CSR. CSR told them not to consider macadamias, as they had little future. When Mel and Tom found that CSR was investing millions into developing orchards, they read between the lines and were convinced that they had found the answer to making their fortune.

In 1967, they bought steep banana land near Lismore, but then moved to Dunoon. There, through their Braham and Hoult Land Company, which became Macadamia Plantations of Australia (MPA), they planted 80,000 trees. Investors were told that money really did grow on trees and macadamias

Tom Hoult, entrepreneur and joint founder of Macadamia Plantations of Australia. Credit: *Australian* (newspaper).

were 'The Money Tree'. MPA advised that, for a 10-acre orchard, an investment of less than $200,000 would buy land and pay costs for the next six years, and deductible expenditure would reduce taxation by $100,000. The appearance of well-managed orchards, with dark green trees and freshly mown grass, attracted new growers.[4] In 1978, MPA built a processing factory, which was soon offering retail packs and continues today as Macadamias Direct. In 1989, Braham and Hoult sold a controlling interest in MPA to the New South Wales Superannuation Board.

The professional promotion of investment in macadamias by Braham and Hoult resulted in the successful establishment of many orchards in the Northern Rivers. The beauty of the countryside, the rich red soils, and the proximity to Pacific coast beaches and resorts attracted growers who developed orchards themselves or used consultants.[5]

Mel Braham, whose entrepreneurial drive pioneered investor-owned orchards. Credit: *Land* (magazine).

The start of ventures by Macadamia Plantations of Australia, Dunoon, c. 1973. Credit: T. Hoult.

MPA factory in 1978. Credit: I. McConachie

In New South Wales in 1979–80, there were 962 hectares planted, of which 326 hectares were bearing and 600 were not bearing. In 1980, the New South Wales Department of Agriculture reported that 147 growers had planted 1,583 hectares, of which 95 hectares were still seedling trees, and that the 282,000 trees cropping produced a total of 400 tonnes.[6]

Part of the macadamia industry's growing confidence was the availability of better-performing varieties. There was ongoing interest in the assessment of commercial varieties, which were predominantly Hawaiian, and locally people were assessing trees and making selections they believed were superior. The Brisbane City Council had planted seedling macadamia trees on nature strips in suburbs such as Windsor, Oxley and Zillmere, and these were assessed by cracking a nut from each tree. Local residents were sometimes interested and sometimes an-

MPA's premium product range, 1979. Credit: T. Hoult, MPA.

tagonistic. No outstanding selections emerged from the hundreds of trees examined.

In Maryborough, Dick Misfeld, an enthusiast with sound science skills, examined many backyard trees. He selected Heilsher, Poacher, Daddow and Goldsmith, each named after the household owner, and planted them with Hawaiian varieties in the suburb of Tinana. Daddow, noted for its robustness, remains a major variety in Australia and Africa. Initially unpopular because it tended to develop an unattractive leaf mottling, it was nevertheless hardy, cropped reliably and had uniform, flavoursome kernels.[7]

Propagation involved improvements in grafting, which had been seen as a technique known only to a few. In 1972, the New South Wales Department of Agriculture published a detailed pamphlet,[8] describing conventional grafting, seed grafting and chip budding. Seed grafting, where a scion was inserted into a germinating nut, became popular, but only surface roots developed and many of these trees were lost to strong winds.

Australian varieties planted in the 1970s were mainly H2 (Hinde) and Own Choice, although many others were trialled.

Hawaiian varieties included HAES 246, 333, 508, 425 and 475, with 660 available in the late 1970s. HAES 425 and 475 were not suited, and 333, while hardy, was found to have a low kernel recovery and was not a high producer. HAES 246, 508 and 660 were the enduring varieties planted during the late 1970s.

HAES 344, a well-regarded Hawaiian variety, was brought into Australia by CSR, who assessed its potential by top-working grafts on to mature trees. It became the variety most planted over the next ten years and in the 2020s it is still being planted.

In 1979, QDPI obtained and released the newer Hawaiian cultivars, HAES 741 and 800.[9] Initially, 741 was recommended for higher elevations and cooler areas but it became one of the main varieties widely planted in Australia, while 800, a large tree, suffered wind damage and had only a modest crop and kernel recovery.

There was a belief that Australia could not achieve the yield performance of Hawaii because of Hawaii's more equable climate and lower seasonal and diurnal temperature differences. After a study tour of Hawaii, Brian Cull, Principal Horticulturist at the Maroochy Horticultural Research Station, pointed out that Hawaiian cultivars were selected for the Hawaiian environment and varieties suitable for the Australian environment were needed to lift productivity. This has taken forty years to achieve.[10]

Near Lismore, Italian immigrant Hugo Baisi took a different approach to orchard layout. In Europe, he had observed that forcing the branches of fruit trees to grow horizontally induced earlier and increased production. For his macadamias, he developed a high-density espalier system, using posts and training wires. This resulted in high early yields, but the macadamias grew too vigorously to maintain the espalier.[11] Hugo also pioneered seed grafting and distilled a potent grappa for unsuspecting visitors.

With support from government departments, consultants and managers, and the example of CSR, orchard practices were refined. The importance of soil health was generally not understood. Herbicides were widely used, particularly glyphosate (Roundup), and bare earth along rows became the standard. The use of inorganic fertilisers, together with soil and leaf analysis, became standard practice. In-

Hugo Baisi adapted Italian trellising systems, seeking to optimise production per hectare. Credit: *Northern Star* (Lismore).

secticides and fungicides and the ability to apply them were adopted, often on a calendar basis rather than a needs basis and often having ineffective coverage.

Tree spacing was controversial and varied from 10–12 metres between rows and 10 metres between trees, to as low as 5 metres between rows and 3 metres between trees. Brian Cull investigated tree

A proto-harvester trialled in early development in the 1970s, before the finger-wheel concept was adopted. Credit: unknown.

This Shaw Dehusker was imported from California in 1975 and, after modifications to suit Australian conditions, became the model for most of the industry. Credit: P. Shaw.

spacing and demonstrated that, when sufficient light was available along each side of a row, enough flowers would be initiated. The distance between trees could thus be reduced to 8 metres inter-rows and 3 metres along rows. By about Year 6, this high-density spacing resulted in a continuous hedge along the rows, which minimised wind damage and increased early yield per acre.[12]

To reduce the cost of harvesting, mechanisation became a priority. Tree shakers were trialled but were not successful. In the late 1970s, mechanical sweepers and harvesters that had been designed for almonds, walnuts and pecans in California were brought into Australia. Crop handling advanced, and an imported Shaw, spring-loaded, pressure-plate dehusker was copied and became the standard. Silos with fans replaced drying racks.

The economic benefits of irrigation were debated. Orchards in higher rainfall areas were not irrigated, but these orchards produced more variable crops.

Johnson Farm Management set up one orchard at Coffs Harbour in 1970, and by 1989 about fifty orchards, containing a total of 30,000 trees, had been established as a retirement plan for a number of Canberra-based public servants from the Australian Defence Force. Later, a small factory, Nambucca Nuts, was established.

Research

At the Sandy Trout Food Preservation Laboratory of the QDPI, Rowland Leverington continued his applied research into macadamia technology, including varietal assessments, crop-handling practices, quality, shelf-life, and processing and packaging. Richard Mason was responsible for most research in these fields from the 1970s to about 2000. Ray Bowden continued applied research into aspects of quality. The QDPI Department of Marketing assisted the industry by providing contacts and market and trade information.

The expanding industry faced major crop losses caused by insect pests, such as macadamia flower caterpillar, fruit-spotting bug and macadamia nut borer. A priority for research was pest and disease control, and during the 1970s, CSR employed five entomologists to develop pest control strategies. David Ironside from the QDPI reported that there were 200 insect pests that could potentially damage trees, flowers and nuts. Six pests were serious enough for control measures requiring high-pressure spraying of specified insecticides. David developed an assessment protocol to monitor both live and parasitised macadamia flower caterpillar eggs, which led to the introduction of integrated pest management.[13] Fungal diseases were also studied by plant pathologists, with trunk canker (*Phytophthora cinnamomi*) and husk spot (*Pseudocercospora*) considered the major problems.

The AMS, in conjunction with processors, marketers and researchers, finalised and presented as an attractive brochure the grades and specifications of macadamia raw kernels.[14]

Marketing

In 1975, Nutta Products, a Brisbane company that produced and marketed a large range of table nuts and margarines under the Daffodil brand, became a processor and retail marketer of macadamias.[15] In the first year, the company processed less than 20 tonnes of NIS, which provided 4 tonnes of kernel. Marketing even this small amount of kernel created an awareness of macadamias and left unfilled demand.

Nutta Products' Marketing Division launched 50-gram foil packs of roasted and salted macadamias on a quota basis, which created such demand that, in a few months, about 20,000 of these packs were sold at 50 cents each. In a few years, Nutta Products were selling 200,000 packs per year.

People

Reg Young was an electrical engineer who 'retired' to a property on Mill Hill Road, Montville, where there were 5 acres of massive seedling trees. In the 1950s, this property had been owned by Eleanor Dark, the Australian historical novelist, and here she wrote her last book, *Lantana Lane* (1959).

An example of processing standards in the 1970s. No image sorters were initially used to remove shell and grade kernel. Credit: I. McConachie.

Norm Greber was awarded a Medal of the Order of Australia in 1993.[16] Reg Young became a close friend of Norm Greber, and in 1976 he established a 'model' experimental orchard on 14 acres at Beerwah, where, for the rest of his life, he deeply considered all things macadamia. Reg invented, built and modified machinery, experimented with tree culture, and assessed new cultivars. Describing himself as a recluse who liked people, he wrote profusely as an answer to insomnia. His thought-provoking papers on root systems, tree spacings, nut drying and light requirements stimulated discussion. At the entrance to his farm shed was a large sign with a message derived from the *Rubaiyat of Omar Khayyam*.

Myself when young did eagerly frequent
The Macadamia Society and heard much argument
About it and about; but evermore
Came out by the same door as in I went.

By 1985, Reg's largely experimental orchard was being assessed annually. A significant part of this orchard included local selections mixed with Hawaiian varieties. After eight years, the accumulated crop yield of the three Australian cultivars, Daddow, Mason 97 and Own Choice, led the Hawaiian varieties 660, 333 and 246, confirming the potential of local varieties (Mason 97 had a low kernel recovery and was discarded).

Henry Bell immigrated from New Zealand in 1956, married Alison, and grew pineapples near Beerwah. A few seedling macadamias, planted on his lower farm in the 1920s by Henry Weyer, captured his imagination. In 1961, after removing the pineapples, he planted an orchard with alternate rows of macadamias and citrus, hoping to minimise insect pest transfer. He named his orchard 'Hidden Valley Plantations', and in 1974 he built a very small factory there.

When Henry planted Hawaiian varieties, he sometimes allowed a branch to grow from below the graft and found that at times this branch produced more nuts than the cultivar. He crossed the Norm Greber hybrid selection of Renown, which had originated from Walter Petrie, with another Greber variety, Own Choice, and planted the results as a trial.

Henry was an innovator, inventor and visionary. When he passed away in 2008, his pioneering passion for plant breeding and willingness to exchange information were sorely missed. His son David inherited the family intellect and vision, and he with Alison successfully continued Henry's work. Alison passed away in 2022, but David continues with innovative macadamia breeding and application of new technology.

Dr George Gray was a leading medical specialist based in Melbourne, whose tax bill was half his earnings. He moved into macadamias, encouraged others to do so, and led creative but legal approaches to tax-effective orchard development. His eccentric character and brilliant mind made him a colourful industry asset. His orchards and nurseries in Lismore were managed by Jack Wilson and later by Jack's son Kim, who expanded George's ventures into Mackay. In the 1980s, George Gray part-owned The House of Macadamias, a small factory in Melbourne. He passed away in 2013.

Influential early growers were the Viola family. The grandfather Tony Conte, on observing wild local trees, selected nuts and established a small *M. tetraphylla* orchard. With his daughter Maria and son-in-law Aldo, they established a larger orchard at Binna Burra near Bangalow. Their sons Ash and Bill became early nurserymen, orchardists and promoters.[17]

Professor Richard (Dick) Hamilton from Hawaii and Dr Bill Storey from California were much respected, popular, experienced, competent global advisors. On their visits to Australia, they made recommendations that assisted the developing industry.[18]

Craig Paton, a commercial artist, became entranced with macadamias from 1975. He had a small orchard, a processing plant and retreat at Numulgi, near Dunoon, which supplied the family confectionery factory in Melbourne. His flair was in package design and promotion, and his chocolate-coated Paton Macadamias became a major duty-free store item in airports throughout the world. He claimed to provide macadamias from the tree to the mouth, and also sold macadamia honey, stating on the label that it was so rare that customers should buy only one jar so others could enjoy it. Predictably, jars just disappeared off the shelves. Paton's four sons continued the business until it was acquired by the Dymock Group of book retailers in 2010, and has since been resold.

In 1971, two gifted entrepreneurs, Bill and Lyn Taylor, purchased a 22-hectare property on the Bruce Highway at Woombye and developed a most successful agri-tourist attraction. It featured a fibreglass 'Big Pineapple' with a viewing platform, and its internal walls displayed the subtropical fruit and agriculture of the region. Apart from a large shop and restaurant, the development included a working farm, a children's farm, an animal zoo and a train ride. The 'Sunshine Plantation' was supportive of the Australian Macadamia Society and held an annual

Macadamia Nut Week. One year, a mature tree was transplanted from Gowen's 'Sahara Farms' orchard to the front of the Big Pineapple. In 1980, the Taylors' combined promotion with the AMS comprised industry displays, competitions, media reports and a Miss Macadamia Competition, and featured Mike Walsh, a popular TV personality. The 'Sunshine Plantation' became recognised as a major Australian tourist attraction. On 12 April 1983, Prince Charles and Princess Diana visited the 'Sunshine Plantation' and rode in the Nutmobile.[19]

From 1976, some of the smaller growers in the Northern Rivers were C. Johnson, F. Crawford, K. Frederickson, P. McCarthy, House With No Steps, A. Howard, G. Watts, W. Hoskins, M. Ivan and H. Baisi. Calvin Ogimori from Hawaii was engaged by Lee Black to develop an orchard west of Lismore.[20]

Larger orchards, from 20 to 60 hectares, were mainly managed for absentee owners. Some owners were the Pinter family, David Macrae of Pacific Plantations, Lee Black, originally from Hawaii, J. Davis at Canaiba, R. Margueles, G. Gray, N. Buckley, Drs Church and Miller, the Queensland Syndicate, G. Wilks and R. Tjoing. Others were J. Middleton, K. Smith and A. Stapleton. The largest orchard, at 288 hectares, was Braham and Hoult's Macadamia Plantations of Australia.

In Queensland, apart from CSR and 'Sahara Farms', growers included N. Richards, C. Hes-

Nutmobile with Prince Charles and Princess Diana at 'Sunshine Plantation', visiting CSR's MacFarms factory, April 1983. Credit: CSR Ltd.

elwood, E. Cottam, V. Buch, T. Grant, H. Bell, B. Morgan, B. Mason, G. McCullough, W. Collard, R. Young, D. Treadwell, D. Misfeld, D. Whittaker, H. and E. Howard, and the Haylock family. Most of these orchards had fewer than 1,500 trees.

As told elsewhere, Australia Post declined to issue a stamp depicting the macadamias, stating they gave priority to Australia's native flora and fauna![21]

The decade ended in an upbeat mood. The prestigious magazine *Woman's Day* wrote positively about the macadamia.[22] Clive James, the iconic Australian writer, described the macadamia as the 'perfect nut'. The first person to import them into the Britain, he asserted, would make £1 million.[23]

Chapter 11

AUSTRALIA, 1980–1990

Growing pains

The price to the grower goes up by the stairs but comes down by the elevator.

(Phil Zadro's comment to a meeting of growers, July 1989)

Industry overview

By 1980, production of NIS was approximately 1,400 tonnes, with CSR's orchards, orchards in New South Wales and independent orchards in Queensland contributing about a third each. CSR had 490 hectares but 90 per cent of orchards were owned by individuals. There were three processors, and 50 per cent of the kernel produced was exported.[1] The industry, co-ordinated and supported by the Australian Macadamia Society, was receiving positive publicity. Farming macadamias was finally being recognised as a business that incorporated growing, harvesting, processing and marketing. The early 1980s were difficult years, because seasonal factors reduced the crop and its quality. Production was growing rapidly, however, and soon led to concern that supply would exceed demand.

In 1982, the Hon. Mike Ahern, the Queensland Minister for Primary Industries, stated that the in-

dustry was expanding at the rate of 10 per cent a year. Some factories were struggling to pay growers promptly, find finance for expansion, and market the kernel. Price to the grower had risen from a low of 80 cents per kilogram in the mid-1970s to $1.80 in 1980, to $1.95 in 1985, to $3.95 in 1988 and then had fallen to $2.50 in 1990.[2] The resultant price increase for kernel, together with increased production in Hawaii, Australia and South Africa, caused global kernel buyers to reduce or stop buying, in the expectation of a price fall.

This fall occurred in 1989–90 and threw most Australian growers and processors into turmoil. Processors were unable to sell their kernel and world market prices fell by at least 30 per cent. Some processors reduced and delayed their payments to growers. Despite some outrage and threats of legal action against processors, plain speaking by the AMS President, Keith Ainsbury, persuaded most growers to accept the situation, to tighten their belts and to be sure to survive.

Left. Joyce orchard, Gooburrum, Bundaberg. Credit: D. Joyce and AMS.

At the start of the 1980s, growers were often paid a flat rate, irrespective of moisture content and the percentage of sound and unsound kernel. This practice discriminated against growers with high standards. At the request of the AMS, Keith Ainsbury, Richard Mason and Ian McConachie developed uniform standards for NIS sampling and evaluation, which, when adopted, helped growers to compare offers for their crop. These AMS sampling and assessment procedures have continued almost unchanged.[3]

NUTS ABOUT MACADAMIAS

Macadamia Maid. A regular feature at Macadamia Nut Week held at 'Sunshine Plantation' in most years of the early 1980s. Credit: *Nambour Chronicle*.

On the Atherton Tablelands, a plateau with a subtropical climate some 60 kilometres west of Cairns, in north Queensland, Macadamia Plantations of Australia, consultant Berry Spooner and local and investor farmers established large macadamia orchards. The trees grew prolifically, but, over the next twenty years, most of these orchards were lost due to excessive rain, tropical cyclones, and declining productivity.[4]

In 1983, at Tiaro, south of Maryborough, Hawaiian Macadamia Plantations Pty Ltd offered an innovative 'agri-investment', whereby the public could purchase and own parcels of fifteen trees in a large, managed orchard. The total cost, $8,050, was to be paid over six years. The estimated yields of 15 kilograms of NIS at Year 6 and 45 kilograms at Year 10 were many times higher than industry averages. The NIS price to the grower was advertised as $12.50 per kilogram, which was actually the kernel market price. The real NIS industry price was only $1.80 per kilogram. The net income for fifteen trees was predicted to be $2,928 at Year 6 and $14,660 at Year 10.[5]

Following a successful launch, the company and its associated Consolidated Macadamia Plantations Pty Ltd continued to offer the scheme. Investors rushed in, but, despite AMS warnings and formal complaints to the government, they lost their money and became embroiled in legal actions. For a number of years, Hawaiian Macadamia Plantations' small processing plant and tourist shop at Tiaro successfully promoted and marketed macadamias.

In 1984, cyclonic winds damaged about 100,000 trees in the Northern Rivers, and many trees were lost. Some of the trees blown over had been seed grafted, resulting in weak support roots, so this form of propagation was discontinued. Seed grafting was a propagation method where a germinated seed had the shoot or plumule replaced with a small cutting from the intended variety. Some orchards were insured against wind and fire damage, but the insurers suffered heavy losses and made their policies less generous.

In New South Wales in 1985, there were 281 growers, mostly in the Lismore area but also in Brunswick, Coffs Harbour, Macksville, Tweed, Illawarra, Gosford and the Hunter regions. Production from 175,000 bearing trees was 1,000 tonnes of NIS.

In Queensland in 1986, there were 252,999 non-bearing trees, while 375,000 bearing trees were producing 5,670 tonnes of NIS.

CSR had established three large orchards at Baffle Creek, north of Bundaberg, in the 1960s, but no commercial macadamias were grown in the Bundaberg and Childers districts in the early 1980s. CSR had also purchased a large grazing property at Yaamba, north of Rockhampton, and named it 'Macfields', but although by 1988, 303 hectares of this property had been cleared for macadamia planting,[6] the CSR era had by then come to an end with the sale of its Macadamia Division to Lance O'Connor, representing a syndicate, and to Tom Modisette from the US RAM Group.[7]

The economy of the Bundaberg region was dependent on sugar, an industry that had developed over the previous one hundred years with highly mechanised farms, five sugar mills, efficient transport systems and government-supplied irrigation water. The prime, almost flat farmland, although ideally suited for macadamias, was considered sacrosanct, and it was almost impossible to secure entry. However, sugar prices are globally cyclical, and during market downturns in the 1980s, cane farmers were planting vegetables and crops such as aloe vera to diversify and supplement their incomes. In 1987, during a downturn in sugar prices, Lincoln Doggrell, who had been an entomologist and manager with CSR, and Ian and Jan McConachie from Gympie formed Macadamia Farm Management Pty Ltd, purchased cane farms for investors and developed them as modern macadamia orchards.[8]

The Pearce and Carroll families, Bruce Moore, Dave and Geoff Chivers, and David and Ann Lynch were among the new growers in Bundaberg, and some farmers, including cane growers, realised the potential and started planting macadamias.[9]

Phil (F.C.) Zadro, a businessman and macadamia grower from the Northern Rivers, saw Bundaberg's advantages and purchased one cane farm, then an-

Example of a young, high-standard orchard in the Northern Rivers, c. 1980–90s. Credit: I. McConachie.

CSR imported a US almond sweeper for use on larger orchards from the late 1970s. Baffle Creek, c. 1985. Credit: I. McConachie.

CSR imported a US almond harvester for use on larger orchards from the late 1970s. Baffle Creek, c. 1985. Credit: I. McConachie

other. He has continued to purchase land and plant macadamias, so that the Zadro family is now the largest grower in the world. A legendary figure, Phil came from Italy to Australia as a teenager in the late 1940s with 2 shillings to his name and little command of English. Through hard work, informal study, sound business skills, self-belief, imagination and lateral thinking, he successfully established a number of companies in the construction industry.

In the late 1970s, seeking a retreat from work pressures, Phil was looking at rural properties near Lismore with John Wilkie Snr, when he saw young macadamia trees. Finding they did not require significant labour and thus entailed minimal trade union dealings, he established 'Victoria Park' near Lismore in 1979 and ploughed his later profits into more and more orchards.

Many of Phil's orchards are in the Bundaberg district, where, with support from the Macadamia Processing Company (MPC), the Zadro family established a large, state-of-the-art factory, Pacific Gold Macadamias, in 2011.

The Zadro family continues to establish orchards, including in a new environment at Emerald in central Queensland, where high temperatures are common in summer. In 2017, the Zadros owned 5,000 acres of macadamias in Australia and 1,000 acres in South Africa, producing a total of 7,500 tonnes of NIS. In 2018, an even larger factory was opened in South Africa.

Phil Zadro's long-term influence on Australian and global processing and marketing began in 1983, when he and three other growers engaged Peter Fusarelli, a young engineering student who had grown up with macadamias at Dunoon, to design and build an experimental processing plant. The New South Wales Macadamia Processing Company Pty Ltd commenced in a shed at Blue Hills, Goonellabah, and processed 53 tonnes of NIS in its first year.

Always a visionary, Phil recognised the benefits of inviting other growers to join the company. In 1985, the plant was relocated to South Lismore with four shareholders who were macadamia growers. This factory was small but technologically advanced, particularly with its separation of shell from kernel. Its growth was rapid, and in 1987 it changed its name to Macadamia Processing Co. Limited (MPC). On 14 acres at Alphadale, east of Lismore, the company in 1987[10] built an attractive, modern, co-operative factory, which expanded on the site and in 2018 was the largest processor in the world. In 1987, MPC processed 543 tonnes of NIS, increasing to 1,632 tonnes in 1990.

Farming

By the mid-1980s, many large orchards began to use expensive sweepers and harvesters from the USA, including the Florey Range and the Savage. These were designed for almonds, walnuts and pecans. Most smaller or newer orchards were harvested by hand, which was costly and labour intensive. However, sweeping and blowing nuts away from trees caused erosion of orchard floors and damage to the health of trees and soil. Researchers considered that maintaining orchard floors should be a priority, but initially many growers did not see this as being important.

In Australia, a competition to invent a practical harvester led to the display of a prototype, based on stiff broom bristles that held the nuts. In the USA at the same time, a finger-wheel machine was invented to pick up golf balls from driving ranges, and this was developed as a nut harvester. Known as the Bag-a-nut, this hand-operated machine had rotary plastic or metal fingers, which rolled over the ground and trapped nuts, which were then ejected into a storage bin.

Crop protection usually took the form of spraying of pesticides or fungicides on a calendar basis. Researchers learnt the biology of pests and diseases, monitored pest levels, determined thresholds of damage, and introduced some programmes that were the precursors of Integrated Pest Management (IPM).

As the industry grew, previously minor pests and diseases became more of a concern. The loss of crop to rats was highlighted when infra-red cameras showed that a rat could gnaw through a shell in eight seconds. An understanding of rat ecology, baiting and orchard hygiene led to a significant reduction in damage.

Felted coccid, a scale insect that reduces tree vigour, was difficult to control because it grew on the underside of leaves. Some varieties of macadamia harboured husk spot, a fungal disease, that causes nuts to drop before they mature. Control procedures were introduced and then refined to use a range of fungicides as required each year.

Environmental issues, such as noise and the use of agricultural chemicals, became associated with 'right to farm' challenges by local residents. Dealing with these issues required open communication, tolerance and goodwill, meeting the needs of all parties and adopting higher standards of farming.

Taxation

A major influence in the economics of a long-term crop like macadamias was that, where a grower met the guidelines for primary production, taxation laws allowed most costs to be deducted from other income. New growers with high incomes from other sources could reduce their tax liabilities until the orchards were profitable. Large commercial orchards were being developed for farmers and absentee owners. Smaller orchards were often planted by owners with no farming background, who sought a rural lifestyle that would provide an income when they retired from city life. However, the Australian Taxation Office (ATO) was monitoring the emphasis on tax benefits and schemes to reduce the capital value of nursery trees that was at best creative. The ATO signalled a closer assessment of the taxation status.[11]

Research

After Norm Greber showed how to propagate in the 1950s, almost all trees were grafted and a few were grown from cuttings. The choice and benefits of different rootstocks were trialled in nurseries. Seed nuts of *M. tetraphylla*, usually from 'Rous Mill' or the *M. integrifolia* variety H2, were used, and the hybrid

Beaumont (HAES 695) gained popularity.

In 1980, Dr Russ Stephenson and Calvin Winks of the Queensland Department of Primary Industries (QDPI) commenced plant breeding, working with Tim Trochoulias, Ross Loebel and others from the New South Wales Department of Agriculture. A plant breeding programme that followed from earlier selection trials became the largest industry research project and, later in the decade, it required significant voluntary and statutory levy funds.

In the early 1980s, through the goodwill of the University of Hawaii, thirty-eight cultivars were imported by the QDPI and planted out in five orchard trials in New South Wales, southern Queensland, Rockhampton and the Atherton Tablelands. These trials involved forty-four selections, some under code names, including commercial and promising experimental Hawaiian cultivars and Australian cultivars such as Daddow, as well as A4 and A16 (neither of which had yet been released). The resultant selections were to change the industry. HAES 816, 842 and 849 showed promise and became widely planted. In March 1989, A4 and A16 were protected by Plant Breeder Rights, the first macadamia plants in Australia to be awarded this protection, and released by the breeders, Henry, Alison and David Bell. Both these varieties were hybrids, which cropped early and had high kernel recovery for the time.

At Knoxfield, Melbourne, the Horticultural Research Institute reported on nine varieties planted in 1976 and assessed in 1983 and 1984. The climate in Victoria proved unsuitable for commercial plantings. A similar CSIRO environment suitability trial, planted in 1976 at Dareton near Mildura in north-western Victoria, was assessed through the 1980s. Many of these trees survived but did not thrive, and they also proved to be unsuited to the temperate climate.

Industry advances through research projects included improved crop protection, erosion control, irrigation, handling, drying and the use of Ethephon (marketed as Ethrel) to cause the nuts to drop from the tree. At processing level, projects to research quality, define and adopt control procedures, and improve shelf-life through barrier packaging were undertaken and quickly implemented.

Selection of superior plants and varieties had been ongoing. From the first commercial plantings in the nineteenth century, growers and nurserymen had sought thin-shelled nuts and trees that produced large crops. From 1948, the Queensland Department of Agriculture and Stock had a small plant-breeding programme, which led to a field trial at the Glass House Mountains in 1963. While this trial provided information, no new cultivars emerged that stood the test of time. The importance of irrigation in producing consistent large crops was emphasised by trials at the Maroochy Horticultural Research Station.[12]

Understanding the genetic make-up of macadamias began in 1987, when Dr Visanthage from the CSIRO isolated isozymes, which are indicators of the plant's germplasm. In the 1990s, CSIRO started an ongoing breeding project, based on plant science principles.

In 1984, there were fourteen proposed research projects seeking funding, and twenty continuing projects funded with small amounts of money. State governments, seeing the potential of macadamias and wanting to support industries that helped themselves, undertook much of the research.

The First Australian Macadamia Research Workshop was held at Newrybar, New South Wales, in 1983, and the Second Macadamia Research Workshop was held in 1987 at Marcoola on the Sunshine Coast. Organised by Dr Russ Stephenson, these events featured advances in nutrition, soil and leaf sampling, marketing, cultural management, quality from farm to retail, and health research. As well, they determined future priorities.

A detailed economic analysis was undertaken in 1984 by the Australian Government, which provided a positive independent assessment of the industry and gave confidence to all aspects of its further development.[13]

Processors

Processors in the early 1980s included CSR's Mac-Farms of Australia, Braham and Hoult's Macadamia Plantations of Australia, Nutta Products trading as Meadow Lea, and the Macadamia Processing Company.

In the mid-1970s, David Macrae had developed 'Pacific Plantations', a large, innovative orchard, at Brooklet, near Bangalow in the Northern Rivers. In the late 1980s, he built an attractive factory, using low-temperature, heat-pump driers. By 1995, 'Pacific Plantations' had established a nursery and were marketing kernel and exporting NIS. They provided services to growers and entered into a joint-venture processing operation in China.

In Queensland, Suncoast Gold Macadamias Australia Limited, an innovative co-operative of eighty growers, was established in Gympie in 1985, and a factory was built to service its members. Ian McConachie was Chair, Peter Zummo and then Jim Twentyman were Managers. Suncoast Gold led the industry in international quality assurance accreditation.

American Deane Stahmann from New Mexico came to Australia with a vision to develop a pecan industry. At Gatton in Queensland and then near Moree in New South Wales, he established Stahmann Farms Inc. In 1982, he built a pecan processing plant at Toowoomba, and in the late 1980s expanded this to process macadamias. At first, they provided contract cracking services, but later they purchased, processed and marketed the nuts in their own right. After thirty years of processing macadamias, Stahmanns have also become large growers and become part of Stahmann Websters.

Stan Fenner, Stahmann Farm's innovative engineer and manager, designed, built and managed their processing plant and led them through implementing world-class quality assurance. A friend to many, he was affectionately known as 'Stainless Steel Stan' because of his insistence on using only the highest grade materials. He passed away in middle age in 1999.

There were other high-standard processors, including Agrimac International, Ozmac, Nambucca

A comparison of Australian and Hawaiian retail packs in the early 1980s. Credit: unknown.

Keith Ainsbury, leader, visionary and AMS President. Credit: *Queensland County Life*.

Macadamias, Henry Bell trading as Hidden Valley Plantations and Nutworks.

In most years, competition for NIS resulted in growers being wooed for their crop. This resulted in processors being forced to pay such high prices that they remained undercapitalised and struggled to finance their operations and expand.

People

Kerry Packer (1937–2005), one of Australia's wealthiest men, inherited control of the media giant Consolidated Press Holdings. A colourful, controversial businessman, he was also the founder of World Series Cricket. Influenced by Harry Chester, an employee and mentor who recognised the potential of macadamias, Packer developed several orchards in the Northern Rivers in the early 1980s. After Harry's death, Packer's interest in macadamias waned and he sold the orchards.[14]

Associate Professor Cathy Cavaletto, a Food Scientist from Hawaii who was acknowledged as the lead global researcher into all aspects of quality, undertook her sabbatical leave in Australia, continuing research into aspects of quality from which Hawaiian and Australian technology benefited.

There were many others who, through their skills, belief and efforts, played a major role in the advances of the time. Andrew Burnside managed many orchards, chaired the R&D Committee for ten years and served on the AMS Board. Others who in their various ways advanced the industry in this decade were, in New South Wales: Rob Baigent, Cliff James, Tim Trochoulias, Ross Loebel, Kim Wilson,

Henry Bell with one of his selections. Credit: A. Bell.

Ian Skinner. From Queensland, Calvin Winks, Ted Davenport and Frank Rivers were the major innovators, along with researchers based at the QDPI Maroochy Research Station, including Russ Stephenson, Paul O'Hare, Eric Gallagher, Brian Cull and David Ironside.

Early attempts to conserve macadamia genetic potential

Norm Greber repeatedly emphasised both the breeding potential and the long-term importance of conserving Australia's flora, particularly the macadamia. Calvin Winks advocated this from the early 1980s and both he and Dr Margaret Sedgley from CSIRO sought international support. A division of the Food and Agriculture Organisation of the United Nations, the International Board for Plant Genetic Studies (IVBGR) in Italy, recommended conservation of the wild genetics through a seed bank. Then, following a submission, they advised that macadamias did not meet their funding criteria as they believed that macadamias, while a food crop, and the wild resources, were not under threat. In 1989, the AMS sought interest in locating and surveying wild macadamias in their rainforest habitat.[15] Despite support from the AMS and industry, little direct conservation occurred for several years.

Initially found in 1982, some 160 kilometres north of *M. integrifolia* habitat, a new macadamia species – later to be named *M. jansenii* – was recognised in 1988. It was both endangered and as rare as the Wollemi Pine, with initially only twenty-three plants found, all in a fire-prone rainforest creek.

The end of the decade

During the decade, confidence was high as both plantings and crops increased.[16] Ross Loebel from the New South Wales Department of Agriculture reported that, in 1985, there were 2,000 hectares of macadamias in the Lismore district, with half not yet bearing. Other significant plantings were in the Brunswick and Coffs Harbour shires. Ross believed production would quadruple from the present 1,000 tonnes of NIS.[17] Dr Russ Stephenson provided a detailed report on the industry which boosted confidence.[18]

CSR's selling of their orchards and factory allowed entrepreneurs into the industry with plans to secure a dominant role. To secure NIS, they offered increasingly higher prices, at a time when production of NIS was increasing. Then in 1989, there was a large increasing inventory of kernel and buyers responded by initially limiting their replacement orders. Almost overnight the price fell and the marketers responded by both actively seeking new markets and being forced to drop prices.[19] This will be expanded on in the next chapter.

It had been a decade of consolidation, growth and confidence, where all sectors of the industry could feel proud of the advances. The 1980s ended on both a low and a high. There was a belief that an industry with unlimited potential was in the making, but the market had collapsed and the industry had to address its now certain growth, create markets ahead of production, and obtain viable prices for the growers. In 1990, the industry was both wiser and poorer, and facing challenges.

CSIRO botanists finding a wild macadamia in Gold Coast hinterland in late 1980s. Credit: I. McConachie.

Chapter 12

AUSTRALIA, 1990–2000

The ups and downs on the way to leading the world

No compromises, no ifs or buts,
This is the eminence of nuts,

...

Enclosed in tight pectoral shells,
A perfect pampered pearl-drop dwells.

(Dr Len Green, from 'Aussie Nuts', 1993)[1]

Industry overview

At the beginning of the 1990s, the macadamia industry had an intact but battered optimism. Rapid increase in the crop, inadequate marketing development and a strong US dollar, which reduced the value of Australian sales, had led to a market downturn in the late 1980s. The industry was divided over how to address increasing production and large, unsold stocks.[2]

Production in 1970 had been about 40 tonnes of NIS, which increased in 1980 to 1,400 tonnes, and in 1990 to at least 8,000 tonnes. It would rise to 20,500 tonnes in 1995, and 30,400 tonnes in 1998.[3] It was estimated that the crop would reach at least 30,000 tonnes by 2000.

The NIS price was $4.00 per kilogram in the late 1980s, but the AMS was warning growers that a long-term price, averaging $2.50 per kilogram, was more realistic. In 1991, NIS prices averaged $1.50 per kilogram, $3.00 in 1995, and $2.80 in 1998.[4]

High prices had caused an upsurge in plantings,

Left. Keith Ainsbury, leader, visionary and AMS President. Credit: *Queensland County Life.*

and by 1990, tree numbers had rapidly increased. When new plantings declined as a result of falling NIS prices, nurseries that had expanded to meet demand were severely overstocked. Grafted trees could be sold for as low as $2.50 each, compared to the previous price of $10 to $12.[5]

Most growers understood the complexities of supply and demand, and accepted that they could have some control of costs and optimise production but could not influence price. Morale was low and tempers were often frayed.

Growers who were concerned about selling their crops often left NIS on the ground until a buyer could be found, which compromised quality. Some growers reacted by cutting back inputs, such as fertiliser, which was counter-productive.

In July 1990, an AMS meeting was held to outline the steps needed for recovery from the downturn, and a plan was developed for both short-term and long-term market development. Although some growers said that they would not accept a price reduction to clear stocks, the President, Keith Ainsbury, told them forcibly to 'lump it or leave the industry'. Keith's strong leadership and respect was much missed after he passed away suddenly in Malawi in 1999.[6]

As a result of the crisis, three of the largest processors formed a joint export marketing company, Macadamia Producers of Australia Limited. During the time delay for the Australian Government to approve this across-industry initiative, the demand for kernel increased. Two members of the joint company Macadamia Producers of Australia withdrew, and the Macadamia Processing Company Ltd (MPC) acquired the marketing company.

Two outstanding pioneers of the industry passed away during the decade.[7] Industry Patron Norm Greber, who had just been awarded an OAM (Medal of the Order of Australia), passed away in 1993, aged ninety-one.[8] Arthur Lowndes, who led CSR Ltd, passed away in 1994.

The AMS in 1990 introduced Quality Initiatives, which had a large effect on raising the quality of kernel by rewarding higher-quality NIS and penalising lower-quality product.[9] The AMS continued to promote the Australian Macadamia Industry Coordination Scheme for a single-desk kernel market.[10]

Dumping actions

Part of the recovery process involved selling accumulated stocks of kernel, some of which were becoming aged, and at whatever price could be obtained. The adjustment of prices to global buyers had a detrimental effect on Hawaiian processors, who considered that Australian marketers were 'dumping' kernel. They sought to have a tariff imposed on Australian macadamia exports and lobbied the US Government for a Senate Enquiry. Bryan Raphael from MPC and Ted Davenport from the AMS flew to Washington and capably represented the industry.

In 1992, the US Government finalised its hearing, but it took six months for the decision to be handed down. While the Senate Enquiry established that a technical breach of US dumping laws had occurred, the US marketers were considered

to have acted unreasonably. The Australian industry undertook not to 'dump', and no penalties were imposed.

In 1998, the US Government held a second dumping enquiry, but Australia was able to demonstrate that no sales below the cost of production had occurred.

For many years, the ability for high-income earners to reduce their tax commitments by claiming a deduction for most establishment and operating orchard expenditure from their total income had been a significant factor in attracting investor-growers. In 1996, the Australian Taxation Office amended the tax laws so that most pre-income expenditure had to be capitalised and later deducted. This initially slowed new investment in orchards, but the industry adapted and plantings continued.

By 1995, 9,000 hectares of macadamias had been planted in Australia, of which 5,500 hectares were commercially bearing. There were 600 AMS members, of whom 500 were growers. *M. integrifolia* together with its hybrids comprised 98 per cent of plantings, and the main varieties were Hawaiian selections. The crop totalled 19,000 tonnes of NIS. Yields averaged 3.4 tonnes of NIS per hectare, which was considered highly productive. In 1995, there were nine processors and other buyers who purchased and contract processed NIS. Most processors were accredited to the International Quality Assurance standard of ISO 9002.[11]

In 1995, 1,700 tonnes of Australian kernel were sold on the domestic market, and 1,000 tonnes were exported to the USA, 630 tonnes to Japan, 359 tonnes to Hong Kong, 165 tonnes to mainland China, 900 tonnes to Europe and 200 tonnes to other countries. Marketers who exported benefited from Australian Government Export Market Development Grants. About one-third of global macadamia production was consumed in manufactured foods, while two-thirds were sold as roasted-salted nuts and in confectionery.

In 1993–94, 1,500 tonnes of Australian NIS had been exported to China. By the end of the decade, this trade had expanded, and China had become a major buyer, grower, processor and marketer. NIS could be imported into China at lower rates of duty than kernel, because cracking in shell created employment for Chinese women, who initially used hammers and sorted by hand. From 1994, at least 20 per cent of Australia's crop was being exported to China as NIS, reducing supply to Australian processors.

NIS prices fell from $3.00 per kilogram in 1997 to about $2.30 in 1999. Growers became concerned about the sale of their crops, which several processors addressed by offering long-term contracts, usually with conditions. The softening market and a 40 per cent crop increase in 1999 put pressure on processors and marketers alike. To some extent, all processors were undercapitalised and found it difficult to receive the whole crop and make timely payments.

A major factor in this downturn was the liquidation of The Peninsular Group, a company formed in the 1980s by Brian Findlay and Winson Woo, who first purchased a macadamia orchard near Bangalow in the Northern Rivers. In 1991, anticipating the potential of the Chinese market, they promoted macadamias as a luxury food and exported increasing amounts of kernel and NIS to China. In that year, they supplied China with 17 tonnes of NIS and marketed kernel to Hong Kong, Malaysia and Japan.

In 1996, The Peninsular Group purchased Mac-Farms orchards at Maleny, Bauple and Baffle Creek and the 'Sunshine Plantation' factory from the Arnott's Biscuit Group. They presented themselves as major players in the industry, but their rapid, ambitious growth led to serious undercapitalisation and, unable to pay growers, they were placed in receivership in 1998. This crisis hampered the recovery of demand and prices into the 2000s, and affected growers did not receive payment for one or two crop seasons.[12]

Stahmann Farms Inc., a major grower and processor of pecans, entered the industry by offering contract processing services and became a leader in technology and quality.[13]

Farming

The earlier decline of the dairy industry in Queensland and New South Wales made suitable land more available. 'Right to farm' issues, arising from the influx of neighbours in farming districts, continued to be debated. These issues included noise, the application of insecticides and fungicides, and the control of spray drift.

By 1997, there were 3,100,000 trees in the ground, 40 per cent mature, 30 per cent at the point of bearing, and 30 per cent not yet bearing. Of these, 95 per cent were Hawaiian varieties.

Yields had always been variable due to the vagaries of nature, the performance of varieties and differing standards of management, making crop forecasting imprecise and planning difficult. Yields were measured by tonnes of NIS per hectare at 10 per cent moisture content. While the industry average was 3.0 tonnes of NIS per hectare, orchards varied by up to 60 per cent above and below that figure.

Mechanisation

The 1990s was the decade of mechanisation, raising standards and applying research. Following on from the Bag-a-nut finger-wheel hand-pushed harvester, in 1990, Andre Rau from Ballina invented a tractor-mounted finger-wheel harvester named Maca-Picker. Then Jack Lennard from Port Macquarie developed and manufactured three sizes of harvesters – the MacMaster, the NutNabba and the Brunt. These machines transformed the industry, reducing the cost of harvesting from an average of 60 cents per kilogram to 10 cents. Similar harvesters are still being used.

An economic analysis by Tim Reilly, Ross Loebel and others from the New South Wales Department of Agriculture in 1993 assessed mechanical harvesters. Their report determined that the full cost – including labour, operating costs and overhead – of using a medium-sized harvester varied between 4 and 24 cents per kilogram of NIS. By contrast, hand harvesting varied between 40 and 80 cents per kilogram of NIS.

Processors

By the early 1990s, Macadamia Processing Company Limited, trading as Macadamia Magic, had become the largest processor in Australia, handling about 25 per cent of the crop.[14] Inspired by entrepreneur Phil Zadro, the Company set professional standards, focused on market development, and was seen as a sound, conservative leader. Bruce Mason was Chair, Larry McHugh was Factory Manager, Bryan Raphael was General Manager and Art Beavis led marketing.[15] The Board consisted of prominent industry people, including Bob Dunlop, Michael Pinter, Jim Grose, Ian Skinner and Noel Buckley.

Macadamia Plantations of Australia Pty Ltd had always punched above its weight with innovation and promotion, and in 1995 it became part of Consolidated Foods Australia Ltd. A new, value-adding factory was established at Convery Lane, Wollongbar, New South Wales, and retail production was expanded under the Pacific Gold label. Graeme Hargreaves was General Manager and Judy Grainger energetically led marketing.[16]

Agrimac International Enterprises Pty Ltd, a compact, efficient processor, which largely marketed bulk kernel, was established in 1993 and achieved quality assurance certification in 1996.[17] Its founders, Keith Ainsbury, John Wilkie and Graeme Fleming, were joined by Quality Manager Darren Burton, who was to play an ongoing role in management and marketing.

Crestnut Products, later to trade as Nutworks, commenced in a farm shed in the Sunshine Coast hinterland in 1995. Keith Ryan expanded the basic processing factory to make a range of products. Relocating to Yandina as Nutworks, it included an attractive tourist shop, and apart from processing NIS into kernel had a value-added section producing a large range of retail products. This benefited from its location opposite the tourist attraction of Buderim Ginger.[18]

In 1994, Suncoast Gold Macadamias was accredited to International Quality Assurance ISO 9002 standard,[19] closely followed by Stahmann Farms, and then by most other processors.

Research

At farm and factory, ongoing research into quality, crop handling and drying, with practical applications, was being led by the Queensland Department of Primary Industries (QDPI), the S.A. Trout Food Preservation Laboratory and its successor, the International Food Research Institute in Brisbane. After leading a team of researchers at the QDPI, Richard Mason moved to the University of Queensland at Gatton. Also at Gatton, Dr Tim Kowitz advanced the understanding of NIS deterioration, and developed handling and drying protocols that became industry standard practice.

Long-term research into the effect of the environment, nutrition and water on the productivity and quality of macadamias was conducted by Dr Russ Stephenson, Dr Rob Baigent and Tim Trochoulias for the Queensland and New South Wales governments.

With an ongoing emphasis on quality, an expanding crop, strong competition for NIS and the needs of new growers, the publication 'Industry Uniform Procedures for Sampling and Assessing NIS' was updated.

Sometimes growers were critical of the accuracy or lack thereof of NIS assessments that determined payments. Although the small size of each test sample limited accuracy, the overall procedure averaged out, and did not favour either the grower or the buyer. An example was that a semitrailer load of 20 tonnes of NIS had four separate tests which were averaged so that any error related to the sample size was minimised. The AMS introduced annual meetings of Quality Control Managers in order to standardise procedures, and 'round robin testing' ensured consistent results between factory laboratories.

Macgroup meetings, introduced and led by the AMS Development Officer Kim Jones, were forums for discussion and technology transfer. As part of these meetings, the QDPI in 1997 launched MacMan, a farm management and data recording system, which later included farm recording software, best practice groups, industry benchmarking comparative performance analysis and videos to illustrate new procedures. Shane Mulo and Paul O'Hare led this innovation with support from the New South Wales Department of Agriculture and with input from the University of Southern Queensland and the AMS. To emphasise quality management in all aspects of growing, the QDPI developed the Code of Sound Orchard Practices (COSOP), which became the handbook for all progressive growers, and then the MacSmart information system.

The Horticultural Research and Development Corporation (HRDC) supported research into more effective spraying equipment and the life-cycles of pests and diseases.

Statutory levies

In 1991, AMS members voted to support the adoption of a Statutory Levy to raise funds to invest in research and marketing.

In 1993, this resulted in $220,000 being available for R&D, $188,000 for promotion and $24,000 for administration of the peak industry body, the AMS. In 1994, growers voted for a 4 cents per kilogram levy, with 2 cents each for research and promotion – an initiative that became a major growth factor.

A Strategic Planning Workshop was held to decide goals and priorities. The resulting plan, 'A Vision for the Future', included a SWOT (Strengths, Weaknesses, Threats, Opportunities) analysis, and covered funding, R&D priorities, promotion, new products, marketing, market research, technology transfer and communication.

In 1996, the National Residue Testing Survey, an Australian Government-managed project, was introduced. Samples of kernel representing all districts were collected and analysed for fifteen farm chemicals, both registered and unregistered, and in some years for heavy metals. Full compliance with Australian Standards was reported, and minute levels of other chemicals detected were below Maximum Residue Limits (MRL). These results gave confidence to international buyers. A new economic analysis of

the market was undertaken by the QDPI, which supported the industry marketing programme.[20]

In 1997, the levy was increased to 7 cents per kilogram, and the promotion component was more than doubled. There were 3,250,000 trees and the crop was continuing to increase. In 1999, a softening market and a large increase in the crop resulted in the levy remaining at 7 cents per kilogram, but 1 cent was transferred from R&D to promotion.

The end of a decade

The 1990s were a time of growth and consolidation for the industry but ended a little subdued. The industry had successfully addressed the marketing and price crisis in the early 1990s, was aware of the potential of its product but recognised that demand must remain ahead of production.[21] The AMS provided leadership and managed the research and promotion professionally.[22]

Many small orchards disappeared as a result of economies of scale, a reduced return per kilogram of NIS and the increasing need to mechanise. In the 1970s, an orchard of 1,500 trees was considered capable of supporting a family, but by 2000, at least 3,000 trees were required. There were approximately four million trees planted by 2000, producing 35,000 tonnes of NIS.[23]

Australia had become the largest producing country in the world and led in research, quality and promotion. Other countries, such as South Africa, Kenya and Malawi, were developing commercial industries but relying largely on Australian knowledge. Hawaii was no longer expanding its macadamia industry.

At the end of the decade, the AMS was twenty-five years old, and the industry relied on its leadership. All sectors of the industry supported working together, but strong competition for NIS production existed. Research and promotion ensured a sound, progressive industry that led the world and assisted global growers, processors and marketers.

Chapter 13

AUSTRALIA, 2000–2010

Consolidation

Even to this day, no native animal species and only one plant species,
the macadamia nut, have proved suitable for domestication.
There are still no domestic kangaroos.

(Jared Diamond, Top 300 Jared Diamond Quotes, No. 20)[1]

Industry overview

At the start of the new millennium, Australia was not only the largest producer of macadamias in the world, but it also led in quality, innovations, promotion and bulk marketing. Approximately 3.8 million trees had been planted, of which almost 3 million were bearing and NIS production had reached 35,000 tonnes. Some in the Australian industry were concerned that the rest of the global industry was benefiting from research and marketing funded by Australian grower levies.[2]

Issues the industry needed to address included high stock levels, undercapitalisation of processors, a decline in Asian markets and price competition from overseas producers.[3]

Carry over of NIS and kernel from 1999 resulted in loss of quality. It was a mixed blessing that, due to harvesting difficulties after heavy rain, the 2000 crop estimate of 35,000 tonnes fell to 29,000 tonnes. In 2001, production increased to 34,000 tonnes compared to 24,000 in Hawaii, 11,000 in South Africa, 9,000 in Guatemala, 6,000 in Kenya, and 5,000 in other countries.[4]

From 2000, overseas producers, particularly in South Africa, dropped prices sharply, affecting the Australian recovery.[5] Stocks built up further, kernel purchasers reduced buying, the exchange rate rose, and price to the grower fell. Despite ongoing plantings, a widespread drought from 2002 to 2004 reduced the crop, which at least resulted in prices firming. In 2004, production had increased by 10,000 tonnes to 43,700 tonnes.

Based on an AMS Census, in 2003 there were approximately 150,000 trees in the mid-north New South Wales coast, 2.9 million trees in the Northern Rivers, 340,00 in southern Queensland, 500,000 north and south of Gympie, 1 million in Bundaberg, 150,000 in north Queensland and 70,000 in Western Australia. Some 65 per cent of these trees were bearing. NIS prices averaged $2.20 in 2000, increasing to $3.80 by 2005.

Left. A typical, high-standard orchard at Bundaberg, about 2008. Credit: I. McConachie.

Prices to the grower in 2000 averaged $2.12 for a 29,500-tonne crop of NIS increasing to $3.60 in 2005 with 35,500 tonnes of NIS and, by 2007, with an increased crop and sluggish demand, prices to the grower had fallen to $1.50 with a crop of 41,800 tonnes before steadily recovering to $2.65 by 2010 with 35,500 tonnes of NIS. Andrew Heap, the General Manager of the AMS, summed up this downturn by saying that consumers were forced to choose between 'bread on the table and nuts on the balcony'. Demand and prices were still low in 2007, but the industry then focused on creating new uses and seeking new markets.

AMS Industry Development Officer Patrick Logue reported that in the two years to 2007, 320,000 trees had been planted in Bundaberg. By 2009, there were 1,500,000 bearing trees in the Bundaberg district producing 6,600 tonnes of NIS and 800,000 not yet bearing.[6]

Towards the end of the decade, demand for kernel had increased, and, with the help of a weaker Australian dollar, stocks were reduced to a comfortable level. However, the cost of growing and processing, particularly with Australia's high labour costs, was encouraging overseas expansion as these countries could benefit from available information and were aware of their lower costs. In 2008, Australia exported a record 7,237 tonnes of kernel and 7,900 tonnes of NIS, but this boost was followed by the 2008 Financial Crisis, which resulted in a global recession.

Australia Post took thirty-three years after implying that macadamias were not a native plant to finally, in 2011, issue a 60-cent stamp intended for standard letters.[7]

Farming

During the decade, social and environmental factors, such as noise and residential development, continued to be debated, and the term 'climate change' was increasingly heard and assessed. Although varieties of macadamias planted varied from district to district, across Australia, 741, Daddow, A4, A16,

A38, 344, 842 and 816 became the most popular.

A decline in the health of trees as they aged led to an awareness of the importance of bio-active soil health. This involved encouraging the growth of beneficial micro-organisms by applying manure, mulch and compost, as well as nitrogen, phosphorus and potash fertilisers. In simple terms, soil health can be summarised as being related to the level of organic carbon.

Farmers recognised the need to maintain a sound orchard floor. Roots exposed by blowing and sweeping during harvesting, and the damage caused by erosion through rain events, became more evident. To rebuild the soil profile, compost could be spread under the trees and soil removed could be replaced. Blowers to move nut-in-husk (NIH) for easier harvesting were modified to reduce soil movement and gradually less damaging harvesting practices were adopted. Shade-tolerant ground-cover such as Sweet Smother Grass was often planted to cover bare earth.

To accelerate nut fall, Ethephon (marketed as Ethrel), a chemical that released ethylene to promote ripening and thus nut-fall, was increasingly applied. As its use was refined over the years, harvest time and costs were reduced and the crop from late-falling varieties could be removed before trees flowered again. This had the added benefit of interrupting the transfer of pests and diseases from the current crop to the next flowering.

Assessments were conducted into the application of insecticides and fungicides, and the control of spray drift. The environmental effects of the use of Endosulfan as the main industry insecticide were being questioned, and the government applied conditions that would eventually cause it to be withdrawn.

Sigastus weevil and then lace bug became emerging insect pests, first in north Queensland and then throughout the industry.

Husk spot (*Pseudocercospora macadamiae*), a fungal disease, also became an increasing problem. It is evidenced by husk lesions, which cause the nut to fall prematurely. Susceptible varieties, particularly A16 with its very late nut fall, spread the disease from the previous crop to the new nutlets. Dr Andre Drenth and Dr Femi Akinsanmi developed strategies for the suppression of husk spot. These rely on alternative applications of different fungicides, which slow the development of the disease so that maturity is reached before the fungi cause nuts to fall.

Most growers dried and stored NIS on their farms with varying efficiency, but this risked deterioration. Dr Tim Kowitz's research on crop handling has resulted in most processors and buyers accepting crop as soon as it becomes available, in order to optimise quality.

On the Atherton Tablelands, some growers of mainly the 344 variety that was planted on deep, well-drained soil noticed trees with extremely vertical shoots, which produced few flowers and nuts. This was named Abnormal Vertical Growth (AVG) or Tall Tree Syndrome, and it spread further south. After much research and observation, M.C.M. Zaheal and Dr Akinsanmi suggested that the cause is a soil-borne bacterium, and that resistance varies with different cultivars. No practical control has been found, apart from planting tolerant varieties, and the causes are still not well understood.

In 2006, Cyclone Larry caused extensive damage to all orchards on the Atherton Tablelands,[8] with several orchards damaged beyond restoration. Losses varied from 25 per cent to 95 per cent, and although most orchards were restored or replanted, the local industry was severely impacted.

Research

Research priorities determined by the industry in 2001 included crop forecasting, food safety, health benefits including antioxidants, evaluation of NIS and control of the fruit spotting bug.

The assessment of a large number of bearing Hawaiian and Australian cultivars from grower trials was led by Dr Russ Stephenson.

Plant breeding expanded into an essential long-term project, taking approximately 30 per cent of research funds. From 1996, Dr Craig Hardner led this project at CSIRO, applying cutting-edge, quantitative genetics and Dr Cameron McConchie contributed as part of his macadamia research. In 2007, Dr Hardner resigned from CSIRO and took a position at the University of Queensland, where his skills and passion for macadamias continued. Dr Bruce Topp of the Queensland Department of Primary Industries (QDPI) then led the plant breeding project.

Dr Chris Searle, a tree crop researcher and consultant based in Bundaberg, worked at CSIRO and then at the QDPI, before leading global advances on aspects of macadamia culture and providing ongoing services to the macadamia and avocado industries.

In 2005, Richard Mason retired after almost forty years as a Food Research Scientist at the QDPI and the University of Queensland. Richard had led many projects related to quality and published over forty research and extension papers on macadamia processing and handling.

Entomologist and plant scientist Dr Henry Drew brought science-based information to growers with his entomological skills and the application of environmental and effective controls. Sadly, he died in a tragic motoring accident in 2009.

During this decade, about thirty R&D projects in the categories of cultural practices, crop protection, varietal improvement, technology transfer, post-harvest and market information, were undertaken.

Funds were also provided to maintain MacMan, the QDPI and New South Wales Agricultural Extension Services that assisted growers to record their management practices and results and to raise their farm productivity and improve their nut quality.

Access to MacMan and MacGroup meetings, AMS *News Bulletins,* Newsletters, Annual Meetings and Field Days resulted in ongoing technology transfer, as the results from projects were made available to growers and the industry.

Kim Jones, the AMS Industry Development Manager, was responsible for supporting and reporting on all research projects, liaising with industry regulators and applying new technologies. His resignation left a big gap in the industry, although he subsequently purchased and expanded Cropwatch Laboratories to provide NIS evaluation and shelf-life testing.

Levies

The Statutory Levy continued and resulted in benefits that allowed ongoing research and expansion of the market. The total levy was 7 cents per kilogram of NIS, increasing to 7.8 cents in 2004, with a Residue Survey Levy of 0.2 cents.

Management of levy funds had been in the hands of the Horticultural Research Development Corporation and the Australia Horticultural Corporation. In 2001, these bodies merged into Horticulture Australia Limited (HAL). The Australian Government's ongoing support in matching R&D projects almost doubled available funding. As an example, in 2001–02, total levies of $2.1 million were collected, which, together with matching government funding for the research levy, meant that almost $2 million was allocated for research with almost $1 million allocated to marketing and promotion.[9]

Projects to ensure quality from farm to consumer were a priority. These included crop handling and drying, sampling and testing, laboratory accreditation, roasting defects, shelf-life, supply chain and quality control.

Wayne Prowse from HAL ably led the marketing programme until the end of 2009, and then the AMS was able to persuade HAL to transfer Lynne Ziehlke to lead the Domestic and International Marketing programme for macadamias – an arrangement that no other industry has been allowed. These programmes created awareness, developed new concepts and markets, expanded demand and maintained the image of macadamias.

Processing

In 1999, 2,610 tonnes of NIS were exported to China, increasing to 8,147 tonnes in 2001–02,[10] and it became obvious that China would become a major buyer, marketer and grower of macadamias. NIS became the preferred export to China, because of its lower rate of import duty compared with kernel. This disadvantaged Australian processors, but the export of macadamias as NIS increased.

Initially, Chinese workers cracked the nuts with hammers, then with a machine like a modified, old-fashioned sewing machine which used a strategically positioned chisel to open the shell. 'Pacific Plantations', one of the large growers and innovative processors, became leaders in exporting NIS and having it processed in China, where a large number of staff cracked each nut with a hammer.[11]

Processors then developed machines that sawed through a section of the circumference of the shell, after which the nuts-in-shell were soaked in brine, sometimes with flavour added, and roasted in shell. This innovation allowed macadamias to be sold as nut-in-shell, but in an 'easy-to-open' form, where each retail pack included a small tool to open the shell.

There were ongoing changes in the processing sector. A new entrant to the industry, Macadamia Industries Australia (MIA), purchased Macadamia Plantations of Australia (MPA) in its entirety. Later MIA was resold and the new company traded as Macadamias Direct. The Macadamia Processing Company (MPC), through their Australian, grower-owned Macadamia Magic and International Macadamias subsidiaries, continued their solid, conservative expansion to become the largest global processors and marketers. Other processors included Agrimac International, Suncoast Gold Macadamias, MacAz, Nambucca Macadamias, Pacific Plantations, Goldmacs, Stahmann Farms, MWT and Nutworks.

In a first for the industry, Suncoast Gold Macadamias in 2002 entered into a renewable energy project with Ergon Energy by establishing a co-generation plant, which burnt macadamia shells. This reduced greenhouse emissions and generated enough electricity to power the plant and return sufficient to the grid for 1,200 homes.[12] However, economies of scale and technical problems caused this green energy project to be discontinued.

Over 90 per cent of processors and exporters

of NIS formed an advisory body, the Data Handlers Group, who collected reliable market and stock data to assist in forecasting. This group became the Australian Macadamia Handlers Association (AMHA) in 2008.

People

Traditionally, macadamias had been grown on the rich basaltic soils of the undulating plateaux of the Northern Rivers, but the Dorey family successfully established highly productive orchards on the coastal flood plains at Newrybar, east of Lismore. Tim Dorey and his wife Lorna had the vision, and in 1989 their sons, Col, Ken, Ron, Ray and Mark, planted their first macadamias on 'Plantation Lorna' and adapted farming practices to local conditions. Second-generation and third-generation farmers, they studied the expanding macadamia industry in the hills to the west and planted a trial among their

sugar-cane in 1990. Despite scepticism from growers familiar with the red, basaltic soils on the nearby tablelands, crop performance in terms of tonnes of NIS per hectare was up to twice the industry average, and kernel recoveries were higher. The Dorey orchards have been sustainable, and their success led to the establishment of other orchards from Ballina south towards Grafton, where more reliable rainfall and gentle slopes were advantages. Practices included improving drainage and mounding the often high organic content soils. Now in the third generation of farming in the district, the Doreys are among the industry leaders in productivity and quality.

In about 2000, Andrew Seccombe and his son Noah established a large orchard near Yarrahapinni, south of Coffs Harbour. Led by John Forsyth and Ann Verschuer, the book retailer Dymock Group followed, planting 75,000 trees over the next ten years.

In 2002, John and Mary Walter pioneered

'Plantation Lorna', orchards of the Dorey family – pioneers of coastal lowland farming   showing their high, industry leading productivity in 2020. Credit: J. Dorey

macadamias at their farm near Emerald, in central Queensland. Competent, visionary farmers, they realised that they had ideal slopes and ample water for irrigation, but had to address high summer temperatures, which regularly exceed 40°C.

Most macadamia orchards had been owned by individuals or families, but this began to change in the late 1990s, when HAIG (Hancock Agricultural Investment Group), known as Hancock's – a large, successful, US-based company – purchased two mature orchards in the Northern Rivers, and then expanded with further investments in New South Wales and Bundaberg.

Syndicated farming, where investors could purchase a financial interest in large-scale agriculture, offered taxation benefits and the lure of future returns. In 2007, Rural Funds Management, an agricultural property trust, established Macgrove, a managed investment scheme, which allowed investors to participate in farming macadamias. Macgrove developed two properties at Bundaberg, where an increasing number of local farmers, many of whom had been growing sugar-cane and vegetable crops, were switching to macadamias. Macadamia Farm Management also established a number of orchards for investor owners.

Jack Gowen, one of the AMS founders, as well as the organisation's long-term President and visionary, passed away in 2003. His drive, belief and leadership did much to develop the modern industry.[13]

In the Bundaberg region, which offered the benefits of flat to gently sloping land and available water, almost half a million trees were planted during the first half of the 2000s decade. Local farmers were experienced with mechanisation and many had long-term, proven farming skills.

By 2006, 'Hinkler Park', led by Phil Zadro and his family, had planted over 400,000 trees on 1,300 hectares, and their bearing trees were producing about 2,500 tonnes of NIS. After purchasing a large sugar-cane farm on the main road from Childers to Bundaberg, they planted another 400 hectares of macadamias and the site became a showcase for the whole industry, as it stands out on both sides of the highway. This led to the establishment of their large-scale, modern processing plant at Bundaberg, Consolidated Nuts.

Phil Zadro, Adrian van Boyen (Operations Manager of 'Hinkler Park') and Dr Chris Searle, researcher, consultant, authority and visionary, 2016. Credit: I. McConachie

In the early 1990s, Col and Pat Sheppard and their son Garry planted on 400 hectares some 140,000 trees at Winfield, north of Bundaberg. This led to Goldmacs, a state-of-the-art factory, being built in the orchard. Sheppards sold the orchards to Macadamia Industries Australia in 2007, and later the factory was sold to Capital Commercial, a division of Hinkler Park.[14] Now Garry and his wife Andrea manage and develop orchards in the Gympie district.

Peter Kermond took over management of his family's orchards, became a processor and marketer at Macadamia Exports Australia at Cooroy and developed a reputation of providing super service.[15]

E.R. (Ted) Davenport, who led CSR in developing their Macadamia Division and then managed the AMS, retired to a boating and caravanning world in 2001.

A successful different approach to marketing macadamias was modelled by Anthony and Teena Mammino, from Childers, who widely sold their macadamia gourmet ice-cream to tourists.[16]

Many researchers had been and were contributing to the industry. Dr Craig Hardner led the plant breeding programme, assisted by Jodi Neal and supported by the research of Dr Cameron McConchie.[17] Tim Trochoulias applied research as a horticulturist from 1964 with the New South Wales Department of Agriculture until his retirement late in the decade.[18]

Macadamia Conservation Committee and Recovery Plan

A few industry members started thinking about where macadamias originated and realised the need to protect them in their native rainforests. By the 1980s, they had started to team up with researchers to find out where the wild trees grow and how to conserve them. This led to development of the National Macadamia Germplasm Conservation Program, implemented by the Australian Macadamia Society, CSIRO, Horticulture Australia Limited and the University of Queensland. As a result, cuttings of three species from eighty-four wild population sites were established at Alstonville, Caboolture and Tiaro. Planted in 2000 and 2001, the resultant trees and their genetics were to be permanently conserved and available for research.

All four species are listed as threatened under the Commonwealth's *Environment Protection and Biodiversity Conservation Act 1999*, which provides for conservation plans to be developed. A Macadamia Conservation Committee was formed in 2004, with the initial goal of preparing a Threatened Species Recovery Plan and having it adopted by the Australian Government. The Committee was initially led by Ian McConachie, with John Gillett, Michael Powell, Ken Dorey, Kim Jones, Alison Shapcott, Annie Keys and Sam Lloyd.

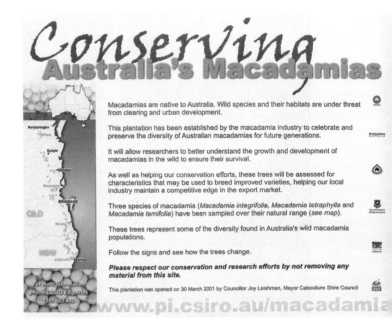

The first Macadamia Conservation Arboretum 2007 planted at CREEC, Burpengary. Credit: I. McConachie.

Searching for *M. jansenii*. Dr Alison Shapcott and Gidarjil Rangers in a rainforest gully at Bulburin National Park. Credit: A. Shapcott.

M. ternifolia leaves and typical beautiful pink flowers. Credit: J. Vaughan.

The Macadamia Conservation Committee prompted the AMS, supported by HAL, to commission Glen Costello, Michael Gregory and Paul Donatiu to prepare a draft Recovery Plan. The plan details the known habitat and threats to wild trees, lists actions to conserve them and their habitat, and presents an implementation plan and budget. In November 2009, the Australian Government formally adopted this as the Southern Macadamia Species Recovery Plan.

On 9 August 2006, the AMS established the Macadamia Conservation Trust (MCT) to collect and hold funds in trust to spend on the following objective: 'to support and conserve Australian wild macadamia trees in their native habitat and in all other ways'.

The following year, the MCT became a Registered Environmental Organisation, enabling donations to be tax deductible, and on 9 September 2007 the MCT was formally launched at the Big Scrub Rainforest Day at Rocky Creek Dam.

A baby wild *M. integrifolia* emerging through the rainforest. Cascades Creek, Amama. Credit: I. McConachie.

The end of the decade

The industry Strategic Database listed production at the end of 2010 as 55,000 tonnes of NIS or 17,875 tonnes of kernel, with a surprisingly high orchard productivity of 3.85 tonnes of NIS per hectare and an average price to the grower of $3.20 per kilogram of NIS. The Australian industry was valued at $176,000,000 at farm gate and $250,000,000 after processing.[19]

The decade 2000–10 was another characterised by ongoing advances and growth in plantings, but disturbingly, production did not expand in proportion to tree numbers. It was a decade of solid growth where marketing an increasing crop and competing with lower-cost producers, such as South Africa, was an ongoing challenge.[20] Apart from contending with unfavourable weather conditions, the industry was still learning about tree physiology and how to manage soil health and the orchard floor. There were many research projects which advanced knowledge and helped growers both to adapt to new requirements and to improve all aspects of tree culture and crop handling. The AMS and HAL, through industry levy funds, addressed the creation of ongoing demand and maintaining or enhancing the image of the nut. But the vagaries of nature, competition to secure NIS, increasing costs of processing and marketing kernel, together with inevitable market cycles, were a constant reminder that life wasn't meant to be easy.

Chapter 14

AUSTRALIA, 2010–2023

Macadamias are a global crop

The continent that had promised the world so many new and useful creatures and products had hatched after a century of British incubation, little more than the odour of eucalyptus oil, cages of bright budgerigars, wattlebark and the macadamia nut.

(Geoffrey Blainey, *The Story of Australia's People*)[1]

Industry overview

By 2010 there were 650 commercial growers, with a total of six million trees of which two million were not yet bearing. NIS sales to China had increased from 1,726 tonnes in 2013 to 16,608 tonnes in 2017, before falling slightly in 2018 and rising again to a record level in 2023. Demand from China was the main driving force for prices to growers to rise and rise to unprecedented levels. From a low of $1.50 in 2007, prices had risen to over $6.00 during the last years of the 2010s and then came a price decline. Average production of NIS per hectare had risen slightly to 3.05 tonnes or 1.04 tonnes of kernel.

Although not used in Table 4, from 2016 an NIS moisture content standard of 3.5 per cent was adopted internationally, which is equivalent to being fully dried NIS, ready to process into kernel. The variation in annual production was, apart from new plantings starting to bear, almost entirely due to environmental factors, which illustrates that current farming practices struggle with the forces of nature. The increase and then collapse in NIS prices is explained within this chapter.

Left. CSP Plantations' mature Bundaberg orchard during harvesting. Credit: AMS Ltd.

Table 4: Australian production and sound kernel recovery price, 2010–2023

Year	Production tonnes of NI at 10% moisture content	Price average at 33% SKR at 10% moisture content
2010	35,500t	$2.65
2011	28,500t	$3.10
2012	40,000t	$3.20
2013	35,200t	$3.00
2014	43,600t	$3.83
2015	48,300t	$5.09
2016	52,000t	$5.51
2017	46,000t	$5.62
2018	52,900t	$5.81
2019	46,600t	$5.83
2020	50,300t	$6.65
2021	55,200t	$5.42
2022	56,800t	$3.57
2023	51,900t (48,500t at 3.5% MC)	$1.80

Source: AMS/ AMHA Facts and Figures Australian Production, 2022.

The 'Australian Macadamias Strategic Plan 2009–2014'[2] summed up the position of the industry entering the new decade. The objectives were to understand markets and improve supply chain efficiency, grower profitability and industry leadership.[3]

Low prices to the grower in the first few years of the 2010s meant that Australian plantings slowed, averaging about 40,000 trees per year compared to 200,000 per year in the previous decade. But the increasing demand for NIS in China and price offered resulted in a growth in plantings and higher NIS prices from 2013. Despite this, by 2016, Australia had lost its mantle as the largest producing country in terms of both plantings and crop to South Africa, and by 2023 also probably to China, whose production and planting reports were varied.

During the 2010s, the nature of investment and ownership in the industry was changing. Although there were still many smaller-scale growers, the viability of individuals developing orchards for lifestyle or retirement was diminishing. Many orchards were established on a scale unheard of twenty years earlier and the growing of macadamias increasingly came into the hands of corporations, syndicated farming and investment funds mainly from the USA and Canada. Orchards were purchased by companies or individuals from China and South Africa, some making Australia their home.

Economies of scale, mechanisation and the continuing higher prices offered for NIS resulted in other forms of agriculture, such as sugar-cane, converting to macadamias. The average market value of an orchard in 2010 was $22,000 per hectare but this was to rise substantially to $200,000 during the decade and then fall from 2021.

In the global industry, Australia had developed a broad international market with a clean, green image. With its high knowledge base and sound quality standards, the industry had benefited from the statutory levy for R&D and marketing, which had led to a higher total kernel recovery and improving post-harvest procedures. However, orchard productivity was not increasing, and Australia's growing and processing costs were higher than those in most other countries.

Farming

In early 2011, widespread floods in Queensland damaged orchards and resulted in crop losses. However, some Bundaberg orchards later benefited from the silt deposited by the floods. In 2013, many farms in Queensland lost between 5 per cent and 20 per cent of their crop, again in floods, and a cyclone with winds of 200 kilometres per hour damaged parts of the Northern Rivers.

By 2018, the Bundaberg region had become the largest growing area in Australia, with 11,300 hectares of macadamias, and in 2020 it produced 20,000 tonnes of NIS. A single orchard near Gin Gin planted over 120,000 trees and over 140,000 trees were being planted north of Noosa.

Alloway Macadamias in Bundaberg had initially planted a high-density orchard of up to 1,000 trees per hectare at 5-metre x 2-metre and 6-metre x 2-metre spacing. This initially generated higher yields per hectare. However, as the trees matured, yields fell and eventually every second row and every second tree was removed. Lessons were learnt and few new orchards adopt this very close planting density. Economic analysis of very high-density plantings followed by later tree removal does demonstrate enhanced returns and is still being trialled. The industry has generally been reluctant to replace mature trees to increase long-term productivity However, some removal of unproductive trees, including those effected by Abnormal Vertical Growth, and replacement with newer varieties is a positive indicator that some orchards or blocks are being modernised.

In northern New South Wales, Discovery Group Australia, owned by Chinese investors, purchased the orchards of Gray Plantations and the Dunoon orchards of Hancock Farms, totalling 680 hectares. South of Coffs Harbour on the central north New South Wales coast, the Dymock Group of companies had established 73,000 trees starting in 1989. John Forsyth and Ann Verschuer with their manager Chris Cook continue to improve productivity and efficiency and 'Arapala Macadamia Farms' is now the most southerly large orchard in Australia. In 2017, they received the Macadamia Grower of the Year Award for a large plantation. In 2010, they purchased Patons Macadamias, the leading producer and marketer of premium packed products designed for duty-free stores, but sold this business in 2020 to focus on their core businesses. The Seccombe family has 23,000 trees nearby at 'Tallowwood', planted about the same time. Together with Dr Murray Hyde-Page's 12,000 trees west of Camden Haven, these three are the most southerly larger orchards in the central east coast of New South Wales, and, despite problems with frost damage, have performed well. At the same time, some of the older orchards on plateaux in the Lismore district have struggled to remain productive. There were several smaller southern orchards: one at Yarramalong near Gosford, 'Hand and Hoe' at Comboyne and Paul Harris at Rainbow Flat nearby.[4]

In late 2015, the Hancock Agricultural Investment Group (HAIG), a US farmland investment company with substantial interests in Australia, announced the purchase of properties in the Bundaberg district with intent to plant 15,000 hectares of orchards to be managed by Macadamia Farm Management. By 2019, HAIG had divested most of its macadamia investments, selling its remaining 664-hectare macadamia orchard at Winfield, north of Bundaberg, to a Belgian company,[5] Finasucre. This company had previously bought Bundaberg Sugar, which had entered the macadamia industry and owned orchards totalling 330 hectares. As part of diversifying from a reliance on sugar, Finasucre continued to purchase orchards and became a major new industry entrant. In February 2020, Finasucre purchased the processing and marketing business of Macadamias Direct at Dunoon in the Northern Rivers.[6]

The low coastal floodplains north and south of Ballina have become major growing areas requiring new practices on often flat, alluvial, poorly drained soils. This initiative had been most successfully led by the Dorey family. Problems in the coastal floodplains may be variable soils, drainage and subsoil acidity,[7] but by 2021, innovative growers had established over 2,500 hectares of macadamias there. 'Boombera Park', a rural investment by Arrow Fund Management, advised that in 2022 they had established 253,000 trees at Lawrence, north of Grafton.[8] The February 2022 floods caused major damage and tree losses in flood-prone areas.

A major new planting area emerged during the decade and into the 2020s. It commenced with Scott Gregson-Alcott from Macadamia Farm Management at Bundaberg working with Trevor Roy, who purchased a sugar-cane property at Owanyilla, south of Maryborough, and developed a 'model' orchard from 2015, which set new standards. The Maryborough or Wide Bay district was predominantly a sugar-cane area supported by the Maryborough, Wide Bay Sugar Mill. However, when the mill closed, many former cane farms with ideal slopes, climate and assured water for irrigation became available for macadamias and other crops. Rural Funds Management purchased 5,000 hectares of farms from the mill, and much of this land is now being planted to macadamias. In 2021, Macadamia Farm Management began purchasing farms for investor owners in the Wide Bay district and by 2023 had planted over 300,000 trees. The Maryborough district will become a major growing and producing area. Hinkler Park had purchased a large cane farm for macadamia development at Glenorchy, just south of the city and on the Bruce Highway, which will become a showcase orchard to all who drive past.

Orchards have also been established in the Mackay district, but most of the trees on the Atherton Tablelands have now been removed due to wind damage, sometimes excessive rain and poor productivity. The twenty-year industry on the Atherton Tablelands is currently represented by one family business, Wondaree Nuts, who grow and market their own nuts, mainly in retail packs. In Western Australia, there was a small industry comprising about 100,000 trees.[9]

Higher NIS prices during the decade resulted in most growers improving infrastructure, farm machinery and crop handling. Many increased expenditure on irrigation and farm inputs. Toward the end of the decade, due partly to the effects of COVID-19, costs were rapidly rising and included fuel and electricity, nutrition, farm chemicals and labour. New technologies, such as satellite, aerial and drone imaging, robotics and advances in mechanisation were increasingly adopted. Drones are being used to survey orchards and identify problems using photography and spectral analysis, which measures characteristics associated with wave and length frequency. LIDAR (Light Detection and Ranging), a new management and research tool, uses aerial or ground 3D imaging to provide information.

The health of trees in many older orchards had been declining, leading to lower productivity, although some well-managed older orchards were still producing in the top tier. A productivity report showed a cross-industry range from 1.0 tonnes to 6.5 tonnes of NIS per hectare. The top fourteen farms in a benchmarking survey in 2017 averaged 4.47 tonnes of NIS per hectare with income of $17,285 per hectare, compared to the middle 50 per cent of farms which averaged 2.53 tonnes per hectare and income of $9,218.[10] Despite the work of competent consultants, the implementation of research and the introduction of new varieties, productivity averaged around a disappointing 3.0 tonnes of NIS per hectare (1.0 tonnes of kernel per hectare). A notable exception, Gary and Julie Davis achieved 2.68 tonnes of kernel per hectare in 2019 on the Sunshine Coast.[11]

Many years of applied research have led to a greater understanding of soil health and the need to increase organic carbon and beneficial microflora for optimum soil structure and nutrient availability. The importance of pollination was increasingly acknowledged, with both European bees and native bees of the *Tetragonula genus* (formerly *Trigona)* used to cross-pollinate macadamia flowers. Verroa mite, which attacks European honey bees, is becoming widely established and may progressively reduce European bee availability. Various techniques for tree canopy management were trialled and adopted. These include selective limb removal, hedging, height reduction and nutritional management to reduce vegetative growth.

Harvesting had been mainly by finger-wheel machines, with a trend to use larger-diameter pick-up wheels. Large machines often had two banks of finger-wheels, one behind the other, to pick up a higher percentage of nuts. The Monchiero range became available, offering self-propelled sweeper and pick-up machines. In one pass, a Monchiero machine could pick up close to 100 per cent of the nuts. Gradually, this and other brands have become more widely used. The use of Ethephon (Ethrel) to hasten mature nut fall has become an essential tool.

More effective crop handling was adopted, often with adaptations of the energy-saving South African Bungay-concept driers for NIS curing. These systems, designed by Dorran Bungay, involved in part applying warm air from under the ceiling of a shed to banks of small, sensor-controlled driers to give efficient curing and maintain the quality of NIS.

Buyers of NIS increasingly accepted unsorted and undried nuts, so that they could better control the grades and quality. The labour cost and inefficiency of sorting nuts led to several solutions. To sort unsound from sound NIS, image analysis machines, formerly known as electronic colour sorters, were developed to detect kernel flaws inside the shell. Independent services, such as the House With No Steps (now Aruma), were available to contract sort and dry NIS.

Pest and disease challenges

A long-known defect in kernels, described as internal discolouration or brown centres, where cell wall damage leads to enzymatic reaction, sometimes resulted in an acrid odour and flavour. Research related the problem largely to drying practices, but control was only partly effective and it remains an industry problem.

There has been an increasing adoption of integrated pest management (IPM) principles, including monitoring and encouraging beneficial predatory and parasitic insects and spraying only when an economic benefit will result. Yet despite this development, previously minor pests and diseases – such as macadamia seed weevil (also known as sigastus weevil), lace bug and husk spot – increased, resulting in the loss of limbs, flowers and nuts. These pests were studied by Craig Maddox and Dr Ruth Huwer at the New South Wales Department of Primary Industries. At the same time, Professor Femi Akinsanmi studied fungal disease at the University of Queensland. Increasingly, biological controls such as *Trichogramma* and *Anastatus* wasps were introduced into orchards. Mistletoe, a semi-parasitic plant spread by birds, reduces tree health and is difficult to detect unless flowering, and major limb removal is required to eradicate it. Increasing numbers of both black and white cockatoos damage the crop when they crack the nuts in their beaks and strip nuts before they mature. Abnormal Vertical Growth (AVG) became more evident as a serious problem, with at least 200,000 trees being affected by 2018. Causes are complex and include moisture stress, particularly in well-drained, deep soils. Early research suggested it was associated with a soil virus, but later research implicates a soil-borne bacterium as part of a multi-factorial problem. Control by limb removal and cincturing the tree trunk annually was partly effective. Varieties resistant to AVG are now being planted in prone areas and affected trees are being replaced.

The increase in macadamia orchards encouraged feral pigs, who relished the nuts, rooted up orchard floors and became a major problem. Trapping, hunting and baiting gave limited control, while expensive, usually electrified, exclusion fencing proved to be the most effective solution. Technological advances such as sophisticated cameras were used to monitor pig numbers. Rat damage required ongoing controls, particularly in orchard floor hygiene.

The traditional rootstock for seedlings in nurseries was the H2 variety. During the decade, Beaumont become popular as an alternative rootstock.

Beaumont is a first-generation hybrid of *M. integrifolia* and *M. tetraphylla*, and has provided greater vigour and earlier production. Availability of Beaumont seed nuts has slowed the uptake of this option. Both rootstocks are grown as seedlings, so the seed female parent is either H2 or Beaumont. The male pollen parent is likely to be a different macadamia cultivar growing nearby. Trials suggest earlier cropping and enhanced production from Beaumont rootstocks but the influence of the male pollen parent has not been investigated.

Windbreaks widely used in the past as an insurance to reduce wind damage are seldom now used, which provides more land for growing macadamias. Windbreaks were often ineffective and competitive, but orchards – particularly those planted close to the coast – are now increasingly vulnerable to cyclonic winds, although these have fortunately been almost absent for many years.

Organic farming often provided a sounder understanding of basic principles, such as using animal and natural nutrients and the biological control of pests and diseases. The industry as a whole benefited and became more 'clean and green'. Macadamias grown 'organically' were in strong demand and attracted higher prices than conventionally farmed product. However, some pests and diseases were difficult to control and few farms were able to maintain strict organic regimes and obtain consistent yields and quality.

Electricity suppliers imposed stricter requirements for clearance and safety under power lines, requiring additional pruning and sometimes tree removal.

From 2009, the industry faced the 'two-headed fish controversy', when the owner of a fish hatchery believed that a genetic defect, which resulted in two-headed fish embryos, was caused by chemicals drifting from an adjoining macadamia orchard. Distortion in the media even suggested a cancer cluster but a State Government Task Force found no evidence that the problem related to either the orchard or the use of farm chemicals. The orchard's managers had documented all operations, monitored weather conditions, used approved farm chemicals, established spray-free buffer zones and notified neighbours. The fish hatchery was eventually purchased and macadamias were planted on the land.

Research

In 2015, the Australian Government restructured the industry-owned Horticulture Australia Limited (HAL) to form the grower-owned Horticultural Innovation Australia (HIA). This is a not-for-profit corporation that manages industry levies for marketing and R&D in horticultural industries, and also provides matching government funds to support research.

HIA introduced long-term, whole-of-industry programmes and required stricter governance. Comprising forty-three horticultural industries, HIA bestowed ownership on levy payer growers who chose to join the corporation, rather than members of the AMS. HIA had capped R&D matching funds to 0.5 per cent of the value of the annual farm gate crop. While acknowledging the professionalism, sound leadership and achievements of HIA, some in the macadamia industry believe that the AMS (which has many grower members who have not joined HIA) should have more input into decisions on levy expenditure. This matter is being addressed. Research and marketing priorities were driven by the 'Macadamia Industry Strategic Investment Plan (2014–2019)', which aimed to increase and sustain productivity and optimise value in all sectors. Some of the HIA staff who supported the macadamia programme were Corrine Jasper, Dumesani Mhlanga, Vino Rajandran and Georgia Shiel, who directed and managed levy funds, which were matched by government funding for R&D. The industry's Strategic Investment Advisory Panel (SIAP) provide guidance.

In 2010, the Macadamia Breeding and Conservation Project, which had commenced in 1996, was transferred from CSIRO to the Queensland Government. It was led by Dr Bruce Topp, with Queensland researchers Dr Mobashwar Alam, Dougal Russell, Paul O'Hare, Dr Craig Hardner, Dr Andre Drenth,

Dr Femi Akinsanmi, Dr Katie O'Connor and, from New South Wales, Dr Ruth Huwer, Jeremy Bright and Craig Maddox. The project is now hosted by the Queensland Alliance for Agriculture and Food Innovation (QAAFI), a research institute of the University of Queensland (UQ), supported by the Queensland Government. The Plant Breeding project has released four commercial varieties: DAF-G, DAF-J, DAF-P and DAF-R. Meanwhile, the Macadamia Conservation Trust has released MCT1, and David Bell has selected new varieties to add to the Hidden Valley A-series, available from 2022: A376, A403, A447 and A538. Genetic research led by Katie O'Connor has determined markers that relate to cropping, in order to shorten the time to make assessments and accelerate plant breeding.

The MacMan programme in Queensland continues to provide training and support. The New South Wales Department of Agriculture, led by Jeremy Bright with Drs Ruth Huwer, Craig Maddox and team, provide information in the form of a Plant Protection Guide, Integrated Pest Management, Growers Guide and Best Practices.

In 2015, the Small Tree – High Productivity Project sought to maximise productivity per hectare by pruning, trellising, and selecting dwarf varieties. The project aims to manage vigour, enhance light interception, manipulate crop load, design tree architecture, select rootstocks and varieties, and assess cross-pollination. Its leader, Dr John Wilkie Jnr, studied the trees' non-structural carbohydrate energy reserves and sought to apply effective light management and tree architecture. The rationale for these cultural advances was derived from the apple industry, which had increased productivity from 10–15 tonnes to 100 tonnes per hectare.

Dr Russ Stephenson, with input from John Chapman from QDPI, had earlier challenged the industry by suggesting that the long-term potential of macadamia productivity was 12–15 tonnes of NIS or 4–5 tonnes of kernel per hectare. This was based on productivity increases in other tree crops, which they believed could be applied to macadamias. The figures were thought to be achievable through advances in breeding, rootstock and cultural management, but its proponents acknowledged that it might take up to fifty years to achieve. This is embodied in Dr John Wilkie's research. The industry is increasingly using this theoretical potential as a goal. The use of drones as a management tool increased.

MacMan, an innovative programme managed by the Queensland Department of Agriculture and Fisheries (DAF), focused on information collection, benchmarking and technology transfer. Led by Shane Mulo, Jeremy Bright, Geoff Slaughter and Paul O'Hare, with other New South Wales Department of Agriculture specialists, it provided farm-level advice in a practical format, including a wide range of online videos. Best practices groups were held in all major production regions to enable growers to compare practices and results and to learn from each other and industry specialists in order to raise productivity and nut quality. The Two Tonne Club was an initiative that challenged growers to double industry average orchard productivity by producing 2 tonnes of kernel per hectare. Some growers achieved this target, but, due to environmental conditions, tree decline and sometimes a low standard of management, the industry average has remained at only half this figure.

Applied research in many aspects of tree culture led to the concept of Integrated Orchard Management (IOM), which focused growers on identifying and addressing their orchards' strengths and weaknesses. IOM embraced the industry's priorities of soil health, orchard floor management, nutrition, tree physiology, irrigation, drainage, canopy management, crop protection, and the need for ongoing research. Adopting IOM increasingly embraced inorganic, organic and biological nutrition.

Dietary studies recognised the importance of macadamias in a healthy, nutritious diet. Macadamia oil contains on average 20 per cent of palmitoleic, a monounsaturated fatty acid known as Omega 7. This fatty acid distinguishes macadamias from all other edible nuts and oils. Some research attributes health benefits, including longevity, to palmitoleic, and it is also a major active constituent in the macadamia oil used in cosmetics. However, some studies identified that there might also be some negative indicators for human health and this constituent has not been promoted.

In 2019, a network of global researchers, the International Macadamia Symposium Committee (IMSC), met in China. The outcome summarised research priorities as:

- breeding and genetics
- current and emerging endemic pests and diseases
- collaborative opportunities
- varietal compatibility
- planting design and density
- light and water
- support for smallholders
- kernel quality, particularly shelf-life.

Processing and marketing

Projects to create demand and position macadamias in the market were based on surveys and statistics. In Australia, macadamias and other tree nuts were promoted under the Nuts for Life Program.

Lynne Ziehlke, who retired from her AMS position in 2020, developed a premium market strategy to release an Australian Brand or Gold Label product that set high standards and could be licensed to marketers. This initiative has not yet been taken up by the industry. Another of Lynne's approaches was to encourage consumers to be indulgent, to enjoy the taste and crunch of macadamia nuts and to value their health benefits. Jacqui Price has capably led this programme since June 2020.

A trend to co-ordinate marketing, reduce competition, and have the volume and grades to provide stability had led to the earlier formation of Green and Gold Macadamias. This marketing venture, led by Brian Loader, initially represented processors in Australia and South Africa and has expanded to include strategic processors throughout the macadamia world. Other marketing companies have been established. These specialise in specific areas or represent specific groups. It is probable that in the short and medium term there will be an increasing range of companies seeking both to expand the market and to maximise return to growers.

Over a five-year period from 2018, the markets for Australian kernel averaged 11,000 tonnes per year. Over the same period, Australia produced approximately 20 per cent of the world's macadamia supply, sold to the markets shown in Table 5.

In addition, Australia sold an average 9,850 tonnes per year of NIS (at 3.5% moisture) to China–Hong Kong importers over this time period.

In 2020, Marquis Macadamias became the trading name and global super-brand for Macadamia Processing Company (MPC), Macadamia Marketing International, Pacific Gold (formerly Consolidated Macadamias) and Global Macadamias in South

Table 5: Markets for Australian kernel, 2018–2022

Market	% total
Australia	32%
Japan	19%
China	10%
USA	7%
Other Asia	7%
Korea	7%
Europe excluding Germany	7%
Taiwan	5%
Germany	6%

Macadamia 'Australia's Gift to the World'

Macadamia belong to the Proteaceae, an ancient angiosperm family whose initial differentiation from ancestral forms occurred in the south-east of Australia 90-100 million years ago. The family is well known for other genera such as Banksia, Grevillea and Hakea.

There are nine species of macadamia, seven of which are found in Australia in two distinct groups. The southern group consists of four subtropical rainforest and wet sclerophyll species endemic to the northeast New South Wales southeast Queensland coastal region.

Habitat and Distribution Summary

All four species are found in rainforest and wet sclerophyll communities in the northeast New South Wales southeast Queensland coastal region. They are genetically closely related and, except for M. jansenii which is known from only a single location north of Bundaberg, all the other species have overlapping ranges.

Current Species Status

All species are endangered due to land clearing apart from Macadamia jansenii, which is listed as 'Endangered' under the Queensland Nature Conservation Act 1992 (NC Act),

These four Macadamia Trees were donated to the Bundaberg Botanical Gardens by the MACADAMIA CONSERVATION TRUST.

'Macadamia – Australia's Gift to the World.' Sign at Bundaberg Botanic Gardens of giftings of the four species.

Africa. This group has become the largest processor and marketer in the world, with a capacity of 35,000 tonnes of NIS per year in 2020. Headed globally by CEO Larry McHugh, who retired in 2023, and in Australia by COO Steven Lee, the company's sound conservative management and long-term vision made Marquis the most influential and stabilising force in the macadamia world. The market collapse from 2021 resulted in concern that Marquis and other marketers did not anticipate the stock build-up and that there was some lack of transparency in its handling of the decline. From 2022, the CEO was Ben Adams and the Board was chaired by Clayton Mattiazzi, who manages the Hinkler Park orchards.

Food safety was a concern for consumers and the industry. Any product exposed to the elements is subject to risk from pathogens, so the industry implemented statistically sound sampling, testing and treatment. This particularly applied to the pathogenic bacteria Salmonella. Suncoast Gold was the first processor to install a pasteuriser to control this risk and reduce microbial levels. Pasteurisation is now widely adopted.

People

Through his family companies, Saratoga Holdings, Phil Zadro with his family are the major shareholder in Marquis Macadamias and in 2018 owned fifty-three orchards comprising 3,642 hectares in Lismore, Bundaberg and South Africa. These orchards in total produced 7,500 tonnes of NIS, making the Zadro family the largest growers of macadamias in the world.[12] Their 400 hectares at Cordalba on the main road into Bundaberg catch the eye of all driving by. Saratoga Holdings had also purchased the Walter family's substantial orchard at Emerald in central west Queensland and expanded it before selling it in 2023.

In 2019, Saratoga Holdings purchased a 623-hectare sugar-cane farm at Glenorchy, south of Maryborough, for future conversion to macadamias. It is the major shareholder in Suncoast Gold Macadamias. Five of Zadro's children and two of his grandchildren are now involved in the business. Annette Fontana is a corporate lawyer, Mark Zadro is in finance and Peter Zadro in management and IT. After playing a major role in its development, David Zadro left the company in about 2016.

In 2023, Phil Zadro, in his early nineties, assumed a quieter role but continues to guide and inspire. In 2012, the AMS acknowledged his vision and success with an International Life Achievement Award.

Larry McHugh, as mentioned above, was CEO of Marquis Macadamias until 2023. A mechanical engineer, he joined MPC as factory and quality manager in 1992 and became General Manager before moving to Macadamia Marketing International. His wealth of experience, sound judgement and managerial skills were an ongoing asset of that group. From 2023 onwards, it was expected that Larry would continue serving the industry in another role.

Henri Bader, an innovative grower near Newrybar in the Northern Rivers, set high standards, encourages other growers and is an industry respected and colourful character.

Marco Prenzel's Australian Macadamia Nurseries at Gympie was one of the several high-standard nurseries propagating trees of many varieties. The standard of healthy, vigorous plants with robust root systems increased and added to strong, uniform orchard development. Many nurseries produced smaller grafted trees, described as 'mini grafts', in about half the time required for conventional grafted trees, at a lower cost. These rely on a high standard of culture to establish in the field.

At Hidden Valley, David Bell continues his father Henry's innovative development in plant breeding and machinery, adding some early developments in robotics. Henry's wife Alison, who brought additional skills to 'Hidden Valley Plantations', passed away in 2021. The Bell family's earlier varieties, A4 and A16, which were protected by Plant Breeder Rights, were followed by A29, A38, A203 and A268. Of these, A16 and A203 continue to be planted substantially. Their new varieties were released in 2022 and are likely to further advance productivity.

High-standard macadamia nursery at Gympie in 2020. Credit: I. McConachie.

Lindsay Bryen, a long-term leader in R&D, thrives in management, industry leadership and consultancy roles. He is widely respected and received the Norm Greber Award for his contributions to the industry.

The Dorey family continue to farm just north of Ballina and in both orchard productivity and sound kernel recovery are at the top of the industry.

In 2018, Henrik Christiansen became Manager Director and a partner in Macfields Farms in Rockhampton, which was established by CSR in the early 1970s.

Alan Coates, who has an encyclopaedic knowledge of macadamias, received the Norm Greber Award for his contribution to consultancy and education services.

Robbie Commens, who led productivity management at the AMS, became manager of a large project on coastal lowlands south of Ballina. In 2022, he was elected to the AMS Board.

Dr Chris Searle, based in Bundaberg, continues to provide consultancy services and his horticultural experience, vision and advice stimulate and lead the industry. He is acknowledged as one of the horticultural global leaders with his accumulated knowledge and lateral thinking in problem-solving.

Chris Fuller, who has twenty years' applied experience in crop protection, is now assisting growers and represents Nutworks to them.

Peter Costi, of Costi Farms, entered the industry in 1998 with the purchase of three former CSR/Peninsular Group orchards at Baffle Creek. An associate of Phil Zadro, he expanded plantings, implemented sound cultural practices, and acquired other established orchards. Over the next twenty years, he purchased large orchards when they had potential and were undervalued, and by 2020 he owned eight farms.[13] In Bundaberg, the increasing crop from Costi Farms and the Hinkler Park orchards resulted in the construction of a large-capacity factory, Consolidated Macadamias, trading as Pacific Gold Macadamias. Having acquired equity in the latter, Peter Costi became a Director of Marquis Macadamias.

In 1997, Scott Gregson-Allcott came to Australia from Zimbabwe. He studied horticulture at the University of Queensland, and from 2008 managed the MACQ nursery in Bundaberg. In 2011, he purchased from Lincoln Doggrell the Macadamia Farm Management (MFM) company that was developing and managing 250,000 trees for investors. Scott's skills, drive and vision resulted in a major growth in planting and mechanisation, and he expanded Bundysort, a centralised crop-handling plant. MFM is a major developer and manages investor-owned orchards in Bundaberg, Maryborough and Gympie. Largely through the standards and imagination of David Harris, also formerly from Zimbabwe, their managed orchards were leaders in tree management, quality and productivity. With partners in Bundaberg, Scott is planning Australian Premium Macadamias, a state-of-the-art processing and packaging factory for both NIS and kernel.

HAIG, which had purchased orchards in both New South Wales and Queensland, sold their Dunoon orchards to the Discovery Group Australia in 2016[14] and their Bundaberg orchards in 2019. In late 2023, they were reported as planning to establish 1,500 hectares of new orchards in Bundaberg from 2024.

Brice Kaddatz was a trusted Grower Liaison Officer at Suncoast Gold Macadamias for fourteen years. He served on the AMS Board, and continues to serve on the Macadamia Conservation Trust Management Committee, where he has been instrumental in making MCT1 available to growers, as well as now providing consultancy services. In 2019, he was awarded an OAM (Medal of the Order of Australia) for his assistance to growers and participation in the broader community.

From 1979, Paul O'Hare served as a horticulturist and extension officer for the QDPI, leading qual-

ity assurance on farms. His knowledge and applied skills helped many growers to raise standards. A specialist in training and applied research, he was involved in the development of MacMan. In retirement, he became Chair of the Macadamia Conservation Trust until 2024. Kim Wilson and Paul O'Hare were both awarded Life Membership of the AMS.

Kevin Quinlan gained his extensive knowledge of macadamias as a researcher and extension officer at the New South Wales Department of Agriculture. He joined MPC as a grower and advisory liaison officer, and, with Jim Patch, advised suppliers to MPC on all horticultural matters. Now back with the New South Wales Department of Agriculture, he leads macadamia horticulture in the Northern Rivers.

The Steinhardt family on Goodwood Road, Bundaberg, were successful vegetable growers, processors and value-adders, trading as Farm Fresh Foods. The business, established by Ron and Marion Steinhardt in 1958, was continued by their children, Kevin, Trevor, Lisa, Janelle and her husband Andrew Gerry. The family started planting macadamias in 2004 and had at least 200,000 trees by 2020. Trading as Macadamias Australia, they developed a value-adding section, established a tourist attraction, and in 2021 built an NIS processing factory. About half of their orchard trees and the new processing plant was purchased in 2022 by Stahmann Webster who are owned by PSP, a Canadian pension investment fund and global agricultural investor which now promotes the industry with its premium image. The Steinhardt family led by Janelle Gerry retained, apart from over 100,000 trees, their new high-standard value-adding factory and tourist facility.

Peter Zummo managed Suncoast Gold Macadamias from 1989, before leaving to develop CL Macs as CEO for David Ng from China. Peter brought an ability to think laterally and innovate, to pioneer techniques to meet or exceed the requirements of the large Chinese market. After purchasing and exporting NIS, the company constructed a high-standard drying, grading and packing factory in Gympie early in the 2010s. Under Peter, the business has expanded its growing and processing operations and has become a model for others to specialise in NIS processing and marketing.

Jolyon Burnett, who led and inspired the industry as AMS CEO from 2008 to 2022, was a strong advocate who influenced and co-ordinated both the Australian and the global industry before his retirement. Jolyon related to all levels of the industry – the public, the international industry and government agencies. He was a most articulate and effective leader co-ordinating all sectors. Following his retirement in 2022, Clare Hamilton-Bate was appointed CEO and has capably led and restructured the industry organisation since then.

Macadamia Conservation Trust

In late 2017, Denise Bond was appointed Executive Officer of the not-for-profit Macadamia Conservation Trust, which had been funded by government grants and donations from the AMS, industry and the general public.

In 2017, the newly released variety MCT1 was gifted to the Trust and protected by Plant Breeder Rights. Royalties from sale of MCT1 have increasingly funded conservation. This privately selected variety is compact, hardy, adaptable and precocious, and usually produces a commercial crop in its third year. It is a prolific bearer of both NIS and kernel, averages 45 per cent kernel recovery, and does not appear to have any undesirable characteristics. Its Australian and international commercialisation and protection is led by Dr Gavin Porter, from the Australian Nurserymen's Fruit Improvement Company (ANFIC), who ensures that all plants are authentic, that MCT1's intellectual property (IP) is enforced and its royalties are collected. These royalties will be used for ongoing conservation of the wild macadamia genetics as part of Australia's threatened flora and for the long-term industry future.

Over fifteen years, the Trust's achievements include the creation of species databases, Recovery Plans, ex-situ arboretums, long-term monitoring sites, research into the geographic distribution of ge-

MCT1, a new generation variety that may advance productivity and orchard economics. Credit: I. McConachie.

netic diversity, and raising the profile of macadamias as an endangered Australian native species.

The Trust's first project was to commission independent experts to prepare a Recovery Plan, which was adopted by the Australian Government – the only such plan prepared by an industry body. This Plan has been updated to 2028 and again endorsed by the Australian Government. Liz Gould, formerly from Healthy Land and Water, ably led the implementation of the Plan with a broad range of education and conservation projects involving local governments. 'Giftings' of display trees of each of the four species were made to botanic gardens. Assisted by a HAL-funded project called WAM (Wild about Macadamias), Dr Michael Powell surveyed wild populations. He defined threats to their survival and the actions required to conserve them, including assessing the likely impact of climate change on species survival. A citizen science programme in 2020, The Wild Macadamia Hunt, requested the public to send in details of very old backyard trees so that samples of DNA could be collected, in part to identify whether they were descendants of lost wild populations. A Gubbi Gubbi Elder, Uncle Russell Bennet, is on the MCT Management Committee and is active in searching for and telling the story of the wild macadamias.

Advances in genetic research were largely led by Dr Cathy Nock from Southern Cross University, who gathered material for DNA analysis and oversaw the first ever chromosome-scale genome assembly for a macadamia (variety HAES 741). Dr Nock built up an important database of genetic markers from cultivars, old residential trees and hundreds of wild trees. Dr Robert Henry from QAAFI also leads genetic research and tissue culture which may help to preserve wild genetics.

The effects of higher temperatures caused by climate change are increasingly being studied, and there is strong evidence that without intervention, by 2050 wild tree habitats and many commercial orchards may not receive sufficient chilling to initiate flowering.

The Macadamia Conservation Trust has continued its goals, including research into *M. jansenii*, the rare species discovered in the 1980s at Bulburin National Park, largely led by Associate Professor Alison Shapcott from the University of the Sunshine Coast and by Professor Robert Henry and his team at the University of Queensland and at DAF. Associate Professor Shapcott has surveyed the population over time and worked with the Gidarjil Development Corporation Land and Sea Rangers to establish habitat restoration plantings to increase the species' area of occupancy. Dedicated naturalists Keith Sarnadsky, Ray Johnson and Carly Sugars continue to search and have found several more remnant populations of this critically endangered species. The outcome from Professor Henry's research was recognised as a model for the future assembly of genomes for other plant species.

Recognising *M. jansenii* as one of the thirty top priority threatened species in 2021, the Australian Government has funded two projects to ensure its survival. 'Insurance' populations are being propagated and planted at secure sites, including the Australian National Botanic Gardens in Canberra. The bushfires that swept through much of the wild habi-

tat at Bulburin in 2019 destroyed about 30 per cent of known trees. Some of these are now re-sprouting, but ongoing threats include human traffic and weeds, such as cat's-claw creeper. Funding to further protect this species has been approved.

The end of the decade

In the early part of the decade, NIS prices in Australia had been low until 2013, when, largely due to demand from China, prices started to increase, reaching unprecedented levels of over $6.00 per kilogram by 2020. These prices led to a global expansion of plantings and high levels of profit for most growers of bearing trees. Many growers used this boom time to upgrade infrastructure, irrigation and tree culture.

In 2013, 1,800 tonnes of NIS were exported to China, rising to 17,000 tonnes by 2017. However, as Chinese production increased, the amount of Australian NIS exported to China declined and in 2020 it fell to 7,700 tonnes. China has rapidly expanding plantings, advanced research and product innovation, and purchases NIS from many other overseas countries. Chinese individuals and companies were also purchasing and establishing orchards in Australia.

The following statistics have been collated from different sources and should be considered as only indicative. The AMS reported that by 2019, there were 7.5 million trees planted on 22,000 hectares. In 2019, the Australian domestic market accounted for 4,046 tonnes of kernel, while the bulk of global kernel sales of 12,284 tonnes went to Japan (2,320 tonnes), the USA (1,595), Europe (1,700), Asia (812), and China (585) as well as 10,910 tonnes of NIS.

During the decade, with an increase in the size of farms, there was a growing trend for the ownership of orchards and processing or marketing companies to change from family to corporate. Agricultural investment companies offered syndicated ownership in orchards, and overseas investment companies began to invest in the industry. The market value of sound, larger orchards increased from about $20,000 per hectare to well beyond $100,000, but then from 2021 settled a little. A restricting factor in growth was the increasing price of suitable land that had potentially available irrigation water.

Due to high prices, many smaller Australian orchards remained most viable, and the industry became over-confident. Sound farmers strengthened all aspects of their orchard during the good years. Many growers naively wanted to believe that the industry had reached a fairly stable plateau and that demand and NIS prices would not fluctuate to the extent they had in the past.

But the boom period for almost all of the decade from 2010 was coming under pressure. While there was increased demand for the product, largely driven by China, it was countered by rapidly increasing global production. Global NIS production increased from about 79,500 tonnes in 2010 to 220,000 tonnes in 2020 and is projected to reach 700,000 tonnes by 2030. China's own production was rapidly increasing, which caused concern that this would reduce their import requirements of both kernel and NIS. Then came COVID-19.

Australia, 2020–2023

Plantings had expanded rapidly during the preceding decade, these newer orchards generally flourished, and standards and confidence were high.

But in early 2020, the COVID-19 pandemic spread throughout the world. Air travel, tourism and travelling almost came to a stop and many traditional macadamia retail outlets were impacted. While travel and trade were returning to normal by 2023, increased global production and reduced sales had resulted in a large build-up of inventory, particularly Style 4 (half-kernels). Based on a 33 per cent sound kernel recovery (SKR), prices to growers in Australia fell from more than $6.00 per kilogram of NIS in 2020 to $5.00 in 2021. In 2022, prices had fallen to between $2.80 and $3.80 per kilogram but worse was to follow. In early 2023, it became evident that there were very large inventories of uncommitted kernel, almost all of it half-kernels. The market reacted and kernel prices fell sharply. Prices to the grower fell further in 2023 to as low as $1.50 per kilogram

'Summerland House', a smaller orchard showing sweeper and harvester, 2021. Credit: AMS Ltd.

of NIS, when the extent of uncommitted stock of half-kernel became known or suspected. Processors and marketers who had unsold kernel were faced with large trading losses and holding stocks of kernel which were becoming aged although stabilised by sound packaging and cold storage.

The larger marketers of kernel received some criticism for both a lack of transparency and for not taking earlier actions. Some processors deferred NIS payment. While there may have been some over-reaction in this downturn, there were concerns that the 2023 crop would be larger than projections and add to the marketing problem, but this did not happen. Industry morale was low and some growers did not harvest their crop. Those with long memories knew the industry had suffered several previous market price collapses where supply exceed demand, but the AMS, the marketing sector and overseas industry had then successfully expanded demand, cleared old and current stocks and, over several years, the market had stabilised and prices firmed.

A new initiative was satellite mapping of Australia's macadamia orchards using sophisticated technology to show a reported 31,878 hectares of orchards. From 2017 to 2021, new plantings in Bundaberg totalled 4,552 hectares and in northern New South Wales, 2,905 hectares.[15] The AMS was forecasting that 8,000 hectares of greenfield new or-

An example of Macadamia Farm Management's Maryborough developments of 300,000 trees from 2020.
Credit: I. McConachie.

chards would be planted from early 2023 to 2025.[16]

Despite the NIS market downturn and with demand and prices for established orchards muted, two large macadamia businesses changed hands in 2022 and 2023. Apart from the Steinhardt family sale to Stahmann Webster, the Manera family in Bundaberg sold their Macadamia Enterprises of 106,000 trees for a reported $71,000,000 to Nuveen Natural Capital, which represents a large US teachers' pension fund.

The market downturn and lack of confidence was eased due to an increased demand for NIS from China and Vietnam in 2023. A record 19,000 tonnes of NIS was exported, although this resulted in a much-reduced production of kernel, so some

marketers had difficulties in meeting existing client demand.[17] While at December 2023 the extent of inventory from 2022 was unclear, there was optimism that recovery would continue during 2024.

The Paradise Dam, built in 2005–06 on the Burnett River, provided a major boost to farming in the Bundaberg district. Concerns about the dam's structural safety due to design flaws led to its capacity being reduced to 42 per cent in 2020, decreasing the available water for irrigation. In mid-2023, the Queensland Government announced a major rebuild to commence quickly, which restored confidence.[18]

Probably partly related to the high prices in the decade up to 2020, global and Australian institutional investors expanded. In 2024, Stahmann Websters owned by the Canadian PSP Group was reported as having 2,051 hectares and Rural Funds Management Group was reported as having nine properties totalling 2,946 hectares. This would be equivalent to over 900,000 trees.[19] In addition, large corporate organisations such as Finasucre expanded with its purchase of the Winfield orchards and Mac Direct processors. Macadamia Farm Management managed orchards for investors, and many others illustrated the changing industry as well.

The World Macadamia Organisation (WMO), led by Jillian Laing, was formed largely in anticipation of increasing production and the need to continually create ongoing global demand. The global crop of 315,425 tonnes[20] of NIS in 2023 was projected to reach 504,000 tonnes by 2027.[21]

The WMO commenced activities in 2022 with a programme to expand consumption in China and India. Eight producing countries joined, providing funding mainly through a voluntary levy on growers. Australian growers voluntarily contributed $900,000 in 2021, but unfortunately membership has since fallen due to the depressed global market prices, which resulted in 2023 in a lapse in membership by China, Hawaii and Brazil. However, the global importance of continuing to push towards WMO goals was recognised by all commercially growing countries. With depressed NIS and kernel prices, the original funding model was not sustainable. In Australia in 2024, processors and marketers through the Australian Macadamia Handlers Association and a co-contribution of industry marketing funds will enable this initiative to continue.[22] The potential of the market in India is massive and tariffs are being phased down or out. There is no commercial macadamia industry in India, but as an example of the potential, the consumption of almonds, all of which are imported, has increased 400 per cent since 2010[23]

A modern, attractive orchard, 2021. Credit: AMS Ltd.

to approximately 200,000 tonnes per year.

In Australia in 2022, much higher than normal rainfall resulted in losses due to flooding but soil moisture levels were optimised going into the 2023 season, giving confidence of a large crop. However, much of Australia's growing regions in 2023 received low rainfall, leading to a reduction in the 2023 crop to 49,000 tonnes of NIS although the quality was high. Water reserves were high in early 2023, but the effect of dry conditions on the 2024 crop were not known when this book was completed. Costs of growing, in terms of labour, fuel, energy and chemicals, had increased over the last ten years.

Recognition of the macadamia was made in 2022 following the death of Queen Elizabeth II, the much-loved monarch who had ruled the United Kingdom and the Commonwealth countries for seventy years. A macadamia named the Queen Elizabeth II Memorial Tree was planted by the Governor of Queensland at Government House, Brisbane, which overlooks the city.

Jacqui Price, the AMS Market Development Manager, in August 2023 summarised the global industry as producing 308,000 tonnes of NIS, where Australia contributed 53,160 tonnes, South Africa about 80,000, and China 56,000. China also imported 51,185 tonnes of NIS, where 14,000 tonnes came from Australia. Jacqui reported that demand was increasing, with macadamias being seen as a functional food and as a snack.[24] The 2023 International Macadamia Symposium held in South Africa provided a September 2023 summary of Australia as having 800 growers, with 41,000 hectares planted and 65 per cent of the trees bearing. They listed Australian and global issues as low prices, high input costs, labour shortages, the need for quality improvement, post-harvest capacity, the growing of markets and the need for collaborative research.[25]

In summary, 2023 was a year when almost all growers operated at a loss and many marketers had high inventories with a significantly decreased value. In May 2024, the International Nut and Dried

The Queen Elizabeth II Memorial Tree with the Queensland Governor, Her Excellency the Hon. Jeannette Young, in 2022. Credit:R. Symmonds.

Plaque on the Queen Elizabeth II Memorial Tree. Credit: I. McConachie.

Fruit Council (INC) Conference forecast the Australian crop at 56,000 tonnes of NIS. Ongoing rain prior to and during much of the harvest resulted in late season insect damage, high levels of pre-germination damage and often lower quality. However, the price increased from the historic lows of 2023, from below $2.00, to over $3.00 per kilogram of NIS at 33 per cent sound kernel recovery (SKR), boosting confidence in 2024.

Going into 2025, the only certainty is change. The rate of growth of global plantings and the certainty of increasing production will dominate the industry for at least a decade. This will put pressure on the Australian industry to market their product as premium quality to attract high prices.

Chapter 19, 'The Future', will address the rate of growth of the Australian industry and delve into the future with its opportunities and challenges.

Increasingly, the macadamia will become known and recognised as the finest nut in the world. That both Australia and several overseas countries will have growth and market cycles is certain. The industry has gone from infancy to adolescence and now embarks on a confident but uncertain adulthood.

CSP orchard, Bundaberg. Harvesting at night, probably so as not to harm both honey and native bee pollinators. Credit: CSP Macadamias and AMS.

Part 4

THE REST OF THE WORLD

Map 4: Countries growing commercial crops of macadamias in 2023

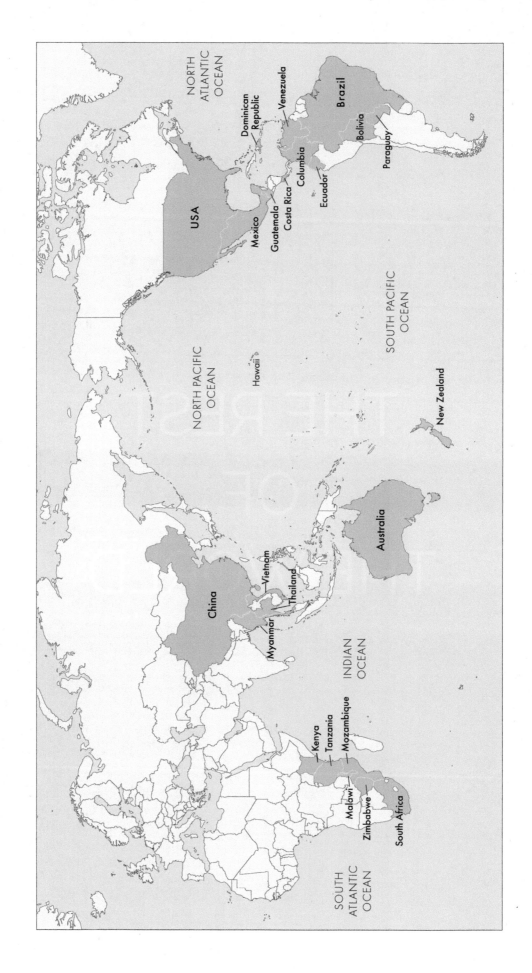

INTRODUCTION

An overview of macadamias
in the rest of the world

So far, this book has told some of the history of the macadamia and recorded the development of commercial macadamias in Australia. Part 4 tells the history of macadamias in the rest of the world (see Map 4). Some of the countries warrant their own book, but space permits only an overview here. What follows is a summary of the defining stages as each country followed its own path to develop a macadamia industry.

There has been such rapid growth from 2010 onwards that the author does not attempt to present the current industries in any detail.

Many people of vision and passion have invested in research and been prepared to risk capital for the dubious pleasure of being industry pioneers. Some of these have established viable industries, some provided optimism, some goals were not realised, with a commercial industry not developing, but in many countries there is likely to be future growth, the scale of which is impossible to project. Global growth in terms of kernel production from the 2020s onwards will be rapid, accompanied by farming, productivity and marketing challenges, but there will also be setbacks and failures.

The global industry is driven by the image and demand for macadamias, with the early 2020s' estimated economic returns often greatly exceeding alternative crops and then, partly because of COVID-19, there was a downturn. In many subtropical country environments, macadamias can be grown by a wide range of farmers, from small-scale, lower-income families, to independent commercial farmers, to large-scale corporations. Macadamias are not readily perishable, they form part of a healthy diet and there is expanding demand for all tree nuts.

From the nineteenth century, macadamias were planted as a botanical curiosity in many countries of the world. There were people who had an interest in new, exotic plants – particularly if they produced such a magnificent-tasting food. So in tropical, subtropical and warm temperate parts of the world, macadamia trees were planted, usually as specimen trees. Macadamias were introduced on this basis to many countries in the late 1800s, but there was little or no commercial interest in them until the middle and latter part of the twentieth century.

Following World War II, there was a global reassessment and seeking of opportunities. It was a time for changes within agriculture, as mechanisation was able to raise efficiencies for many countries. As well, horticultural and processing information was available and pest and weed control chemicals were being developed. Countries had to assess and consolidate existing viable industries, as well as seeking new industries and crops.

During the war, many servicemen passed through Hawaii, either being based there or having their recreational leave there. They saw and tasted macadamias for the first time and enthused over the distinctive flavour and texture. They associated macadamias with the exotic image of Hawaii. Macadamias became an exclusive and different gift, available from no other country, ideal to give to friends, and these qualities created an interest in their commercial potential.

After World War II, Hawaii more fully realised the potential of this crop. Their industry expanded and they made their applied research available to other countries. Macadamias appeared to be an easy crop to grow; based on their performance in Hawaii, there was strong market demand; the nuts sold at high prices. California also played a major part in creating an awareness of macadamias, particularly in the Americas. Australia played little part in the global industry at this stage.

Also through Hawaii, there was the availability of both scion wood and expertise to introduce this crop into other countries. As in many fields, not all pioneers were successful but we should acknowledge, respect and appreciate their vision and entrepreneurial spirit.

FIVE POUNDS OF
ROASTED-IN-THE-SHELL
Macadamia Nuts

•

HAWAII'S BEST-KNOWN
DELICACY

Mac Adamia

PACKED BY

A. G. GREENWELL

KEALAKEKUA, HAWAII, T. H.

Chapter 15

THE AMERICAS

The USA – Central America –
South America – The Caribbean

The USA

Hawaii

> Last Christmas in a weak moment we decided to chocolate coat some of our nuts. This was a mistake as we have been pestered with orders ever since.
> (Walter Naquin, Hawaii, 1949)[1]

From the 1930s to the 1980s, the global macadamia industry was almost entirely concentrated on the Hawaiian Islands. It was their vision, passion, government support, extensive research into horticulture, varietal assessment, processing and marketing that created the image and demand for macadamia nuts. It was fortuitous that Hawaii had an ideal climate for growing macadamias, or the global industry might have been set back for many years. In the 1960s, Dr William (Bill) Storey, commenting on the many pioneers whose names had been lost, said, 'in the accumulation of knowledge who is to say whose contributions were the most valuable and who the least'.[2]

In 1883, macadamias were introduced into Hawaii by William Herbert Purvis, who obtained some nuts in Australia and planted a few trees near Kapule-

na, between Honokaa and the Waipio Valley on the island of Hawaii (the 'Big Island'). He established his own botanical gardens and one of the original trees planted at Kapulena later became a tourist attraction.[3] Then six macadamia trees, of possibly three species, were planted between 1892 and 1894 by the Board of Agriculture and Forestry at Mount Tantalus on the island of Oahu.[4]

In 1892, Robert Jordan was first mate on a vessel berthed in Brisbane, Australia, and he visited a friend, Francis Lahey, who had settled at Hotham Creek just north of the Gold Coast region. He collected a half-sugarbag full of macadamia nuts from the wild *M. integrifolia* trees on or near the property. Robert planted some of these seed nuts at his brother Edward's home at Wyllie Street, in the Nuuanu Valley of Honolulu, on Oahu, resulting in at least six macadamia trees. Most of the trees were lost some sixty years later when the Pali Highway was constructed over a part of Honolulu. One tree remained but it has since died, though a neighbour's tree still remains.[5] One of the wild trees that Jordan collected nuts from in Australia was known as

William Purvis, and Edward and Robert Jordan, who all brought nuts from Australia in the 1880s and 1890s, as well as planting the first trees. Credit: unknown.

the Jordan Tree or the Mother of all Macadamias, but it was lost in a fire in 2020. In 1915, the Annual Report of the Hawaii Agricultural Experiment Station (HAES) included a section on macadamias, reporting on a trial planting of one acre. In 1912, the HAES provided seedlings to farmers in the Kona District on the Big Island, and another 1,000 seedling trees were distributed to Big Island coffee growers in 1918.[6] An Australian seed merchant, Hubert Rumsey, provided 10,000 to 20,000 seed nuts to Hawaii, California and Florida in 1910 and it is possible that the resultant plants became the basis of later selections. From 1925 to 1930, about thirty acres were planted each year but plantings slowed during the economic depression and World War II.

Walter P. Naquin was Manager of the Honokaa Sugar Company from 1916 to 1948 and his introduction to macadamias was seeing boys eating what they called a 'monkey nut'. He found this nut had come from the Purvis Tree and is reported to have paid a dollar for each of 600 seed nuts. He recognised their potential and in 1923 planted several acres, increasing this to 24,000 trees. His first small harvest was in 1926, when he engaged sixty school children to pick up the nuts.[7] His wife Ethel experimented with roasting and flavouring the kernels. Other growers followed and they supplied

their nuts to the Hawaiian Macadamia Nut Company until Otto Hermann and Leo Thevenin designed and built processing and value-adding facilities, trading as Triangle H. Naquin used a rotary steam-driven press to crack the hard shells and at times connected a steam locomotive to provide the energy.[8]

A dehusker designed by Walter P. Naquin in the 1930s. Credit: unknown.

Ernest Sheldon Van Tassel. Credit: unknown.

'Nutridge', Van Tassel's home and orchard at Round Top, above Honolulu, about 1925. Credit: Bishop Estate Museum.

But it was Ernest Sheldon Van Tassel (1881–1943) who deserves to be acknowledged as the 'Father' of the Hawaiian industry. In 1915, Van Tassel came to Hawaii to recuperate from what was probably poliomyelitis, and attended a dinner at Gerrit P. Wilder's home, where he tasted macadamias. He dedicated the rest of his life to them, starting in 1921 by planting 2,000 trees at Round Top, Mount Tantalus, above Honolulu, on a property he named 'Nutridge'. This was to be his forever home, where he lived with a handyman and a nurse, Lillian Jonsrud who, because of limited cash flow, was often paid with shares in his Hawaiian Macadamia Nut Company (formed in 1922). Van Tassel's first trees died as they were planted still in their metal tins.[9] He used cattle for weed control, cats to control rats and held hunting parties to control wild pigs. Other orchards used geese for weed control. Van Tassel successfully lobbied the Territorial Legislature for tax exemptions, which encouraged more orchards. He leased and planted 100 acres at Keauhou on the Big Island from the Bishop Estate, a large philanthropic land trust derived from the King Kamehameha Estate which contributed much to the development of the islands and the industry.[10]

In 1931, Van Tassel established the first processing factory in Honolulu under the brand Vans Mac-

Hawaiian Macadamia Nut Company, cracking and kernel separation, 1937. Credit: unknown.

Drying racks, probably at Hawaiian Macadamia Nut Company. Credit: HMNA.

Value-adding section of Hawaiian Macadamia Nut Company, Honolulu, 1930s. Credit: R. Moltzau.

adamia Nuts, which successfully marketed retail packs in Honolulu, New York and Boston. He had engaged Ralph Moltzau, a young university graduate, to learn and apply sound horticultural management.[11]

Orchards were established on the islands of Maui and Kauai. By 1937, the macadamia industry consisted of twelve commercial growers as well as smaller growers and several processors.

Macadamias were considered most difficult to graft or vegetatively propagate in order to produce trees with consistent desired characteristics. Ralph Moltzau and Bill Storey were university students gaining work experience and in 1928 were given an unwanted broken branch to practise grafting and which was still attached to the parent tree. To the surprise of all, some grafts 'took' but it was another eight years before it was fully understood that the broken limb was equivalent to cincturing or girdling (a process which causes carbohydrate energy to accumulate above the removed bark, keeping the scion or graft alive until it has joined with the root stock). Later Moltzau, Storey and others initially assessed 15,000 seedling orchard trees and helped to define selection criteria. Fortunately, from 1938 to 1941 selected grafted trees were planted in experimental plots, so they were able to be assessed after World War II. By 1948, 65,000 seedling trees had been assessed before five varieties were recommended and released. One of these was Keauhou (HAES 246), selected from Van Tassel's 'Keauhou' orchard, and Kakea (508) from 'Nutridge'. Keaau (660) was released later in 1948, Kau (344) in 1971 and Mauka (741) in 1977.

The availability of these varieties, both in Hawaii and later to the rest of the world, enabled macadamia farming to be based on reproducible, understood tree and kernel characteristics. After World War II, E.T. (Eddie) Fukunaga actively led the plant-breeding programme and he was followed by Dr Phil Ito.[12] Dr Craig Hardner from Australia visited Hawaii in 2014 on a Winston Churchill Memorial Trust Fellowship and clarified the origin of many Ha-

Ralph Moltzau demonstrating the first successful grafting technology, 1927. Credit: R. Moltzau

waiian varieties, showing that they were bred from a very small number of trees from the Australian rainforest.[13] Subsequent genetic research indicates that the original few trees may all have been seeds from a single or small number of wild trees of *M. integrifolia*, probably from Mooloo.[14]

The Hawaii Agricultural Experiment Station was established in 1901, to research and develop agriculture. Its ongoing role in research and its implementation contributed enormously to the commercialisation of the Hawaiian and international industry. The first booklet on growing macadamias, *The Macadamia Nut in Hawaii* by W.T. Pope, was published in 1928 followed by *Processing of the Macadamia* by R. Moltzau and J. Ripperton in 1939. The policy of the Hawaiian Territory and then the USA in making available such knowledge resulted in the development of macadamia industries in many countries. There were several outstanding macadamia

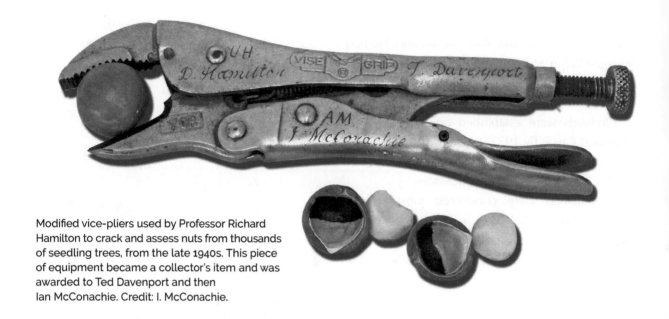

Modified vice-pliers used by Professor Richard Hamilton to crack and assess nuts from thousands of seedling trees, from the late 1940s. This piece of equipment became a collector's item and was awarded to Ted Davenport and then Ian McConachie. Credit: I. McConachie.

researchers, including Dr John Ripperton, Dr W.T. Pope and later Dr John Beaumont, Eddie Fukunaga, Dr Phil Ito, Dr Richard (Dick) Hamilton and Associate Professor Cathy Cavaletto. Dr Bill Storey was another dedicated researcher, who moved to Riverside University in California in 1954, where he continued to support macadamia study.

Largely through the Hawaiian Macadamia Nut Company and Honokaa Sugar, a typical gift for World War II US servicemen to take back home to the mainland was a can of roasted and salted macadamias. This created an awareness and demand which accelerated the expansion of the Hawaiian industry.

Castle and Cooke, one of the 'Big Five' companies, expanded into macadamias. By 1949, they had purchased 4,000 acres of old aa lava uncleared land at Keaau, near Hilo, and commenced planting. The orchard floor was largely crushed lava overlaid with organic matter produced by the native rainforest. By 1956, they were marketing their cans of Royal Hawaiian Macadamias. United Airlines served small packs on some flights, buying four million packs per year, and a survey showed that many passengers chose United because of the macadamias. Castle and Cooke claimed to have created a 'Garden of Eden out of an old lava field'.[15]

In 1948, Mamoru Takitani, later the owner of Hawaiian Host Macadamia Nut Company, commenced planting 12,500 grafted trees on the island of Maui and subsequently established the retailer Hawaiian Host, which became a major retail brand. Many Japanese Hawaiian farmers had trees and continued to plant small orchards on the Kona Coast. The Kona Macadamia Co-operative was established in 1962. C. Brewer and Co. planted macadamias at Pahala

Dr John Beaumont and Dr William (Bill) Storey, 1936. Students together, then researchers at the University of Hawaii, where Bill continued his macadamia research and built on knowledge gained from California.

Retail pack from the 1940s. Credit: unknown. Retail jar of macadamias from the 1950s. Credit: unknown.

on the Big Island during the 1960s, then purchased the Royal Hawaiian Macadamia Company which was renamed Mauna Loa Macadamia Nut Corporation. John A. (Doc) Buyers was Chair from 1972 to 2005, Robert Grimmer was Manager and Masao Nakamura headed orchard development. These were the halcyon days, with tourism the major island industry and macadamias the perfect gift for tourists to take back home, both of which further increased demand. A successful marketing theme was 'Say Aloha with Macadamia Nuts'. Tourists, mainly from the US mainland and Japan, flocked to the islands.

In 1961, the Hawaii Macadamia Producers Association was formed, later to become the Hawaii Macadamia Nut Association (HMNA). It represented all aspects of the industry and, with support from the government through the University of Hawaii and their Hawaii Agricultural Experiment Station, held conferences and research workshops. The HMNA was a body that strongly represented and co-ordinated its members and was a major support for the whole industry.

As the industry prospered and the image of macadamia as a premium nut grew, small and large businesses were attracted, consolidated and led the industry. The larger businesses increasingly undertook their own research and promotion, which tended to weaken the united presentation of Hawaiian macadamias.

The largest orchard in the world at that time was established from 1958 at Honomalino, near where Captain James Cook was speared in 1779. The orchard has had a chequered life and ownership and

Roller crackers used at Royal Hawaiian and Mauna Loa factory from at least the 1950s to the 1980s. These required NIS size-grading and damaged the kernel. Credit: I. McConachie.

is situated in a low-rainfall (750 millimetres/ 30 inches) area, which experiences frequent droughts and where rocky soils made mechanisation difficult. High labour costs, wind damage, funding and legal disputes were all ongoing difficulties. Equity owners were Macadamia Nuts of Hawaii, Lands of Kapua and Honomalino Agricultural Company. Investors in these groups at times included the film stars Jimmy Stewart and Julie Andrews. It consolidated as Mac-Farms of Hawaii, was owned at one time by the Australian CSR Group, changed hands several times and in the early 2020s became part of the Health and Plant Protein Group incorporating Royal Hawaiian Macadamias, which is based in Australia and has now been purchased by the Hawaiian Host Group.

Hawaiian Holiday at Honokaa, Big Island of Hawaii, 1974. Credit: I. McConachie.

MacFarms of Hawaii. Honomalino, Big Island of Hawaii, 1980s. Credit: CSR Ltd.

In 1974, the macadamia factory of Honokaa Sugar became Hawaiian Holiday, owned by Paul and Anita DeDemenico. They launched a range of one hundred products and were superb marketers. Anita was known as 'Mrs Nut', which was the number-plate of her top-of-the-range Mercedes Benz. The couple stated that they 'were making and exporting paradise'.

By the 1970s and 1980s, the image and success of the Hawaiian macadamia industry was legendary and it was the global base of knowledge. The extent of research and its implementation covered all aspects applicable to the commercialisation of macadamia nuts. Research was professional and included botany, breeding, nursery propagation, orchard establishment, tree culture, nutrition, pests and diseases, harvesting, crop handling, quality, processing, value adding, economics and marketing. This knowledge encouraged other countries to consider growing macadamias and assisted in developing overseas industries. The HAES stated that expected yields at Year 20 were 100–150 pounds (40–60 kilograms) of NIS per tree, but this was based on wide-spaced, high-standard orchards. Average yields were much lower than projections and published figures.

Hawaii, apart from the hurricanes and drought which mainly affected the east of the Big Island, had an ideal climate with few seasonal extremes and little diurnal temperature variation range. The macadamia trees received sufficient chilling to initiate flowering, while the benign climate caused the trees to flower and mature their nuts over an extended period, which spread the energy demand on trees and thus minimised stress. For many years, pest and disease impact was low but as orchards became widespread, both native and introduced pests and diseases expanded and controlling them proved difficult. Macadamia Quick Decline resulted in tree death; bug or sucking insects and felted coccid somehow introduced from Australia resulted in losses and added to costs. Many older trees declined, and many trees were replaced.

By 1974, almost all orchards were on the island of Hawaii, with 600 hectares at Keaau, 1,220 hectares at Pahala and Kona, and 400 hectares at Honokaa.[16] While the islands of Maui and Kauai both had over 40 acres planted in the 1930s, more were planted in the 1980s which included 640 acres at Kahului on Maui. By 1990, these orchards were largely lost to winds and poor performance.

Levelling and aa lava crushing at Royal Hawaiian/ Mauna Loa, 1950s. Credit: Mauna Loa Nut Corporation.

Young tree planted in aa lava at Royal Hawaiian, Keaau, 1974. Credit: I. McConachie.

Much of the land was classified as aa lava containing little soil, which had to be broken up, and in high rainfall areas growing was almost by hydroponics. This limited the ability to mechanise, particularly to machine harvest. Moreover, as soil organic carbon fell, it was difficult to maintain long-term tree health.

There were many people who contributed to the advancement of the more modern industry from the 1990s to 2010. For forty years, Alan Yamaguchi, a research horticulturist at the Mauna Loa Group and associated companies, led horticultural research, shared his knowledge and applied it through management on their orchards. David Reitow led the industry association and provided consultancy. Cathy Cavaletto at the University of Hawaii undertook basic and advanced research on all aspects of quality. Rick Vidgen from MacFarms of Hawaii became a spokesperson representing the industry. Mike Nagao and Skip Bittenbender provided extension services from the University of Hawaii. During the 1990s, the major growers and processors were Mauna Loa Macadamias, MacFarms of Hawaii, Hawaiian Host Candies, Hamakua Macadamia Nut Company, Island Princess, Hawaii Macadamia Company and Big Island Candies.

A range of Hawaiian retail packs from the 1980s. Credit: unknown, but likely HMNA.

Dr Gordon Shigeura played a major role in horticultural advances and Dr David Curb's nutritional studies demonstrated significant health benefits from macadamia nuts. Jim Kendrick from Mauna Loa and within the HMNA provided strong leadership.

Debbi Fields, trading as Mrs Fields, had an international chain of cookie stores and established a processing factory at the Kona store. This was very effective as a tourist attraction, but less so at processing nuts. However, it added to the Hawaiian image until the business failed.

While there were market and economic cycles, over time plantings steadily increased. In the 1950s, 200 acres per year was typical, expanding to plantings of almost 1,000 acres per year during the 1980s.

In 1989, the HAES published the direct farm cost of production averaging US 54 cents per pound without overheads, and for most of the decade from 2000 to 2010 prices to growers averaged US 62 cents, which meant that many farms were not viable. During low-price cycles, many orchards were abandoned.

In the early 1990s an increase in kernel coming on the market, mainly from Australia, tested demand as global inventory soared and prices fell. Hawaii believed that dumping was occurring and in 1992 and again in 1998 petitioned the US International Trade Commission (ITC) to investigate and impose restrictions on imports. The ITC did not support the Hawaiian industry's claim and gradually the market expanded, prices stabilised and concerns subsided.[17]

The decade from 2000 to 2010 was an unstable period for the industry. Apart from droughts, the major growers, processors and marketers were subject to mergers, takeovers, resale and some legal disputes which destabilised all sectors. Overall this resulted in low prices to growers, with many unable to sell their crop. MacFarms of Hawaii was on the market and the sale fell through. The Mauna Loa business was sold, then bought by the Hershey Group, then resold.

In 2015, some 8,164 tonnes of NIS were exported to China and only 9,979 tonnes processed in Hawaii. Demand from China resulted in prices to the grow-

Table 6: Hawaii: indicative plantings, production and prices over time

Year	Number of farms	Total planted area (acres)	Bearing planted area (acres)	Production of NIS net (tonnes)	Price per pound (US cents)
1938		1,086		30t (estimated)	
1943		667		190t	10c
1950		1,767	808	378t	17c
1960	240	3,515	2,100	1,308t	15c
1970	275	8,670	4,050	6,600t	22c
1980	480		5,300	16,000t	90c
1990	660	22,600	18,200	25,000t	82c
2000			17,700	25,000t	59c
2010	570		15,000	20,000t	75c
2020	550		16,900	20,000t	120c

Sources: R.A. Hamilton and E.T. Fukunaga, 'Macadamia Nuts', report undated. USDA Macadamia Market Reports.

er doubling by 2020 but their long-term purchases are uncertain. The main processors and growers remaining in the early 2020s are Mauna Loa, Hamakua Macadamia Nut Company, MacFarms of Hawaii, Island Princess, Island Harvest, Hawaiian Host and Hawaii Macadamia Nut Company. COVID-19 affected tourism and resulted in increased bulk sales to the US mainland, although the traditional attractive snack and confectionery tourist trade remains. Despite COVID, tourism remains the backbone of the industry. A major consolidation occurred in April 2023 when the Hawaiian Host Group, parent company of Mauna Loa, acquired MacFarms LLC which included Royal Hawaiian Macadamias.

The Hawaiian crop estimate for 2024 was 15,000 tonnes of NIS, which has been relatively static for several years. The import of lower-priced kernel was weakening demand for Hawaiian-grown product and representations were being made in 2024 to have labelling or other restrictions imposed on imported kernel.

A different retail pack, 2006. Credit: I. McConachie.

The Hawaiian industry, in terms of size and production, has been broadly static for the last thirty years. While it has an ideal growing environment and strong tourist demand for the product, there have been hurricanes or windstorms, droughts, high labour costs, high land cost and limited ability to mechanise. As a result, the price to growers was often below growing cost, and the industry arguably suffered a loss of direction and domination by corporations whose commitment to the industry was not their main priority. HMNA, the industry body, was not well supported and was unable to contribute to government research, which was reduced. A combination of high levels of unsound NIS from growers, high cost of producing marketable kernel, and pests and diseases, all resulted in reduced new plantings and initiatives. Overall Hawaii's position in the global industry has declined and this is likely to continue.[18]

From being the pioneers who achieved so much and gave so much to the rest of the world, the Hawaiian industry has not been able to sustain the momentum generated from World War II to the 1980s. Despite their considerable strengths, their growth failed as global plantings greatly expanded. The reasons are complex, and possibly in summary the industry became fragmented, arguably the growing sector was not supported and economic viability declined.

The HMNA was not supported to co-ordinate and lead as was the AMS in Australia and SAMAC in Africa. In 2018, the HMNA had only thirty-three growers and three processor members and no levy in place.

The Hawaiian industry will continue as a relatively stable grower, processor and marketer of high-image tourist products, but its days of leading the world are over.[19]

California

Californian macadamias have always been minor in the global scene but their contribution in the past has been far out of proportion to their size. A number of enthusiasts believed that a commercial industry was possible and in 1953 formed the California Macadamia Society. This Society, through its excellent *Yearbooks* which commenced in 1955, became the main way of disseminating knowledge of the macadamia and its cultivation throughout the world. Although they followed different paths, California must be acknowledged with Hawaii as leaders in the commercialisation of the macadamia. Growers and specialists combined to support global development through skill, enthusiasm and information transfer after World War II. Enthusiastic amateurs led by applied researchers in botany, horticulture, management and engineering united to develop the foundations of a commercial industry which, due to factors largely outside their control, did not develop.

In 1879, Professor C.H. Dwinelle from the University of California brought nuts from Australia which came from the Walter Hill Tree in Brisbane. At least two were planted at the Berkeley campus near Strawberry Creek. One tree remains. Before 1890, the Spiritualist Health Colony, a vegetarian community in Orange County, planted some macadamias. In 1888, David Balch at Coronado and in 1900, C.

Left. Plaque for *M. tetraphyllas*, telling the early history of the macadamia in California. Credit: C. Hardner.

Below. *M. tetraphylla* trees commemorating the history of the macadamia in California from 1890. Credit: C. Hardner.

Seymour Faulkner at Santa Paula both planted trees. Nurseries began to supply seedling trees and in 1939 a survey listed 300 specimen trees, which was considered to be an underestimate.[20]

During World War II, American servicemen passing through Hawaii tasted and often brought home macadamia nuts, which created much interest about a possible new industry. After the war, land in California was cheap, avocados and citrus were declining and macadamias were seen as having potential. The first commercial planting was of 200 trees in Oceanside by Robert Todd in 1946, and this property was purchased by Dr Art Sneider, who leased ninety-five trees to Colonel Wells Miller, who became an industry leader for many years.[21]

In 1953, eleven people met and formed the California Macadamia Society (CMS). The following year, membership had grown to 154. Their annual *Yearbooks* were detailed and for the next sixty years comprised a broad spectrum of research, advice and encouragement. Both Californians and a significant global membership used the information in the *Yearbooks* as a reference. Communication between people throughout the world was often through the CMS. Dr Bill Storey from Hawaii joined the University of California in 1955 and passionately supported the industry. Many people contributed their skills and enthusiasm. Because available growing areas were often subject to frosts, the industry differed from Hawaii and focused on varieties of the *M. tetraphylla* species or its hybrids which, growing further south in Australia, were incorrectly believed to be more frost hardy. Dr Beaumont from Hawaii provided selections to assess and one of these, HAES 695, was named after him following his early death. It became widely planted, particularly in South Africa and New Zealand. Cate became a popular variety. From 1970, Paul Shaw, a retired tool-maker from Whittier, progressively developed a blade-type cracker which revolutionised the processing sector, was often copied and continues to be used in the 2020s.[22]

While orchards were small, the skills, experiments, enthusiasm and transfer of knowledge all contributed to a belief that a large commercial industry was possible. In 1975, there were 330 acres planted, almost all bearing and producing about 15 tonnes of NIS. Southern California's winters were milder than those in the north, and rainfall averaged just 10 to 20 inches per year, making irrigation essential.

An orchard considered large in California in 1957. Credit: California Macadamia Society.

As early as 1978, Dr Bill Storey on a visit to Australia reported that the opportunities for a commercial industry were dashed by a population explosion, which created a real estate boom, high labour costs and a substantial increase in water and land taxes. While macadamias were demonstrated to be possible horticulturally, they were not an economic crop in California.[23]

Avocados overcame their disease problems and were a viable, expanding crop while macadamias remained a cottage industry. There were earlier marketing attempts of mainly NIS retail sales until, in 1973, the Gold Crown Macadamia Association led by Alva Snider, a grower from Fallbrook, was established. It was a co-operative formed to purchase, value add and market the crop and has continued into the 2020s.[24]

Tom Cooper from the Nut House entered the industry in 1969, establishing a nursery offering grafted trees and consultancy services, and over the next thirty years he provided 15,000 trees to California, Florida, Mexico and Central America. His vision was to hold a World Macadamia Symposium in the early 1980s,[25] but he was ahead of his time.

Tom passed away in 2009. Dr Gary Bender, a Farm Advisor from the University of California, provided advice for many years. Jim Russell, though paralysed and confined to a wheelchair, together with his wife Barbara was a grower. Jim was also President of the Society and an industry spokesperson and tireless promoter for thirty years. He was active until his death in 2020.

The 2019 crop was 150 tonnes of NIS. The industry had slowly declined in size, due to the age

Gold Crown Co-operative retail NIS pack, 1965. Credit: California Macadamia Society.

Orchard at Vista in 1974, showing typical California growing environment. Credit: I. McConachie.

of orchard owners, the extremely high land prices, limited availability and high cost of water and the resulting fact that many orchards had become real estate. Now remaining orchards are mainly in San Diego County, North County and Santa Barbara and in 2020 totalled about 150 acres. Most of the crop is sold at farmers markets or through Gold Crown.[26] The California industry remains as hobby farms and the lofty goals of the 1950s have not been achieved.

Central America

Costa Rica

The image and expansion of the Hawaiian industry after World War II and early plantings in Central and South America created a wide interest for macadamias as a potential crop. The first known tree in Costa Rica was planted in 1948.[27] During the 1950s, the Costa Rican Government at their Turrialba Research Station led by Dr Hestes Barbes and Edilberto Camacho brought in *M. tetraphylla* seed nuts from California for rootstocks. They followed up with both Californian and Hawaiian scion wood, mainly HAES 246, 333 and 508. Growers established small plantings which were closely observed.[28] As part of US aid to developing countries, Hawaii assisted Costa Rica with new crop opportunities.

Orlando Rojas is from a well-known local family and in 1968 planted a small orchard of Hawaiian varieties and his family became long-term promoters and enthusiasts. In 1977, Macadamias de Costa Rica (MCR) was formed and over four years established 530 hectares at two sites. MCR built a small processing plant and then a larger commercial factory in 1985. By the late 1980s, kernel was being sold on world markets.[29]

The company American Flower, led by the Thomas family, entered the industry in 1993 and later built a processing factory. Michelle Cloutier in the north established a small orchard, then a processing plant producing tourist-oriented retail products trading as Rio Frio Macadamia Farm. In a small but effective way, she promoted macadamias for about forty

Older orchard in Costa Rica held by Macadamias de Costa Rica, in hilly terrain, 2001. Credit: I. McConachie.

Small, efficient factory in Costa Rica in the 1990s. While health regulations required staff to wear face-masks, it was really to stop chatter. Credit: I. McConachie.

years. Alex Murray was an early enthusiastic grower.

One report in 1988 stated that there were between 700 and 1,000 farms, almost all in the foothills above the Atlantic and in the Central Valley, comprising 9,000 acres of which 3,500 acres were bearing. Twenty years later, almost all had been abandoned.

Led by an enthusiastic promoter, Alfredo Volio Perez, investors from Costa Rica and Swiss bank clients established orchards totalling 600,000 trees in the northern Miravalles and Tenorio volcano districts as well as a modern processing plant. They later acquired the troubled MCR.[30]

The Hacienda Juan Vinas family had established orchards and with other growers established a processing factory, Sol Caribe, in 1990. This is prob-

ably the only processing facility still operating in the 2020s.[31] Plantings increased in the 1980s and 1990s. In 1982, there were 160 growers with 5,000 hectares of trees and in 1992, there were 730 growers with 8,350 hectares with over half not yet bearing. These growers included MacFarms of Hawaii, who purchased a large orchard but later abandoned it. The annual crop of wet-in-shell (WIS) nuts averaged 2,000 tonnes or 250 tonnes of kernel, most of which was exported.[32]

Initially supported by horticulturists from Hawaii, the growing environment was believed to be suited to macadamias. With a humid environment, high rainfall of between 2.5 and 5 metres per year and fertile soils, the trees established well. Trees generally cropped reasonably well when first bearing but sound kernel recoveries were low, averaging less than 20 per cent, with unsound kernel averaging 8 per cent. Most Hawaiian varieties planted adapted poorly. There were serious insect pests and fungal diseases, high levels of immature kernel, variable quality and most plantings were growing in a marginal environment.[33]

To address the poor performance, many consultants were engaged and wide-scale research was undertaken, but productivity remained very low. Many trees declined and progressively died. Probably a combination of an unsuitable growing environment with excessive rain, tropical rather than subtropical temperatures, pests and fungal diseases and variable management all resulted in orchards being lost. An example was the 35,000 MCR trees at Turrialba and the factory which were all abandoned in about 2000. The larger orchards at Miravalles and Tenorio, consisting of 600,000 trees, were also abandoned and became cattle ranches.[34]

Over the last twenty years the industry has declined, but with remaining orchards now reasonably healthy and productive. Macadamias in Costa Rica continue as a small and relatively stable industry with one processing plant. In 2021, there were no records of any new plantings and approximately 150,000 mature trees remained producing an estimated 800 tonnes of NIS.[35] The 2024 crop was estimated at 500 tonnes of NIS and reported as now expanding.[36]

Guatemala

The California Macadamia Society in their 1985 *Yearbook* reported that in 1931, Jorge M. Benitez gave ten trees to Pedro Cofini and two to the Retana farm, both near Antigua, and that the trees were thriving in 1985. There was a single tree planted in Guatemala City in 1932. In 1958, Jorge Benitez planted a large number of trees in order to graft them. He later also sent trees to El Salvador and Honduras. After World War II, in Guatemala as in other Central and South American countries, there was an interest in macadamias due to the establishment of a successful industry in Hawaii. Also, through US Government aid programmes, Hawaii was supporting new industries in many countries. So the genesis of the Guatemalan industry goes back to the early 1950s. Seventy years later, Guatemala, despite having early failures and less than ideal growing conditions, has become a successful and expanding producer of kernel in world markets.

During the 1950s at the Chocola Experimental Farm, *M. tetraphylla* seed nuts from California for rootstocks and cultivars from Hawaii were introduced. In 1966, Robert C. Axtell from California reported that about 2,000 plants had been established at Chocola. Many of these trees were transplanted to other sites for assessment.[37]

The major industry of coffee growing was subject to horticultural risks and often low prices, so there was a reason to diversify. Interest in macadamias was high, which led to a number of small orchards being planted during the 1960s. The orchards in the northern part of Guatemala were unsuccessful, but from the early 1970s rapid expansion commenced in the middle and south. This was mainly led by Agronomionles Internacionales de Hawaii, a subsidiary of C. Brewer and Co. from Hawaii, who planted 1,900 acres from 1972. Storms, marginal site selection, poor drainage and poor management led to about 1,600 acres being abandoned. In 1978, there was an attempt to recover some of this. Starting in 1973, John G. Smith planted 60 acres adjoining C. Brewer and this orchard was of a higher standard and provided confidence to new growers for the next thirty years. Fritz Rosengarten and his syndicate planted 200 acres. Fritz had planted trees at Patzulin after selling land to C. Brewer and was paid in part with 12,000 grafted trees, which were planted and performed satisfactorily. All varieties were Hawaiian.[38] Cardamom and coffee were typically planted in conjunction with the macadamias. Coffee prices were low during the 1970s, resulting in macadamias partly replacing them. From the early 1980s, cropping commenced and was processed for several years in Costa Rica and Hawaii. There were about 2,000 hectares planted by 1985.

The Valhalla Project has, for the last forty years, been a visible, worthy undertaking whose goals are reversing global warming, providing education, undertaking reafforestation and assisting Indigenous people to develop sustainable agriculture. American Lawrence Gottschamer managed macadamia orchards in Costa Rica in 1975, before moving to Guatemala where he conceived and developed his vision with his Guatemalan wife. In 1984, Valhalla obtained access to the diversity of California's germplasm and Lawrence and his wife selected seedlings from which, over the next thirty years, almost

500,000 trees were released to regenerate forests, act as carbon sinks and provide income for local people. Valhalla still continues as a small eco-tourism promoter and attraction.[39]

In 1994, a Guatemalan family trading as Central American Trading Company acquired Industria Guatemalteca de Macadamia SA, the remnants of C. Brewer's orchards and factory. Since then, they have become the country's largest grower, processor and marketer. Based at Rio Bravo and trading as Mayan Gold, they have undertaken applied agronomic research, addressed the many problems in adapting macadamias to the environment, and expanded and modernised the factory. They have been successful in becoming a global producer and marketer. They are also major growers and marketers of cardamom and coffee.[40]

Progressively over the last forty years more than a hundred small, cottage-sized growers together with larger growers have steadily expanded the industry, raised standards and generally produced kernel of acceptable quality. The cottage-sized and smaller orchards are owned mainly by Guatemalan Indigenous people of Maya origin, who also provide most of the workforce on the larger orchards and factories.

The orchards in Guatemala are picturesque and attractive, often set close to active volcanoes. Rainfall varies from 1.6 to 8 metres per year. Developing a major industry there has been challenging. Despite deep fertile soils, orchards face severe windstorms to which they lose on average 3 per cent of the trees each year. Just one hurricane in the early 2000s resulted in 80,000 to 100,000 trees being lost in one day. Elevation is critical to productivity and many Hawaiian varieties have not adapted well. Local selected varieties have performed better. Most orchards are on steep broken slopes and in relatively inaccessible areas.

Visitors inspecting many of the orchards have described it as 'an adventure'! Mechanisation is limited due to the broken slopes and transport costs are high. A range of varieties from Australia have been assessed recently. Sound kernel recoveries have averaged about 20 per cent, with mainly immature and insect-damaged kernel averaging 10 per cent. Reducing losses has been an ongoing priority. Most orchards are grown in combination with coffee. Buyers of NIS from smaller growers usually collect the harvested nuts from the farm weekly to maintain quality. The relatively stable government has been a factor in providing confidence to the industry,[41] which consolidated during the 1990s, led largely by

A Guatemalan orchard, showing typical lush but difficult terrain. Credit: M. Tapernoux.

the example of Industria Guatemalteca de Macadamia. Nearby Costa Rica has by comparison been unable to establish a strong, stable industry.

From the first commercial industry crop in 1981 of 50 tonnes of NIS, the industry had increased by 1991 to 773 hectares of bearing trees in a total of 2,800 hectares with a crop of 1,429 tonnes. By 1998, 2,208 hectares were bearing out of a total of 3,400 hectares and a crop of 2,745 tonnes of NIS.[42] In 2018, 2,000 hectares were cropping out of 3,000 hectares planted. By 2020, the crop was 14,200 tonnes of NIS – the equivalent of 3,000 tonnes of kernel.[43] Mayan Gold is the major processor, but there are other processors and buyers of NIS, largely for export to China. J.I. Cohen, a diversified local company, has become a major grower and processor from the mid-2010s. With high prices for NIS from 2013 to 2021, many coffee trees were replaced with macadamias. The International Macadamia Symposium in September 2023 reported the Guatemalan industry as growing 18,118 hectares with 45 per cent at full production, with 186 growers producing 15,850 tonnes of NIS in 2022.[44]

In 2005, Alan Yamaguchi, an experienced horticulturist from Hawaii, reported that orchards at middle elevation and receiving high levels of sunlight were performing well, and that the main factory and quality goals were of international standard.[45] Cliff James from Australia has advised the industry on a number of occasions.

In the early 2020s, the industry with about 1.5 million trees had both small and large growers, with two main processors, smaller factories and buyers of NIS for export. While there is limited domestic consumption, all industry participants have persevered, set high goals and are increasingly working together to address productivity and promoting their country's product.[46] The 2024 crop was estimated as 15,000 tonnes of NIS, making Guatemala the sixth largest producer in the world.

Mexico

Mexico's interest in macadamias stemmed from California and the California Macadamia Society in the 1950s. Grafted trees were available, there were a number of individual and small plantings to assess and there were both grower enthusiasts and food connoisseurs.

There are records of macadamias being trialled and small plantings on the west coast, particularly in the Baja Peninsula which became part of Mexico in 1952. The growing environment there was generally dry and harsh, and unsuited to macadamias.[47]

In 1959, 135 bare-rooted trees were sent to Hualahuises in the state of Nuevo Leon. Many of these trees survived but did not prosper.[48] At the same time, fifteen trial varieties plus several hundred seedlings were grafted and planted at Rancho Rio Vista in the state of Veracruz. The climate appeared suited to macadamias and the trees established well. There were a number of plantings of individual trees and small orchards. The trees planted at Rancho Rio Vista were assessed several times during the 1960s and 1970s, for their varietal performance and bearing potential. In a report from November 1962, the trees were inspected and some were producing their first crop. The owner, who was passionate about macadamias, was absent for most of the year and the local staff gave priority to the coffee, bananas and citrus as that brought in their main income. However, the macadamia trees were well established and had potential. To a significant extent, the Mexican Coffee Institute promoted the growing of macadamias, particularly when there was a downturn in the coffee industry and concern for its future. This led during the 1960s to growing and variety assessments and commercial plantings during the 1970s. Many of these orchards were later abandoned and unstable government caused loss of ownership,[49] but by the early 1980s about 500 tonnes of NIS was being produced and processed in small factories.

Considerable interest was generated and between 100,000 and 150,000 trees were planted from the late 1960s, mainly in the state of Chiapas. Horticultural knowledge was limited and only a small number of these trees were managed to a sound standard. The largest orchard contained 5,000 trees. At that time, it was a fledgling industry with little direction. There was scant knowledge about sound orchard management, or how the nuts would be processed and marketed, or of industry economics, and the majority of these trees were abandoned and lost.[50]

During the 1990s, interest in the industry increased and from older orchards there was some commercial production. In the early 2000s, a sound assessment was made of the state of the existing industry including varietal and district performance. Samples were taken and tested for quality and some seedling trees assessed to identify any that were suited to the local environment. This resulted in a small but expanding industry being developed. While there are now some sound orchards, the majority are of a low standard, many trees are not grafted and most growers do not practise sound nutrition and crop protection. Usually, macadamias are planted in conjunction with coffee, maize and cattle grazing. Three local varieties have been selected that show promise: Albercan, Huatusco and A527.[51]

An orphanage, Oaxaca Mission, was established in Baja County in 1981 and macadamias were planted to create work experience and income. In 2015, there were 3,000 bearing trees planted on 10 acres where visitors could assist in farm work.[52]

In 2006, there were about thirty-five growers with 30,000 non-bearing young trees and 140,000 trees over six years of age. NIS was produced, with most of the kernel sold within the country. Production increased from 500 tonnes of NIS in 2003 to 750 tonnes in 2006 and 150 tonnes of kernel. In 2011, plantings were 1,292 hectares, almost all bearing,

and produced 2,122 tonnes of NIS. By 2012, the industry was centred at an elevation of around 1,000 metres, where between 600 and 1,000 tonnes were produced.[53]

Over recent years, Finca Kassandra has planted 350 hectares of coffee and macadamia in total and is offering an extension service to support others in the industry. Finca Kassandra is a large coffee and macadamia organisation in Veracruz. In 2015, they built a modern factory trading as Mexadamia.[54] This company appears to be committed and have high standards.

A report to the Eighth International Macadamia Symposium in China in 2018 stated that there are 1,043 hectares planted in Mexico, comprising 77,500 trees not yet bearing, 121,500 trees from six to ten years old, and 160,000 trees more than ten years of age. Production was 1,140 tonnes and bearing orchards averaged 1.7 tonnes per hectare. In 2021, there were 100,000 non-bearing trees and 300,000 bearing trees. Thomas Nottebohm at the INC Congress in 2024 reported an estimated crop in 2024 at 500 tonnes of NIS and expanding. Mexico appears, in the medium term, likely to become a modest commercial producer and marketer of macadamias.[55]

South America

Bolivia

A number of South American countries have Japanese immigrants who have brought different agricultural approaches with them, including an interest in macadamias and a belief that their home contacts could assist them to market their kernel. Japanese immigration commenced in Bolivia in 1955 and a few macadamia trees were planted from seed. In 1965, the Cooperativa Agropecuaria Integral Society of Yapacani (CAISY) was formed to promote agricultural production, initially rice, soya and chicken. The CAISY co-operative would grow to lead the Bolivian macadamia industry. In 1967, a macadamia specialist was sent from Hawaii with 150 grafted trees which were planted for assessment. In 1971, the variety 333 was introduced. Following several years of assessment, the most promising plants for the region were identified, giving rise to the Bolivian selections of B6, B8 and B14. However, over the next twenty years there was little commercial development.

In the 1990s, John Thomas from Costa Rica supplied scion wood as it was believed that Costa Rican varieties would also be suited to the Bolivian environment. He supplied 30,000 grafted trees as well, mainly to small growers, but most of these orchards were later abandoned.

By 2000, about 450 hectares had been planted in about fifty orchards in the Santa Cruz area, with about 100 hectares producing 5.3 tonnes of NIS that year. Most of the plantings had replaced citrus, which was suffering from overproduction.[56] Tree density was quite wide, at an average of 204 trees per hectare. The varieties planted were the newer Hawaiian selections and Bolivian selections. There was much optimism and yields were projected at 25 to 30 kilograms per tree at 30 per cent total kernel recovery. A report in that year projected there would be 1,500 hectares established by 2005.

A small processing plant was built in 2000 and was achieving about 20 per cent sound kernel recovery. The older trees were producing about 25 kilograms of NIS. Production grew from 5.3 tonnes in 2000 to 23 tonnes in 2001, 80 tonnes in 2002 and approximately 120 tonnes in 2003.

Ikuo Nishizawa from CAISY advised that in 2006 there were 94,000 trees planted, of which half were bearing. NIS production was averaging 120 tonnes, with 30 tonnes of kernel. Nuts were being sold to a small local market with most of their production exported to the USA. However, the Eighth International Macadamia Conference in China in 2018 reported an annual crop of only 30 tonnes.

In 2021, Philip Lee estimated there were 100,000 trees under six years old and 250,000 bearing.[57] This suggests that production of NIS could be between 1,000 and 2,000 tonnes, but this estimate has not been confirmed.

The original Santa Cruz orchards were in the general area of latitude 17.05 at an elevation of 270 metres, with average annual rainfall of 1,850 millimetres. Macadamias were also promoted as an alternative crop for diversification in upland regions.[58] Now most orchards are at a higher elevation.

Brazil

As occurred in many countries, it is likely that nuts or specimen trees were introduced to Brazil earlier than the first recorded date, but records commenced there in 1931 when Mr Dierberger acquired some seedling trees and planted them at his 'Citra' haçienda at Limeira in São Paulo state. Macadamias did not attract much interest at the time, because of expensive land prices and the dominance of citrus.[59]

The Instituto Agronomico de Campinas (IAC) in 1955 imported seed nuts from Hawaii and the resultant plants were shared with other research stations in Limeira, Ribeirão Preto and Campinas. Sometime after this, the IAC published performance data from the 'Shangrila' orchard in Bauru, São Paulo state. Trees from the experimental stations were assessed and selections made for local varieties. Two that appeared promising were Campinas A and Campinas B. Others were IAC 4-12B, 4-20 and 9-20. All the promising selections are preserved in the Brazilian

Macadamia Varieties Clonal Garden.[60]

Macadamias were seen as having a good deal of potential due to the performance of individual trees, lower land prices in the 1960s and 1970s, favourable climate, availability of water, the ability to mechanise and low labour costs. During the 1960s, several small orchards were planted largely as trials.

In the 1970s, Dr R.A. (Dick) Hamilton from Hawaii studied the industry and considered that macadamias had potential in a number of Brazilian states. He considered that the risks were low, the climate was suitable and water available for irrigation if required. However, he found little interest in planting a new long-term crop where markets did not exist in that country.

In 1974, the writer met Mr Takitani, the owner of Hawaiian Host Candies in Hawaii. Mr Takitani was convinced that Brazil could produce macadamias as efficiently as Hawaii and for a lower cost, and he was planning to develop a large orchard there to supply his candy business. He asserted that by the late 1980s, Brazil would become the largest macadamia producer in the world. The name of his macadamia enterprise was Bahia Agroindustrial and in 1976 he planted about 100 hectares of Hawaiian varieties at Taperoa, near Valença in the state of Bahia.[61] Pedro Luis Blais de Toledo Piza reported that Mr Takitani's manager and business partner, Mr Maekawa, had been in poor health and passed away in the 1990s. There was no one interested in purchasing the orchard, which was abandoned and probably no longer exists.

In the late 1970s, there were about twenty eleven-year-old M. integrifolia at the Experimental Station at Cruz das Almas, Bahia. They were not impressive and would not have encouraged commercial plantings in that area. Most of the research was conducted at the Instituto Agronomico at Campinas in the state of São Paulo. Dr Mario Ojima was the Senior Researcher. Then varieties from Hawaii, and some Australian and Brazilian selections, were trialled. Farmers generally considered macadamias to be high-risk and not to provide the returns obtained from growing oranges.[62] However, in the late 1970s and early 1980s, small commercial orchards were planted in the states of Minas Gerais, São Paulo and Bahia. At this time, commercial orchards in the Campinas district of São Paulo totalled fewer than 10 hectares.

South of Campinas, the Dierberger Nursery had grown and planted about 10,000 grafted trees. They kept to themselves and had limited association with the rest of the industry, apart from selling grafted trees. A small orchard near São Mateus was based on trees that came from California, so these were probably M. tetraphyllas or hybrids.[63] Then in the 1980s there were further plantings in the states of Espírito Santo and Río de Janeiro. These used Hawaiian and Australian varieties and adopted those countries' processing and growing technologies. Throughout the country, there was a modest increase in the rate of new plantings.

Pedro Piza, a sound historian as well as Manager at Queen Nut, recorded that from the early 1990s, plantings expanded including in the states of Goiás and Parana. Production in 1990 was 22 tonnes of NIS from 400 hectares of total plantings. This had increased in 1995 to 974 tonnes of NIS at 10 per cent moisture content, which produced 195 tonnes of kernel. There were approximately 4,000 hectares of younger plantings which were not yet bearing. Production steadily increased and in 2000 there was a crop of 2,254 tonnes of NIS, yielding 485 tonnes of kernel. Plantings continued and the industry steadily expanded. In 2006, there were 180 growers with 714,000 trees producing 3,250 tonnes of NIS.[64]

The first processor in Brazil was established in 1985, increasing over ten years to four larger processors and a number of smaller ones. The major processors in the 1990s were Tribeca in the state of Río de Janeiro, Queen Nut in the state of São Paulo and Coopmac in the state of Espírito Santo. Processing plants were of a high standard and most equipment was manufactured in Brazil. Most factories achieved international quality assurance accreditation.[65]

Few orchards are irrigated and overall horticultural standards are below those of Australia, South Africa and Hawaii, but this is being addressed. Pests include ants, species of trigona bees, stink bug and tropical nut borer. Some diseases have been identified, including *Lasiodiplodia pseudotheobromae* (a fungal disease). Harvesting commences in February and most orchards are harvested weekly by hand to minimise nut borer damage. Sound kernel recovery is low by international standards but has increased from 18 per cent in the 1980s to 20 per cent and now 25 per cent in higher-standard orchards. There is considerable variation in the performance of individual orchards, but this applies to all growing countries. Over 90 per cent of the industry is in the states of São Paulo, Bahia and Rio de Janeiro.[66]

The initial industry body was the Association of Producers of Macadamia Nuts in the State of São Paulo (APROMESP), founded in 1990. It represented the São Paulo macadamia growers and expanded to become the Brazilian Macadamia Society, which now represents and co-ordinates the overall industry. It has 150 growers and processors within the states of São Paulo, Espírito Santo, Minas Gerais, Rio de Janeiro, Bahia, Goiás, Parana and Mato Grosso. Dedicated industry people have led and contin-

ue to lead this Society, but there is no government funding available for research and promotion and currently there are no industry levies. This means that research and promotion must be privately funded and undertaken. A major initiative was the Brazilian industry holding the Third International Macadamia Symposium in 2006, which attracted about 300 growers, processors and researchers from around the world.[67]

Dr Andrew McGregor, in his report 'A Review of the World Production and Market Environment for Macadamia Nuts', in 1991 stated that Brazil was an industry waiting to happen. Dr Skip Bittenbender from Hawaii studied the Brazilian industry in the early 1990s and forecast production of 13,000 tonnes of NIS in 2000, which would be approximately 15 per cent of world production. The industry did not achieve this projection. In 1998, the United States International Trade Commission (ITC) reported that 6,500 hectares were planted and 40 per cent were bearing, yielding 1,600 tonnes of NIS. The report commented that macadamias are relatively unknown to the public within Brazil and 75 per cent of their production is exported, mainly to the USA.

Over the last fifteen years, growth has been steady but not at the rate of South Africa, Australia, China and Vietnam. Global demand and price increases have resulted in increased plantings from

José Eduardo and Maria Teresa (Tica) Camargo, principal of Queen Nut, with Philip Lee, South African industry leader, 2006. Credit: I. McConachie.

Queen Nut, a leading high-standard processor, São Paulo, 2006. Credit: I. McConachie.

2015.[68] In 2017, production was reported as 5,300 tonnes of NIS or 1,378 tonnes of kernel, of which 450 tonnes were consumed in Brazil. It was then estimated that plantings of one million trees would be made between 2017 and 2022.[69]

In 2019, the major processors and marketers were Queen Nut, Coopmac and Tribeca. Queen Nut has 400 hectares of orchards, producing about 1,000 tonnes of NIS, and they were purchasing about the same amount from other growers. The Brazilian Ministry of Agriculture has supported the new industry body, Association Brazilian Nut Castanhas (ABNC), representing nuts, chestnuts and dried fruits. Its President is José Eduardo Camargo, the CEO of Queen Nut and his daughter Beatriz is a global marketer of macadamias.

In 2021, there were 1,200,000 trees not yet bearing and 1,600,000 bearing trees.[70] The International Macadamia Symposium in September 2023 reported 8,000 hectares planted, producing 6,500 tonnes of NIS. The 2024 INC Congress amended the 2023 crop to 7,685 tonnes, but widespread drought has resulted in a forecast reduction to 6,500 tonnes for 2024.

The Brazilian industry will continue to expand, although it is unlikely to become a major global producer. Orchard productivity in terms of premium kernel produced per hectare is low by world standards. There are many reasons for the less-than-expected growth. First, the political and economic climate is probably restrictive. A major impediment has been high interest rates, which were approximately 18 per cent in 2005 with inflation of 8 per cent. This makes shorter-term crops such as coffee, sugar and high-density citrus more attractive.

Leoni Kojetin from the AMS studied the industry in early 2024 and reported that there were 250 growers having variable standards of orchard management, with few being irrigated. While there were thirty processors, most were small and much of the crop was exported as NIS. Domestic consumption absorbs 30 per cent of the crop and middle incomes are rising. With low yields per hectare, cost of production averages $US 1.50 per kilogram of NIS. Interest rates are between 15 and 40 per cent, making new planting difficult to fund and usually requiring inter-row crops. There are no government subsidies.[71]

Competition for land between sugar-cane, coffee, oranges and other crops that are quicker to generate income, low kernel recovery and sometimes low horticultural standards for growing macadamias have been factors in Brazil remaining a sleeping giant in its potential to be a major global producer. However, the industry remains confident and enthusiastic.

Colombia

In the early 1960s, Pat Ganie from a Hawaiian agronomic company visited Colombia and recommended macadamias as a potential new crop. While the coffee association planted a few trees from seeds in the mid-1960s, the first serious plantings were at the end of that decade as part of a crop diversification programme from the Colombian Association of Coffee Growers. They were planted at an experimental station in Paraguicito, in the Armenia region at 1,200 metres elevation. Some 30 per cent of these trees remain and are generally healthy. The research station monitored these trees for many years, which allowed later selections to be made. Dr R.A. (Dick) Hamilton and later Dr Phil Ito from the University of Hawaii visited Colombia over at least a ten-year period, commencing in the 1970s. On his first visit, Dr Ito brought over eighty trees of a range of varieties. Dr Ito's visits and advice were instigated and funded by two sugar mills, Ingenio Monuelith and Centro Tumsco, as well as by both the Uribe family and the Dorronsoro family trading as Del Alba. Ito recommended grafting procedures and detailed horticultural requirements. Most of the grafted trees in the industry came from his introductions. As a result, grafted trees were available to interested farmers in the 1980s, which provided some confidence. The first commercial orchards were established around 1986 and the rate of planting increased from 1990. Macadamias were often interplanted with coffee and most orchards were owned by large coffee growers. Typical sized orchards are between 20 and 80 hectares, with a small number at 200 hectares.[72]

In 1998, an experienced horticulturist from Ecuador, Heinz Gattringer, studied the industry and produced a detailed report. He said there were probably between 1,000 and 1,500 hectares of macadamias planted in traditional coffee-growing areas around Manizales, Armenia, Pereira and Cali. All of these plantations are situated between 1,000 and 1,500 metres elevation. In addition, there are quite a number of small coffee farmers with up to several hundred trees, but these do not contribute significantly to the industry. Most orchard owners have high agricultural and technical standards and many years of farming experience.[73]

Rainfall in the growing areas is between 1,800 and 2,500 millimetres per year. The macadamia trees have a long flowering period with two peaks, around June–July and December–January, which result in two harvest peaks, in February and July. The main varieties are the Hawaiians 344, 660, 246, 800 and 508, plus Beaumont. Newer varieties which appear promising are 788 and 741. Colombians are evaluating their own selections, described as the P series. As in most of Central and South America,

macadamias were grown in conjunction with coffee and often replaced it.

In the early 1990s, following a visit from Dr Ito from Hawaii and Sr Fonseca from Costa Rica, growers formed the Colombian Macadamia Society. However, this group did not obtain enough support and the major family grower groups largely determine industry priorities.

In 2005, an independent orchard trading as 'Coconut Farms' at Manizales was reported as producing 300 tonnes of NIS which was shipped to Costa Rica for processing.

There is only one commercial processor, Del Alba SA, although there are several small cottage processors which largely cater to the local tourist trade. Del Alba have led the industry through all stages of growing, processing, retail packing and marketing. They process other tree nuts, peanuts and snack food, and have high standards. Having focused on developing a local market, they did not export any kernel until about 2000. In 1997, they processed about 200 tonnes of NIS, producing 40 tonnes of kernel. They are also major growers who offer an extension service to independent growers who supply them with crop. The principal shareholders of the company are Jaime and Edgar Dorronsoro and family.

Del Alba SA, Colombia high-standard young orchard of 44,000 trees. Credit: Jaime Dorronsoro, Del Alba.

In 2003, the NIS crop was 550 tonnes, which increased to an estimated 1,200 tonnes, producing 240 tonnes of kernel in 2006. In 2006, there were about 48,000 trees on 400 hectares that were not yet bearing and approximately 96,000 trees on 800 hectares that were over six years of age. The figures suggest a factory recovery of 20 per cent, which while low is typical for most of Central and South America. Since 2000, an export market has been developed mainly to Europe and in 2006 half of the production was exported.

The Eighth International Macadamia Symposium in China in 2018 listed the crop as totalling 880 tonnes of NIS, producing 221 tonnes of kernel. In 2021, there were 100,000 trees under six years of age and 300,000 bearing.[74] The 2024 INC Congress in May estimated production as 1,100 tonnes of NIS, with the crop estimated to be similar to that of 2023, at 1,100 tonnes of NIS. The industry is stable and progressive, with high standards and production should continue to increase.

Ecuador

Macadamias in Ecuador got off to a confused start. In 1974, seed nuts from varieties 246, 333 and 508 were sent from Hawaii to the Government Agricultural Department. They were germinated but all the labels fell off and this also happened to acerola which were brought in at the same time. So for a period, the acerola were grown as 'macadamias'.

Donald Brainard from the USA moved to Ecuador in the 1960s. He had a love of the land, of plants and his fellow human beings. Donald knew what a macadamia looked like and corrected the government officers. In 1976, he was given 400 of the plants and established them on his farm. Some died, some were culled but, as seedling trees, they were all different and only slightly resembled their Hawaiian parents. These appeared to be the only trees planted at that time apart from a few as residential ornamentals.[75]

Matthias Tapernoux was an enterprising young businessman from Switzerland and came to Ecuador

Factory workers sorting nuts, Ecuador. Credit: I. McConachie.

to seek his fortune. He married an Ecuadorian lady, had a family and settled there. In 1995 he received a gift of roasted and salted macadamias from the former Costa Rican Ambassador to Ecuador. Matthias became entranced with the quality of macadamias and dedicated much of his life to establishing a small industry in Ecuador. Initially he bought three grafted trees from Costa Rica and, with his business partner, commenced developing their own orchard at La Concordia at approximately 250 metres elevation, on land which was formerly a dairy farm. They recruited Alberto Arguedas, a pioneer agronomist from Costa Rica, and later Henry Fonseca who developed nursery procedures and horticultural practices and who visited the orchard twice a year. From this dairy, they developed a nursery and then an 8,000-tree orchard, 'Via Lactea'.[76]

Scion wood from Costa Rica was from varieties

246, 333, 508 and Tambor, a local Costa Rican selection. The seedlings used for grafting came from the trees of Donald Brainard. 'Via Lactea' was planted in the early 1990s. Business colleagues, investors and progressive farmers visited the orchard and became interested in its potential. By 1995, 35,000 trees had been planted through the country and the first crop of 1,800 kilograms of NIS was harvested. 'Via Lactea' supplied grafted trees, engaged a horticulturist, and developed more orchards or acted as a consultant on behalf of the owners.[77]

By the late 1990s, the nursery was producing 20,000 trees annually and there were about twenty macadamia orchards in the country. The author of this book was fortunate to be engaged to assist the industry, mainly in commissioning a processing plant and establishing quality procedures.

The country had a major financial crisis in 1999, which resulted in the economy becoming depressed and then largely based on oil exports. The emphasis on agriculture and its exports was stifled.

In 2006, Ecuador had approximately 100,000 trees but half of them had a low standard of tree management. Production averaged 250 tonnes of NIS, producing about 50 tonnes of kernel. Local markets had been developed, with retail packs and bulk sales in the industrial sector.[78]

Ecuador is on the Equator, with a climate modified by the Humboldt Current, and most orchards are situated at 200 to 600 metres elevation. One small orchard owned by Walter Guggenheim is in the Andes foothills near Quito, at 2,450 metres elevation, which makes it probably the most elevated macadamia orchard in the world. The trees are

Removing husk from NIS using rocks, before dehuskers were available in Ecuador. Credit: I. McConachie.

healthy but grow slowly and did not crop until about eight years of age. The rest of the orchards have a subtropical climate but usually a dry season of seven months with little rainfall. Apart from relatively minor pests, such as leafcutter ants, a nut borer and trigona bees, they have a trunk disease causing some mature trees to rot from the centre and be lost.

Matthias Tapernoux led the industry for twenty years with his enthusiasm and skills. But the overall standard of orchard management was poor, the growing environment at best marginal, yields were low and variable, and sound kernel recovery was low. Gradually orchards were allowed to decline and many were abandoned. In 2015, the main processor, Macadamias La Concordia, closed and while small processing facilities continued, the interest in macadamias as a commercial crop has largely disappeared.

At the Eighth International Macadamia Symposium in China in 2018, the crop was reported as being 50 tonnes of NIS. Philip Lee in 2021 reported 40,000 non-bearing trees in a total of 140,000. This was not supported by Sr Tapernoux, who believes that many trees have been abandoned and the commercial industry is declining. Most macadamias are sold at road-side stalls as NIS. There appears to be little prospect of Ecuador becoming a significant producer in the world market.[79]

Paraguay

Clarence Johnson, originally from the USA, came to Brazil and in 1953 moved to the neighbouring Paraguay. In 1964, at the Annual Meeting of the California Macadamia Society, he spoke of the potential for macadamias in Paraguay and encouraged investment there. He was planning to grow macadamias on a large scale.[80] Clarence lived near Amambay, on the border with Brazil. This town is 60 kilometres south of the Tropic of Capricorn, surrounded by virgin rainforest on rich soils, with rainfall averaging 90 inches. He was a large coffee grower and had an experimental farm containing many tropical and subtropical crops, which included seventy macada-

mia trees. Part of the incentive to develop the area was that land costs were $US 3–4 per acre.

In 1964, Clarence planted 200 grafted trees and had several thousand seedlings in his nursery. Many of his coffee plants were lost through frosts, but the latter were not severe enough to affect the macadamias. Some 370 grafted trees were supplied from California.[81]

A neighbour in 1968 planted 700 grafted trees and had 800 in his nursery ready to plant. The clearing and development of this part of the country and subsequent planting, mainly with coffee, created much employment and when the new city of Corpus Christi was founded, an esplanade off the principal street was planted with macadamia and named Macadamia Avenue. Banks were prepared to make loans to establish macadamia orchards and interest in the potential of macadamias was high. The government of the day erected a statue in Corpus Christi of Clarence Johnson to honour his contribution to the area.[82]

By 1972, Clarence had 3,000 producing trees and 4,500 trees not yet bearing. In addition, his nursery had sold over 10,000 grafted trees. Japanese residents and emigrants through the Japanese colony established a nursery and planned major plantings.

The main commercial introductions occurred in 1980, when scion wood produced by the Agronomic Institute of Campinas in Brazil was imported by CETAPAR (Centro Technologico Agropecuario do Paraguay). The government supported research and established a number of experimental trials. In 1989, a Macadamia Nut Cultivation Investigation Group was formed, composed mainly of Japanese growers. In 1992 there were 500 acres of macadamias recently planted, all of Hawaiian varieties and most less than two years old. As in much of South America, Japanese colonists played a part in establishing the industry. In 2002, NIS production was around 90 tonnes, and processed in Paraguay. By 2003, it was reported that there were orchards from Amambay in the north to Itapúa in the south totalling about 300 hectares and containing approximately 58,000 trees. German emigrants and others from the Ukraine also planted small orchards.[83]

At the end of 2001, the Paraguayan Macadamia Society was formed to promote planting macadamias, undertake research and support macadamia growers. In 2002, an International Macadamia Symposium was held in Asunción and many experienced macadamia persons from California, Costa Rica and Brazil attended.[84] In 2006, a nursery in Brazil supplied 40,000 trees, many of which were planted in large orchards by emigrants from the Ukraine.

The International Macadamia Symposium in Brisbane in 2012 projected 1,600 tonnes of NIS from Paraguay by 2020, but this has not been achieved. In 2018,[85] a report listed that there had been 190 tonnes of NIS in 2017 and 90 tonnes in 2018. In 2021, there were about 20,000 young trees and 120,000 bearing.[86]

Venezuela

Like most countries in Central and South America, Venezuela's climate is largely tropical, so macadamias need to be planted at higher elevations to find suitable conditions. In 1971, the government agency FUDECO was formed to develop the central west area. At Barquisimeto, three experimental sites were planted and a nursery of 2,000 trees was established. The resultant small orchards were interplanted with castor oil shrubs. A small orchard was established in 1971. Elevation was about 1,000 metres, with an annual rainfall of 1,200 to 1,800 millimetres. Coffee was the major crop.[87]

Little is known of these trial sites although there are reports that a major flood destroyed the bridge providing access to the area, resulting in the decline of commercial farming there.

In 2001, there were between 100 and 200 hectares planted to macadamias in small orchards. The crop is processed and sold within the country.

The Caribbean

Following the early development of industries in Guatemala, Costa Rica and Brazil, there was widespread interest in macadamias as a new commercial crop. This applied to several of the Caribbean islands in the early 1960s. Overall, however, no commercial industries were developed. Most of the farming land was too tropical, so only elevated areas were suitable. Hurricanes destroyed many tree crops. Increased prices to growers from 2015 resulted in some interest, but no records are available on commercial development.

Dominican Republic

In the Caribbean Sea is the beautiful Dominican Republic, one of the places where Christopher Columbus landed in 1492. It is a poor country and in 1979 was devastated by Hurricane David which killed 2,000 people and destroyed 70 per cent of agriculture and forests.

From 1980, the environmental group Plan Sierra sought to use macadamias to help stabilise the denuded hills, partly with the idea of providing future income for small farmers. But for many years the trees and their nuts were largely ignored, due to the very hard shells.

Alfonso Moreno founded the popular Dominican ice-cream company, Helados Bon, which introduced a gourmet line of macadamia ice-cream that helped to provide a local market for macadamias. For ten years, all local macadamias collected from the forests or from smallholders were hand processed and went into the ice-cream. The project is described as the only sustainable business model promoting social, economic and environmental solutions.[88]

Then in 2018, Alfonso's son, Jesus Moreno, established the country's first macadamia processing factory, La Loma Macadamias, supplied by farms run on regenerative agriculture principles. Jesus also founded the Loma Quita Espuela Foundation to conserve the largest rainforest in the Caribbean.

The Loma Quita Espuela Scientific Reserve was supported by the Minister of Environment and Agriculture and the International Development Bank.

Now there are more than 100,000 orchard trees, more than 200 small growers and 400 hectares of forests restored with macadamias.[89]

Jamaica

Apart from individual trees brought in as exotics, the first plantings were by Charles H. Deichman who, in 1962, planted twenty grafted trees from Colonel Wells Miller in California together with importing about 50 kilograms of seed nuts from Dr Hamilton at the University of Hawaii. The trees and nuts were used to establish a small nursery and, as was typical of much of the Americas, Deichman planted his macadamias in conjunction with coffee. He had problems propagating grafted trees, which restricted industry growth in Jamaica. His goal was to produce grafted trees and make them available on whatever basis was necessary to encourage further plantings.[90] Deichman's orchard was named 'Sherwood Forest'. In 1968 it was reported that Donald Turner, at the town Bog Walk, had planted a signifi-

cant macadamia orchard in 1965. At Newport, Mr and Mrs Michelin had a 35-acre orchard of citrus and macadamias.[91]

There have been no recent reports from this country and it is probable that a combination of wind and low production has caused the orchards to be abandoned.

Trinidad

Seed nuts were supplied from Queensland to Port of Spain in 1877 and a report in 1892 stated that the trees were healthy and fruiting. In 1968, a report on these trees stated that they had poor-quality nuts, so they may have been the true, bitter *M. ternifolia* species.

At the University of the West Indies at St Augustine, a small number of seedlings were planted for assessment. In 1965, seeds were taken from the University of both *M. tetraphylla* and *M. integrifolia* species and supplied to the Department of Agriculture for planting out and assessment.[92]

No further recent reports have been received, so it is unlikely that there is any commercial industry in Trinidad.

Stages of macadamia nuts, left to right: looking backwards from final shelled nuts, to earlier nuts-in-shell, to original nuts-in-husk. Credit: Eliza Powell.

Chapter 16

AFRICA AND BEYOND

Africa: industry leaders – Africa: industries
in development – Middle East and Europe

Africa: industry leaders

South Africa

> Macadamias seem well adapted to several areas and
> high hopes are held for their future and the industry
> should be cautiously optimistic.
>
> (Peter Allan, 1961)[1]

There are many parallels between the development
of the macadamia industry in Australia and South
Africa. In both locations, most of the pioneers were
enthusiasts feeling their way and few reaped the
hoped-for rewards. Similarly, in both, large corporations entered, and there were setbacks as well
as periods of growth, the latter sometimes gradual
and sometimes rapid. By 2015, South Africa had become the largest producing country in the world, to
be overtaken briefly by Australia in 2017, 2020 and
2021. To strengthen their ties to the industry even
more, South Africa's national flower, the protea, is in
the same botanical family as the macadamia.

Left. Rwandan agronomist checking Macadamia sapling
on an industrial farm, Rwamagana District, Rwanda,

*Table 7: Comparative NIS production, Australia
and South Africa*

NIS production, Australian and South Africa

Tonnes Nut In Shell at 10% moisture by year

■ Australia   ■ South Africa

Sources: https://australianmacadamias.org/industry/
facts-figures/australian-production and https://www.
globalafricanetwork.com/company-news/macadamia-nuts-
the-future-of-high-value-crop-farming/.

The introduction of macadamias to South Africa

The first introduction of macadamias into South Africa was in 1872, when two trees were planted at the Durban Gardens. They were labelled *Macadamia ternifolia* but Curator John Medley Wood recorded the fruit as edible, so they must have been either *M. integrifolia* or *M. tetraphylla* (neither tree survives).[2] It has been reported that these trees were provided from Kew Gardens but it is known that Walter Hill, the Superintendent of the Brisbane Botanic Gardens, had been providing seed nuts and trees throughout the world from the late 1860s. In Western Australia's newspaper *Western Mail,* a regular column titled 'The Flower Garden' reported in 1911 and again in 1913 that the macadamia was doing well in Johannesburg.[3] There was a tree growing in the Durban Botanical Gardens in 1915, which could have been from the 1872 introduction. Another tree was planted there in 1932.[4] From an early time, there was a magic that macadamias seemingly wove over their enthusiasts.

In 1931, the Agricultural Research Council (ARC) of the Institute for Tropical and Subtropical Crops established macadamias from seeds brought from Australia. Dr Oosthuisen, who visited Australia in 1933, reported on the wonderful nut there. From the 1930s, mainly single or small plantings were being made.[5]

The Reim family first imported *M. integrifolia* seed nuts in 1935 from Yates Nursery in Australia and, a year later, *M. tetraphylla* from Walter Petrie's nursery. These formed the basis of the industry.[6] Dr E.P. Reim, a former Professor of Engineering at Durban University, with his wife who was a horticulturist, established an orchard of about 1,000 seedlings of very variable characteristics which, when mature, produced about 5 tonnes of NIS. They also selected seeds from the better performing seedlings and sold about 50,000 trees.[7]

The ARC Institute for Tropical and Subtropical Crops started research on macadamias in 1963. A meeting of local interested persons was held at Dr Reim's nursery at Hillcrest in 1966. Pioneer growers

Reim pioneer nursery, Natal, 1962. Credit: *California Macadamia Society Yearbook*, 1963.

Dehusker designed and built by the Reim family, early 1960s. Credit: unknown.

in Natal (now KwaZulu-Natal) were Messrs Gibson, Hilerman, Porter, Pringel and Wyatt. To the north, in what was then Transvaal, pioneer growers were Messrs Hobson, Howard Blight and Bertie le Roux. Dr Joubert undertook considerable research. Further north in Zimbabwe (then Rhodesia) were Ted Tonks, and Arthur and Anne Lane. The ARC imported Hawaiian cultivars and made selections to cross with two hybrid cultivars, Nelmac 1 and 2 released by Dr Joubert in 1973. Nelmac 26 was selected by Bertie le Roux. These cultivars were widely planted from the 1970s. Probably the ARC was influenced by California's preference for the *M. tetraphylla* species and a belief that hybrids would add vigour to the selections. Dr Reim believed that *M. tetraphylla* was the species most suited to South Africa.[8]

Varietal trials, Nelspruit Experiment Station, 1969. Credit: *California Macadamia Society Year Book*, 1970.

Reim family: Christopher Reim, Mrs E. Reim and Dr E. Reim, Durban, 1969. Credit: *California Macadamia Society Yearbook*, 1970.

The Keith family from Stegi in Swaziland also obtained seed nuts from Petrie and another small orchard was planted in 1957 in Soekmekaar. In June 1966, the first South African Macadamia Society was formed at the home of the Reims. It had fifty members by 1969. Mrs Wyatt was the Chair for the first four years.[9]

Dr C.A. Schroeder from the University of California visited South Africa and Malawi in about 1969. He noted the changing emphasis towards crops of higher economic value and with longer storage life. Several horticultural crops appeared to be suited to the growing environment. Macadamias were one of these.[10]

Macadamia pioneers

Bertie le Roux was one of the pioneers whose passion and belief helped to enthuse the industry. In 1957, he was given approximately 100 nuts, thus commencing his lifelong affair with, in his words, the 'Queen of nuts'. In 1961, he expanded his orchard and in 1967, he started grafting local cultivars together with Hawaiian cultivars. In 1969, he had planted the first commercial orchard in the Levubu district. Farmers in the area watched, with some sharing his enthusiasm and planting orchards of between 500 and 2,500 trees. He helped to select Nelmac 2 and believed his selection Nelmac 26 to be an even superior cultivar.[11]

Professor Peter Allan (1930–2016) had been a passionate, dedicated and skilled researcher since the mid-1960s. He became a most respected researcher, providing advice and developing a network of experienced global contacts. Over fifty years, he contributed to and saw macadamia farming develop from a cottage scale to a major successful industry.[12]

Len Hobson, another passionate enthusiast, was first attracted to macadamias in 1968 and maintained this passion until his passing in 2014. From the late 1960s, Len and this author corresponded, met, shared enthusiasm and exchanged many thoughts. Len provided advice and consultancy services, and he promoted all aspects of the indus-

Possibly Bill Wyatt with Dr Peter Allan, industry researcher and enthusiastic supporter. Credit: *California Macadamia Society Yearbook*, 1970.

SAMAC in 1999. Philip Lee and Julie Marjoram. Credit: I. McConachie.

try. From 1965, Len initially had a fruit tree nursery near Tzaneen. After a visit to California in 1968, he came back full of enthusiasm and soon added macadamias to his nursery range. Peter Allan demonstrated the advantages of vegetatively propagated plants compared to seedlings, and Len focused on commercial propagation. Len imported the hybrid variety HAES 695, later named Beaumont, from California – where cuttings developed rapid, vigorous roots – and he promoted this variety as both a cloned rootstock and a commercial variety. In later years, he sought to develop smaller orchards as family retreats towards the East Cape.[13]

An industry develops

In the 1960s, most of the 50,000 macadamia trees in the Natal area and 5,000 in the Transvaal area were seedlings. There was ongoing assessment of cultivars for their suitability. The Windy Hill Wattle Company in Natal (7,500 acres) had over 3,200 seedlings and some grafted cultivars which were subject to extensive assessment.

The Soutpansberg Nut Producers Association was formed in 1973 and, by 1976, 43,000 grafted trees had been planted by its members. With a commercial crop in 1977, the small Levubu Co-operative

was formed. Bertie le Roux described it as an out-shed with tobacco driers and a new cracker.

The South African Macadamia Growers Association (SAMAC) was formed in 1979 after a macadamia and pecan symposium was held in Politsi. In 1990, the name was changed to the Southern African Macadamia Growers Association. When SAMAC was first formed, there were fewer than 100,000 trees planted in South Africa, both seedlings and grafted trees.

In the early days, macadamia farming in South Africa was a hobby venture on a small budget. One of the strengths of Dr Reim and his wife was their ability to attract publicity. Dr Reim built a dehusker and a cracker. Nuts were sundried in trays and then dried again using a paraffin room-heater placed under a stack of trays. The cracking machine had two rollers, each with different flutes. One roller rotated at twice the speed of the other and the gap was adjustable. The kernels were hand sorted, graded and marketed in 100-gram clear packs. Most of the output was sold to health stores. By 1960, their son Christopher had expanded and improved the shelling, sorting and packaging processes. Both Chris and Len Hobson developed successful grafting techniques.[14]

In a detailed 1969 horticultural guide, Professor Allan recommended that the production goal should be 10,000 pounds weight of fresh NIS per acre, which he later amended to 1,200 kilograms of kernel per acre. This was logical and forward thinking, and it took more than forty years for most of the global industry to adopt his sounder measure of production in terms of kernel rather than NIS. He also commented that macadamias had now moved beyond the speculative stage and a commercial industry was a real prospect. In the 1960s and 1970s, it was assumed that the South African production of macadamias would be consumed within South Africa at road-side stalls, tourist areas and retail outlets.

In 1972, two ventures almost simultaneously announced their plans to grow macadamias as syndicated investments on a scale previously unheard of. They were to be grown at Komatipoort, to the far east near the Mozambique border. These orchards were close to each other and belonged respectively to Macadamia Finance and to SA Farm Investments.

Macadamia Finance proposed 400,000 trees and their brochure depicted 'South Africa's most exciting growth investment where money does grow on trees'. The brochure also proposed: 'Darling, what are we going to do with the profits. – Why don't we buy a couple of farms for the grandchildren.' They attracted 3,500 investors.[15]

The SA Farm Investments proposal was for debentures and shares to develop an area in the nearby Komati Valley, comprising 2,028 acres with the inter-rows planted with cash crops to generate early income. The concept was controversial, with some existing growers concerned that the newcomers would dominate the industry.[16]

In retrospect, much of both concepts was flawed. Assumptions were based on limited facts and some creative accounting masked the likelihood that investors would never get sound returns. They extrapolated Hawaii's horticultural technology and yields,

S.A. PLAAS JOERNAAL
FARM JOURNAL

APRIL 1979

Aerial view of a portion of Seekoegat and Squamans. Prominent are the healthy macadamia and windbreak trees with the large storage dam in the background.
Lugfoto van 'n gedeelte van Seekoegat en Squamans. Duidelik sigbaar is die gesonde macadamia en windbreek bome met die groot opgaardam in die agtergrond.

Komatipoort investment report 1979, showing the scale of this investment. Credit: SA Farm Investments.

much of which was not applicable to Komatipoort. District temperatures were often very high in summer and, in places, there were heavy frosts in winter. Expected yields were not realised, cash flow dried up and both ventures collapsed.[17] Possibly with more capital and sounder management, the orchards may have become viable.

About 1978, Leo van Schaik became the manager of the first commercial macadamia factory in South Africa. It may have originated from an earlier co-operative at Levubu that also handled avocados and bananas. Leo's daughter, Alida Kirk Cohen, inherited her father's passion for macadamias. This factory was the first to export macadamias, followed by Zetpro which built what was considered a state-of-the-art factory at Shayandima in the then homeland of Venda.

By the early 1990s, Sapekoe had become a shareholder in the Zetpro factory with the Levubu Co-operative. The factory had the first electronic colour sorters (one of the very early belt models) in a macadamia factory in South Africa. The factory was relocated from Shayandima to the old Sapekoe tea factory at Mambedi River Estate, next to Maclands Estate, when Philip Lee became General Manager in 1996.

In the earlier years of development, macadamias were planted in three distinct growing areas: Levubu, Southern Lowveld and Natal South Coast.

In 1979, the Australian Trade Commission and a 1980 CSR report summarised the industry as having 605,000 planted trees, with 42 per cent Hawaiian varieties and the rest local selections plus *M. tetraphylla* and *M. integrifolia* seedlings. Production was reported as 167 tonnes of NIS. Apart from Komatipoort's two processors, there were co-operative factories at White River and Levubu. Orchard standards were considered sound but market development was barely being addressed. Production steadily increased.

The failure of the Komatipoort ventures did help to shape the development of the industry. Independent unpaid growers met and resolved to form a lo-cal growers' association chaired by Roly Schormann and to build a co-operative, which was processing 600 tonnes of NIS by 1990. Other processors entered the industry in response to increasing plantings and a belief that processing and marketing would be profitable. One innovative small processor was Ed van der Hoek, who only processed Nelmac 2, which he roasted in the shell and marketed himself.

By the late 1980s, tobacco growing was declining and macadamia NIS prices were high. About 300,000 trees were growing in the Southern Lowveld and, though 80 per cent were not bearing, interest, enthusiasm and prospects were strong.

Other early growers in the region were Lund Roetgers, Des Altona, Messrs Pretorius, Degenaar, Duffus, Skar, Mike Graham and Doug Grant. By the end of the 1980s, the crop reached 250 tonnes of NIS.

KwaZulu-Natal farmers had planted mainly *M. tetraphylla* and hybrids in small numbers since the 1950s. Sometimes macadamias were interplanted with bananas on steep slopes, which had been unsuccessful in Australia in the 1930s. Interest increased during the 1980s and by the end of the decade about 28,000 trees had been planted. Because of isolation, a Southern Natal Macadamia Association was formed, to represent and encourage its growers. In time, it became part of SAMAC.

In the early 1980s, an independent assessment of the industry looked for additional suitable growing environments. Excluding Komatipoort, more suitable districts were identified and sound horticultural procedures were increasingly adopted. Pests and diseases (mainly false codling moth, southern stink bug and *Phytophthora*) reduced yields. Export market development was fragmented but was being considered and mainly left in the hands of brokers or agents. During the 1980s, SAMAC provided strong leadership in a similar manner to Australia's AMS, but their levy was voluntary, there were no matching funds, research was limited, with industry promotion and market development almost non-existent.

South Africa's kernel was almost all exported, as the domestic market was small and there were few retail products apart from local markets and roadside stalls where some of the NIS was stolen from orchards. Seeking to dismantle apartheid, many countries led by the USA imposed trade sanctions from 1985 to 1993, which in theory would stifle the export of kernel. In large part, however, kernel was still marketed through third countries or the embargo was not strongly policed. Nonetheless, it did reduce industry growth.

The industry matures

From at least the 1990s, South Africans have shown their entrepreneurial drive, sound farming skills and determination to develop an industry. Strong leadership from SAMAC oversaw orderly development, co-ordination, initiation of research and representation – all of which supported industry growth.

Many people have contributed to this growth. Philip Lee, a gentle giant, initially managed Maclands, led SAMAC and has provided direction and consultancy for more than thirty years. Other contributors include Alan and Jill Whyte at Green Farms (Jill was the first female President of SAMAC, in a country where gender discrimination was then the norm); Dorran Bungay at SAMAC, whose engineering skills have, over thirty years, revolutionised the curing of NIS and established quality, engineering and economically efficient crop-handing and processing systems; and Duncan McGregor from Golden Macadamias. There are many more who provided leadership, research, variety development, advances and examples.

A detailed census was undertaken in 1993, which defined the industry. Maximum yields of 4,500 kilograms of NIS per hectare for irrigated orchards and 2,300 kilograms for non-irrigated orchards were considered achievable. There were 717,000 trees planted, almost 40 per cent not bearing, with 58 per cent in the Levubu district, 35 per cent in Lowveld and 7 per cent in Tanzeen. Some 15 per cent of the trees were *M. tetraphylla*, and about 50 per cent were Beaumont, a hybrid which produced nuts considered to have the processing character-

Maclands Limpopo in preparation and planting stage, 1999. Credit: I. McConachie.

istics of *M. integrifolia*. Production of 5,076 tonnes of NIS was projected to increase to 9,100 tonnes of NIS in 2003. Seven factories serviced the industry.[18]

After the first democratic elections in 1994, land restoration and restitution became government policy. Black African people and tribes who were displaced by apartheid after 1913 and who had not received compensation could claim their land. Most of the farming land was held by Europeans who, in the case of macadamia orchards, had made considerable capital investment. Laws and interpretation were complex, ownership or title was lost by the occupying farmers, and this threatened many existing orchards and future investment. Sometimes where title had been lost, the orchards were leased back; sometimes restitution was paid, but the uncertainty weakened the growth of the industry. In the 2020s, land claims continue and often take many years to resolve. Extensive legal research is required by macadamia farmers. Overall, while it remains a concern, the industry accepts and manages the process. In 2016, 8 per cent of orchards were owned by Black South Africans.[19]

Philip Lee provided an overview of the industry in 1996, reporting that factory kernel recovery now averaged 24 per cent. There was no government mandatory levy but there was a voluntary industry levy equivalent to US 1.0 cents per kilogram of NIS, collected by SAMAC. Farm and factory wages averaged $A 100 per employee per month. Most of the kernel was sold in the USA, Europe and China. There was a strong demand for nursery trees and rapid expansion of orchards. In sugar-cane growing areas, macadamias were increasingly replacing cane.[20] An Australian study group in 1999 estimated that the total cost to grow and process macadamias was about 60 per cent of the cost in Australia.[21]

Beaumont (HAES 695) remained the most planted variety, accounting for 50 per cent of tree sales from 2000 to 2005. Now in the 2020s, Beaumont's modest kernel recovery and low percentage of whole kernel is reducing its appeal. The need to hand harvest because of its late fall has been addressed by the application of Ethephon (Ethrel), which results in complete nut drop.

During the first decade from 2000, annual plantings reached 500,000 trees in some years. Most orchards are in the Limpopo Province, Mpumalanga, KwaZulu-Natal and expanding into the Cape region. By 2005, 5.8 million trees had been planted, exceeding Australia's tree numbers but not its NIS production. Stink bug, droughts, theft, land claims, limited funds for research and export market development all were issues. Positives were the standard of farming and indirectly hand harvesting, which preserved the orchard floor and soil health; quality initiatives including Quality Assurance and Hazard Analysis and Quality Control Points (respectively, QA and HACCP); favourable exchange rates; and having a vision for their future.[22]

In 2003, Philip Lee summarised the industry as having over three million trees, with 24 per cent fully mature, 30 per cent in early bearing, 46 per cent not yet bearing and overall producing 11,500 tonnes of dry-in-shell or factory dried (DIS), which had almost doubled since 1998. There were nine major processing plants, most accredited with international quality assurance systems. Small amounts were marketed in the Republic of South Africa,

South Africa Developments macadamia factory, 1999. Credit: I. McConachie.

with North America, Europe and Asia the main outlets. Production was expected to double by 2007.[23]

Processing facilities used modified blade crackers and had sufficient capacity. Consignments were segregated and processed by supplier batch, which enabled accurate reconciliation and assessment but required expanded storage facilities and sound record-keeping. NIS was not fully dried before cracking, to maximise the percentage of whole kernel and reduce losses, which meant that final drying was required although there is a gradual trend towards fully drying to 1.5 to 2.0 per cent kernel moisture. Processors had been using either dry kernel processing or wet separation systems to separate kernel from shell – both systems having advantages and disadvantages. By 2022, most wet processing had ceased except for chip washing to separate shell from small kernel pieces. The vagaries of weather, pests and diseases, global cycles, land claims, declining yields, electricity disruptions, limited research and market development were all ongoing issues, but they applied to a greater or lesser extent to most growing countries.

As in Australia and other growing countries, South Africa's export of NIS to China became a major outlet for the crop during the 2010s and, later in the same decade, an average of 40 per cent of the South African crop was exported as NIS.

Many individuals and corporations have invested in macadamias on a large scale. Camellia PLC is a UK company that has orchards in South Africa, Malawi and Kenya, with a total of 2,906 hectares of bearing trees and 786 hectares of immature trees in 2022, making them one of the largest growers in the world. They have major equity in the processing company Zetmac.

The decline in tobacco growing and sugar-cane together with reduced replacement of forestry resources (for paper pulp) freed up suitable farming land in several provinces, allowing macadamias and other crops to expand. Growing and processing costs to produce marketable kernel were low by global standards and, together with the high prices for NIS and kernel during most of the 2010s, resulted in macadamias being considered a most viable crop with high returns. Large ongoing plantings continued. The rate of plantings during the 2010s was extraordinary.

Table 8: South Africa tree numbers and production, 1976–2030*

Year	Tree numbers	Dry-in-shell production (tonnes)
1976	15,000	5t
1980	73,000	147t
1990	300,000	2,010t
1993	717,000	5,076t
2000	1,800,000	7,902t
2010	6,000,000	37,990t
2021	18,000,000	53,200t
2030	25,000,000 (estimated)	190,000t (estimated)

Averages calculated from several sources, including SAMAC, P. Lee, WMO.

Wages rose steadily from an average of $A 8.50 per day in 2009, encouraging an increasing use of mechanisation at both farm and factory level. By 2015, wages were $A 15 per day and increasing.

The cost of electricity increased, as did that of purchasing water. Stink bug and other pests increased. Felted coccid was accidentally introduced from Australia. The increased use of mechanical harvesters has probably damaged the biological structure of the orchard floor, contributing to a decline in tree health.

SAMAC has adapted to industry needs and provides leadership and representation that has greatly stabilised their industry. Their goals are to lead world-class research, establish strong relationships with the government and industry, deliver value for money to growers and communicate strongly with all. Promotion of the product and expanding demand had been limited by funding and was largely in the hands of industry marketers and market development initiatives from Australia. For twenty-seven years, the research levy of 5 RSA cents per kilogram of DIS was voluntary. In 2014, a statutory levy was introduced of 23 RSA cents increasing to 62 cents in 2021. This introduction of a statutory levy was a major advance for the industry. Over the first five years, $A 800,000 was collected, with $A 10,000,000 projected by 2027. In addition, South Africa strongly supported the establishment of a global macadamia organisation (WMO) to create demand and expand markets.[24]

Philip Lee cleverly summarised the industry in 2015 as the 'the rise and rise'. By 2022, the major growing districts were:

- the Limpopo Province, with 20,000 hectares, but facing issues of drought, old orchards and varieties, pests and low total kernel recovery;

- the Mpamalange Province, with 26,000 hectares and a rapidly continuing expansion; and

- the KwaZulu-Natal Province, with 24,000 hectares and ongoing issues with water availability and wind damage.

There are smaller growing areas in the Cape Province and Northwest Province.

Investing in the future

The South African industry benefits from long-established research and its application to orchard practices to optimise productivity and kernel recovery, quality initiatives and some market development. Mark Penter at the Agricultural Research Council has taken a leading role in applied research into understanding and applying quality principles.

Phil Zadro from Australia through his family company consolidated his position as the largest grower in the world by establishing orchards in South Africa from the early 2000s. This resulted in the establishment of Global Macadamias, later to be renamed Marquis Macadamias Africa, a consortium-owned processing business which established a factory at Lows Creek in 2008 and a larger factory at Alkmaar in 2020 with an initial capacity of 15,000 tonnes of NIS but designed for twice that. The Whyte family – Alan, Jill, Alex, Grahame and Cairine – had three factories which they merged into a large new factory

at White River, designed for a capacity of 25,000 tonnes.

Philip Lee in his professional, detailed and sound *Macadamia 101* (2021) summarised the industry in the early 2020s, which included projections and assessments. The high planting rate was strongly influenced by NIS prices and from 2007 to 2021 tree numbers increased by 240 per cent or thirteen million trees. Yields declined from 3.14 tonnes of DIS per hectare in 2018 to 2.3 tonnes in 2021, where a goal of 4 tonnes per hectare is a realistic target. Pests and diseases were estimated to have cost the industry R3 million per year. Half of the trees in the country are Beaumont, with A4 and 816 making up another 30 per cent. Demand and the exchange rate resulted in Rand terms of averaged kernel prices in $US rising from $US 6.55 per kilogram in 2007 to $US 18.80 per kilogram in 2018. Farm gate prices expressed as kernel increased from R26 in 2007 to R243 in 2020. In the same period, processing costs increased from R20 to only R48, which included

Phil Zadro inspecting plantings, c. 2011. Credit: Hinkler Park Plantations.

substantial increases in costs such as electricity, labour and transport.[25]

From 2013 to 2021, exports of NIS to China averaged 40 per cent of the crop, but as China's domestic production increases, this is likely to decrease, requiring a major expansion in kernel markets. The domestic market within South Africa is small. Kernel production in 2026 is estimated to be 55,000 tonnes of kernel or about 160,000 tonnes of NIS. South Africa currently produces about 44 per cent of global production, which is expected to remain constant in the medium term. The value of the South African crop was reported as R32 million in 1996, increasing to R4.2 billion by 2018.

Philip Lee states that the 2020 to 2022 fall in global kernel prices was due not to oversupply but to the effects of the COVID-19 pandemic and the war in Ukraine. Inventories increased particularly with Style 4 or half-kernels. The role and success of the World Macadamia Organisation (WMO) to promote the product, expand existing and develop new and emerging markets, is essential. Continued funding for the WMO will be a major challenge because of the price collapse to the grower. At orchard level, productivity must be increased largely through breeding of precocious and high-yield varieties, possibly as dwarfing types combined with higher planting densities and application of advanced horticultural technology.

The ACAMOZ 'Report on Mozambican Macadamia Value Chain' (2021–22) summarised the South African industry as having 700 farms planted on 45,000 hectares and averaging 1.6 tonnes of DIS per hectare. Plantings were continuing at 400 hectares per year. Average cost of kernel production was $US 4,000 per hectare. This was compared with Australia, averaging 2.7 tonnes of DIS per hectare at a production cost of $US 9,000 per hectare. Their stated conclusion that South African costs per tonne of NIS are similar to those in Australia should be questioned, but most reports list production per hectare as being similar to Australia.[26] Current inde-

South African typical high-standard orchard. Credit: SAMAC.

pendent reports are of 1,000 growers and fourteen processors in 2020.

At the IMS Symposium in September 2023, it was reported that 70,000 hectares were planted, with 68 per cent in full production, an 80,000-tonne NIS crop in 2023 where 36,000 tonnes were exported as NIS to China, and a projected 180,000-tonne crop in 2034.[27]

Jill Whyte from Green Farms in May 2024 assessed the current crop as 92,000 tonnes NIS at 3.5 per cent moisture content, a major increase from the previous year. Quality was satisfactory but the amount of NIS exported was 70 per cent of the crop in 2023, reducing the amount of kernel available to established markets and local processors who had expanded their processing capacity. Marketing through e-commerce was expanding and assisting NIS export sales.

South Africa overtook Australia as the largest producer in 2015, but due to environmental factors its production rates have not increased as rapidly as was projected. Overall, South Africa deserves to be commended for the sound development of an industry that is certain to expand and where the cost of kernel production is low by global standards. Macadamias have been an outstanding success story in South Africa and their horticultural standards, applied research, quality emphasis, and ability to plan and execute on a large scale will ensure their future. Even excluding future plantings, production is certain to increase massively until at least 2030. But by 2023, the high prices which had held up to 2022 had fallen dramatically, global inventories were high and recovery was seen as likely taking several years. With a 2024 crop increase forecast of 15 per cent to 92,000 tonnes, quality is sound and the crop will increase steadily. With 70 per cent of the crop delivered to China as NIS in 2023, the Republic of South Africa factories are operating below capacity and unable to satisfy the demand for kernel.[28] The rapid growth in plantings is likely to slow until the industry consolidates and demand is in balance with supply. However, there is optimism that the industry will continue to expand.

Kenya

It is difficult to put Kenya into perspective. There are, in effect, two industries. There are up to 150,000 smallholders whose family livelihood is at least partly dependent on small macadamia plantings. Many of these have a low number of mainly seedling trees, of mixed species with variable quality, and this has created many problems that are slowly being addressed. The horticultural standards of their trees and the ability to protect crop and sound kernel quality is sometimes poor. Then there are commercial orchards of increasingly large size whose managers, over time, have adopted better horticultural and crop-handling procedures, and generally produce sound marketable kernel.

After early struggles and challenges, few would have anticipated that Kenya would become one of the larger macadamia-growing countries of the world. This is a proud achievement and the government is committed to supporting smallholders and their families as well as to securing export earnings from kernel and now NIS sales.

Kenya straddles the Equator and has a typical tropical climate, so macadamias are suited only to the elevated areas where high temperatures are moderated and there is sufficient winter chilling to initiate flowering. Macadamias were initially grown at higher elevations, usually in the Central and Eastern Highlands where coffee and tea were already grown. From 1995, they expanded to parts of the Western Province and Rift Valley Province of between 1,450 and 1,900 metres elevation, with rainfall of 1,200 to 1,700 millimetres and on largely suitable soils.

The first report of macadamias in Kenya was in 1926, when a tea planter obtained seeds to plant for both fruit and shade. These nuts may have come from Herbert Rumsey, the Australian seed merchant, who from 1911 dispatched both small and large consignments of NIS to many countries together with instructions on germination and how to grow the seedlings. There is no record of the outcome of these seeds.[29]

In 1946, Robert Harries – later known as Bobs Harries – brought in six nuts from Australia which resulted in five trees being planted at his coffee estate at Thika, near Nairobi. He believed in the potential of macadamias and founded a company, Bobs Harries Limited (BHL), which sought to pioneer an industry. Initially he imported *M. tetraphylla* seed nuts from Australia and in the late 1960s he brought in 10 tonnes of *M. integrifolia* seed nuts from Del Monte in Hawaii. He received initial support from the colonial Kenyan Government, which was encouraging diversification into a range of cash crops including coffee, tea and sisal. In 1963, when Kenya gained its independence and became a republic, he planted 8,000 trees.[30] In 1967 and 1968, BHL proposed to sell 825,000 seedling trees to small-scale farmers in the Central Province, where the farmers were informed that the nuts produced would be a source of cooking fat. The Kenyan Government through the Minister of Agriculture in 1967 launched a campaign to plant 1,000,000 trees over small, separated farms totalling 8,000 acres. When these trees commenced bearing from 1971, there was no market and half the trees were lost.

It appears that politics and personalities intruded and in 1971 the government forced the closure of the BHL nursery, which resulted in about 400,000 young seedlings being lost. BHL investigated the establishment of a processing plant and received government approval for it to be funded from Germany, with equipment supplied by the Meyer company in the USA. The government again intervened and required planning for the factory to cease, and there is some doubt that the factory was ever built. In the early 1970s, farmers who had bearing trees had no market and many of the trees were either abandoned or removed.[31] It would be 1983 before it could be considered that a commercial industry had commenced in Kenya.

The origins of the Kenya Nut Company (KNC) date from 1971, when Mr Yoshiyuki Sato from Japan visited the National Horticultural Research Station at Thika and learnt about macadamias. He prepared a feasibility study of macadamia processing and was supported by investors from Japan and by the Japanese confectionery group Meiji, which provided both technical and financial support. In 1974, a basic factory in a small shed was commissioned. The supply of nuts from the large number of small-scale farmers was arranged through the Co-operative Farmers Societies. There were eighty-five such Societies and 450 primary processing factories for coffee in macadamia-growing areas. As coffee and macadamias were harvested at about the same time, farmers brought their nuts into the factory with their coffee and these nuts were collected every two weeks and sent to the KNC. Processing commenced in 1975.[32]

By 1975, there were five small, commercial macadamia orchards, BHL's 8,000 trees and between 70,000 and 80,000 small-scale farmers usually having only upward of ten trees each. The amount of crop actually harvested and sold from the small farms was inversely related to the profitability of coffee, so often nuts were left on the ground for up to three years. There were no fertilisers applied and there was no crop protection. Nuts generally had high levels of insect damage and mould, and, as expected from seedling trees, the kernel recovery was variable and low. Often much of the kernel produced was inedible or very low grade.

Eventually in 1977, the Kenyan Government established the Kenya Agricultural Research Institute (KARI), recognised the importance and potential of macadamias as an additional cash crop to supplement coffee and tea, to become a foreign-exchange earner and an income generator for their small-scale farmers. Growing macadamias was intended as a poverty reduction plan, to support the rural economy. The macadamias were usually interplanted with coffee.

In 1977, the Government of Japan, through the Japan International Co-operation Agency (JICA), provided horticultural and other advice plus finance to rehabilitate the industry. This led to a formal agreement which, with extensions, lasted into the 1990s and was the beginning of the Kenya Horticultural Development Program. The Kenyan Government agreed to a macadamia monopoly with a Japanese company. As a result, the Kenya Nut Company marketed almost all of its kernel to Japan, mainly to Meiji and mainly to be used in confectionery. After considerable lobbying, BHL was permitted to process and market kernel only from their own crop. Most of this kernel was sold locally through PJ Products. The KNC was required to purchase nuts from all small growers and the price was not related to sound kernel recovery, so there was no incentive to supply fresh nuts or sound quality. The Horticultural Crops Development Authority actively promoted the industry and encouraged its expansion to other regions. KARI has, through its Macadamia Research Station, developed several varieties and provided nursery advice, grafted trees and extension services.

Part of this included an assessment, by Japanese and a West German horticulturist, of Hawaiian commercial varieties as well as the large number of seedlings, most of which were of the *M. tetraphylla* species. Preference was given to local selections and six or seven of these were made which were considered more suited to Kenyan growing conditions. While it was proposed that all seedlings would be top-worked within two years to the new varieties, which were either of the *M. integrifolia* species or hybrids, little happened. Estimates varied of between 400,000 and 800,000 trees having been planted by 1979. The varietal selection programme, particularly in the early stages, had flaws and the published projections of performance were not credible.[33]

In 1980, KNC processed 980 tonnes of NIS. Mr Kunihiko Yamaguchi undertook marketing in Germany from 1987 through Nutfields.

Up to the 1990s, 95 per cent of the crop came from small-scale farmers. While the concept of supporting and encouraging the very large number of small-scale farmers to plant viable crops was both commendable and necessary, there were many problems that have taken years to address. The final quality of kernel is largely determined by on-farm practices and it is difficult for small-scale individual farmers to meet sound standards. A mixture of *M. tetraphylla* and *M. integrifolia* species, particularly when many are seedling trees, made it much more difficult to produce and market uniform kernel. Buying practices in some seasons resulted in immature nuts being harvested and sold. Sound kernel recovery averaged 15 per cent, although by the 1990s it had increased to 20 per cent. Kernel quality from small-scale farmers was often suitable only for confectionery.

People who addressed these problems during the 1980s and 1990s included:

- Mr Takitani from Hawaiian Host in Hawaii, who in 1977 was seeking involvement in the emerging Kenyan industry;

- Mr Yoshiyuki Sato, the President of KNC;

- Mr Kunihiko Yamaguchi, a marketer who from 1987 played a major role in steadily addressing the issues and developing a sound industry;

- Mr Masahiro Shiota, with his engineering and cultural skills as well as much enthusiasm, was a leader who contributed through KNC and has recorded much of the history of the Kenyan industry. Later in Japan he provided services through his company Hardnut International.

The retail brands Nutfields and Out of Africa became well known. Industry statistics have been difficult to obtain and often varied from source to source. Table 9, while incomplete, provides some information to show the growth of the industry from the early 1990s to 2022.

In the early 1990s, most of the 350 tonnes of kernel produced per year was still being sold to Japan. The Kenyan Government supported the domination of the market by Japan and considered that it strengthened ties between the two countries. This joint venture started in 1977 between the Government of Kenya, the Government of Japan and the KNC was to last for twenty years. From the early 2000s, Mr Pius Ngugi ably led the KNC, which by that time had 200 hectares of orchards and 460,000 nursery trees.[34]

Table 9: Kenya industry overview, early 1990s – 2022*

Year	Hectares planted	Number of trees growing	Tonnes of NIS/ year	Tonnes of kernel/year
Early 1990s		400,000 to 700,000	2,500t to 3,000t	350t
1994	5,600 (of which 5,160 on small-scale farms)	969,000		430t
1998			4,632t	
2000				1,080t
2003	8,300 (of which 7,300 bearing)		7,300t	
2005			10,800t	1,585t
2006			12,500t	2,500t
2006	Statistics for trees grown by 100,000 small-scale farmers	900,000	4,000t	800t
2010			10,000t	
2021			39,750t	
2022			40,000t	

* Data in this table has been collated from the following sources: International Macadamia Symposium Report, China, September 2018; Ten Senses Africa, 'The TSA Story'; online: https://tensensesafrica.com/the-tsa-story/; Camellia PLC, website November 2022, see https://www.camellia.plc.uk/; A. and D. Spurling, 'Macadamias in Malawi', *California Macadamia Society Yearbook,* 1971.

The Kenya Nut Company's position of control was not well accepted by some, and in 1983 BHL tried to enter the purchase and processing of macadamias but was not successful. Regardless of the monopoly, KNC has been a leader and stabilising force for almost fifty years. The Kenya Farm Nuts Co-operative Society (KFN) was established in 1992. This co-operative lobbied the Kenyan Government to be allowed to act as a second processor and purchaser of NIS. Their factory was built at Muranga, 150 kilometres north-east of Nairobi in Central Province, and was supported by MacFarms of Hawaii. They attempted to secure crop by paying a cash advance, which changed the long-term buying structure. KFN failed as a company but was later re-formed and now continues as Equatorial Nut Co. Ltd.

Of the 5,600 hectares planted to macadamias by 1994, over 5,000 hectares were on small farms with ten or more trees (occasionally up to 100). The few larger orchards were owned by KNC (271 hectares) and BHL (28 hectares), with 41 hectares on other commercial orchards.

During the 1990s, sales commenced to Germany with decreasing amounts to Japan, which was sourcing kernel from other countries. The industry continued to expand and the Kenya Nut Company provided strong leadership. Through Mr Shiota, KNC had established a large nursery reputed to contain 460,000 trees, which was producing up to 150,000 grafted trees per year and planting 2,000 hectares of orchard in its own right. Kenya's 2006 production of 12,500 tonnes of NIS produced 2,500 tonnes of kernel – an improvement in sound kernel recovery, although some of the crop was rejected because of low quality. The industry projected 20,000 tonnes of NIS and 4,000 tonnes of kernel by 2011. Apart from the KNC, there were six major processors in 2007 plus smaller ones and exporters of NIS. The processors included:

- Mt Kenya Nuts;
- Agricultural Commodities Co. Ltd;
- Earthnut Co. Ltd;
- Saw Africa Ltd;
- Global Nuts Co. Ltd.

Linton Park Africa, which is now part of the Camellia PLC Group and through Eastern Produce, have since established a factory at Kakuzi.

Partly due to a global market downturn in 2006, increasing quantities of NIS were exported to China, although quality variability limited sales. The Kenya Nut Company has been processing 3,000 tonnes of NIS from their own farm, with 4,000 tonnes of NIS supplied through the collection centres.

Support to macadamia farmer societies from the Kenya Horticultural Development Program and the Kenya Agricultural Research Institute has resulted in the supply of grafted varieties and assistance in top-working seedling trees. This has, in turn, reduced the amount of low-grade, often *M. tetraphylla*-based NIS being supplied.

In the first decade of the 2020s there were more than 700 collection points for small-scale farmers. The industry has been restricted by financial constraints, slow adoption of research and an increasing need to address pests and diseases. In 2006, an average price was $US 0.42 per kilogram of NIS and the average macadamia income was $US 16.70 per farmer. While kernel from Kenya in the past was generally considered to be of low quality and often weakened the global image of macadamias, this has changed in recent years and samples and consignments of Kenyan kernel for export have been of acceptable quality. Lower-quality kernel is used in confectionery products and for local retail sales.

The modern structure of the industry commenced in the 1990s with larger-scale, corporate-owned orchards with grafted trees and higher standards of horticulture. Also, through government support and applied research, many local farmers established orchards of between 100 and 2,000 trees, and many small-scale farmers adopted sounder procedures resulting in an improved kernel quality.

In 2006, much of the crop was rejected because of low quality. After some years of disappointing quality, the industry reached further lows when, in 2009, the government banned export of NIS in order to support local processors. This resulted in claims of exploitation by small-scale farmers unable to find an outlet and fair price.[35]

There were at least 100,000 small-scale farmers producing 70 per cent of the crop in the early 2020s, including 500 growers with more than 1,000 trees.

A detailed report at the Eighth International Macadamia Symposium in China in late 2018 summarised the Kenyan industry as benefiting from a stable government which supports farming and processing within the country. The ban on export of NIS from 2009 remained until the 2020s, despite significant leakage to China, and was finally lifted in early 2023 although this is being challenged. Also there has been the emergence of medium to very large orchards which includes Kenya Nut, Eastern Produce at Kakuzi, and Wondernut covering some 2,500 hectares.[36] Plantings have been extending to new suitable ecological areas. Over ten years, the number of processors has grown from five to twenty-six.

A major advance was the formation of the Nut Processors Association Kenya (NutPAK), an alliance of eleven processors, all accredited to HACCP. Another advance was the formation of Ten Senses Africa Ltd in 2006, a fair-trade certified group who have increasingly been successful in improving incomes for small-scale farmers, eliminating middlemen, exploitation, and engaging with growers. They have developed a strong advisory, sustainable and co-ordinating relationship with farmers.[37] As an example, prices to smallholders rose from an average of 30 Kenya shillings per kilogram of NIS in the 1990s to 200 Kenya shillings in 2018, although prices have fallen from 2022. While Kenya is reported as the fourth largest producer in the world, that is in terms of NIS and not the amount of edible kernel.[38]

In 2021, yields were reported as averaging 30 kilograms of NIS per tree but sound kernel recovery was still low at 13–19 per cent and delivery moisture content was high at 20–30 per cent. Losses to theft, pests and immaturity are high.

In 2022, the UK company Camellia PLC had 767 hectares of bearing trees of a total of 1,356 hectares, producing 492 tonnes of NIS. Their factory at Makuyue trading as Kakuzi PLC was commissioned in 2016 and markets under the Maclands brand.[39]

One report in 2022 described the industry as being wholly organic, with pests controlled by smoking and molasses traps, which may have referred to the small-scale farmers. There were in 2022 more than 150,000 growers and a clear roadmap for economic development of small-scale Kenyan farmers. These farmers still contribute much of the country's crop, and are essential to the wellbeing and income of families, but in some opinions, governments' policies and laws have resulted in guaranteed prices, guaranteed acceptance of crop, sometimes leading to low quality. Many small-scale farmers' trees are now up to fifty years old and often a mix of two species that should be kept separate. The farmers have limited ability to purchase and apply crop protection chemicals and nutrients, and with a long supply period from harvesting to processing it will be difficult to market an acceptable product. The high-quality kernel is generally exported, with lower grades sold within the country. The global fall in NIS prices from 2022 is a further disincentive to maintain sound standards.

Macadamias are the most important nut crop in Kenya. With Kenya producing about 10 per cent of the world's production of kernel, macadamias have great potential to reduce poverty due to the high returns achieved up to 2021 and the relatively low inputs required. The IMC 2023 Symposium listed production that year as 41,650 tonnes of NIS on 31,000 hectares, with 65 per cent of trees bearing and with more than 200,000 growers.[40] At the INC Congress 2024, in the Roundtable Recap, Graeme Rust, CEO of Kenya Nut Company, estimated the crop at 46,000 tonnes of NIS, although very heavy rain could affect quality. The low prices and the amount exported to China were issues being investigated by the Kenyan Government. Significant plantings from 2015 will now increase their future crops.[41] Despite the global market downturn in the early 2020s, the crop forecast for 2027 is 63,000 tonnes of NIS at 3.5 per cent moisture content[42] and 90,000 tonnes by 2032.[43]

However, sound kernel recovery averages only 15 per cent, which implies that much kernel produced would have low sensory qualities, unattractive appearance and reduced shelf-life. Despite wages being low by global standards, the cost of processing this NIS would be high, with the resultant kernel of variable quality. The industry practice of paying for NIS irrespective of kernel recovery and quality continues and is a disincentive to improvement. Global purchasers of kernel and NIS are becoming more discerning, which will make Kenyan product difficult to sell. Global oversupply and the fall in NIS and kernel prices from 2022 will be a challenge, as it could reduce income and confidence, and it will be difficult to restructure their industry to

African nut-in-shell inspection belt, including Kenyan product. Credit SAMAC.

meet market requirements. Despite this, the industry is expanding rapidly, confidence is high, poverty is being addressed, small-scale growers are seen as exemplars for Black economic development, and, while there are major quality and other problems, the industry's future seems assured.

Malawi

In 1932, twenty trees were planted from seed nuts supplied by Australia's Rumsey's Seeds and in 1942 there were plantings at the Bvumbwe Research Station and in private gardens. Very small but commercial orchards from Bvumbwe were established at Kumaladzi near Thyolo in 1953.[44] Also, in 1950 William MacFarlane Oddy planted a small orchard in the north. In 1953, *M. tetraphylla* seed nuts were brought from South Africa.[45] Sir Malcolm Barrow, owner of the Naming'omba Tea Estate in Zimbabwe, encouraged the planting of macadamias in both countries. Edward Tonks from Zimbabwe in 1968 provided ten mother trees which were rapidly repropagated so that, by 1972, 15,000 grafted trees were reported to have been distributed. Initially the emerging industry was led by Andrew and Daphne Spurling, who were enthusiastic promoters and provided much information through the Bvumbwe Research Station.[46]

The Commonwealth Development Corporation (CDC), formed in 1948 and owned by the UK Government, played a major role in funding and developing British colonies. The CDC in the 1970s supported the establishment of 37 hectares of macadamias at Kawalazi near Mzuzu in the north, expanding these and adding more orchards at Mapanga, near Thyolo, and at Ngapani Estate.[47] The CDC supplied a range of clonal material to Sable Farming Estates near Blantyre, which they made available to the industry.[48] In the early 1980s, the Tropical Products Institute (TPI) from England studied macadamias and developed a prototype processing plant and procedures for the Malawian industry.

By 1968, some tea and tung oil estates were establishing macadamia nurseries and in 1973, the Malawi Tung Oil Growers Association included in its charter all tree crops.[49] Also in 1973, twenty small-holders planted several hundred trees in the Central Districts.[50] The Malawian industry has been developed largely by big overseas (mainly UK) corporations as part of their tea and sometimes coffee estates. Early large growers were Naming'omba Tea Estates Limited, Eastern Produce Estates, Kawalazi Estates and Sable Farming. The Malawian Government supported crops like macadamia due to their export-earning and poverty alleviation potential, and evidence that they were suited to much of the country. With low growing costs, particularly for labour, a local industry had a competitive global advantage. In the 1980s, the government required that the tea estates diversify into other crops for least 20 per cent of their acreage, with macadamias as one option. Insect pests, mainly green stink bug and nut borer, resulted in losses of up to 80 per cent and sound kernel recovery varied from levels as low as 4 per cent, although industry average increased over twenty years from 14 to 30 per cent. Australian entomologist David Ironside was engaged by the government at Bvumbwe Research Station from 1989 to 1994, to study insect pests and diseases and to develop control procedures. Wayne Hancock, also from Australia, provided horticultural advice. By the early 1980s, seventeen tea estates were growing macadamias.

David and Sue Emmott were early pioneers, leading Naming'omba Tea Estate macadamias and promoting and assisting industry development in the 1980s. They planted 200 hectares at Thyolo and converted a tea factory to process nuts using water flotation to separate shell from kernel.

By 1993, there were 2,200 hectares planted to 460,000 trees and producing 315 tonnes of kernel, with claims the country was then the third largest producer in the world. Other corporations such as British African Tea Estates were also planting macadamias. The Malawi Tree Nut Growers Association, later to become the Tree Nut Growers Association of Malawi, supported the industry both in horticultural matters and in co-ordinating and representing the industry. The large corporations had skilled but inexperienced macadamia agronomic staff, and, supported by the Bvumbwe Research Station, developed an understanding of horticulture and productivity.[51] The author of this book was engaged by the government in the early 1990s to commission a new processing plant at Kawalazi and to provide advice to other sectors of the industry. I would like to think that my advice to the corporations and the government was sound, but in retrospect I did not understand the importance of encouraging smallholders and instead was influenced by their challenges of producing quality kernel.

Over 80 per cent of employment was in agriculture and over 90 per cent of all food was grown by smallholders. Most lived at subsistence levels growing tobacco and maize, the staple food. Tobacco demand had fallen and maize was facing horticultural problems. The government implemented ongoing strategies for smallholders to support, assist, encourage diversification, and protect and provide funding. The African Development Bank provided some finance. Macadamias were seen as one of the ideal replacement crops which, after say five years, would provide food to families and hopefully a reliable income. The National Resilience Strategy 2018–2030 defined how this would apply. Obtaining foreign exchange was important.

Smallholder macadamia plantings during the 1980s and 1990s were developing slowly, but due to the sometimes uncertain market demand, time to reach maturity, variable yields and price fluctuations, confidence and commitment declined. Smallholders found it difficult to fund their plantings, faced losses through insect damage and, apart from small quantities exported as NIS, the kernel was mainly sold and consumed locally.[52] Some of the problems they faced were the availability of both sound trees and extension services, funding, pests and diseases, and variable, often very low yields. A benefit of growing macadamias was that it provided employment and income for females.

The situation improved in 2004 when David Emmott founded the Neno Macadamia Trust, which supported the establishment of seven co-operatives

Examples of high standards in processing, even in 1992. CDC Kawalazi. Credit: I. McConachie.

through the Highland Macadamia Co-operative Union Limited, to co-ordinate collection, receival, drying, grading and delivery services.[53] By 2020, there were 3,500 macadamia smallholders growing a total of 1,500 hectares, ranging from 0.1 hectares to more than 4 hectares.

Industry plantings doubled during the 1990s. By 2012, there were 5,224 hectares on commercial estates and 56 hectares on small-scale farming blocks, totalling over 1 million trees and with a crop of over 7,000 tonnes of NIS. New plantings were in Mulanje, Mchinji and Ntaja districts.[54] Concerns at this time were climate change that was contributing to drier seasons; increasing costs for crop protection; and the distance to ports for shipping.

Camellia PLC, a large UK company trading as Linton Park and Eastern Produce Malawi, had by 2006 established large orchards at Nasonia, Kumadzi, Thyolo South and Masauve, totalling 1,231 hectares.[55] In 2021, there were seven processing plants with a projected capacity of 13,000 tonnes of NIS per annum. All have International Quality Assurance Accreditation. Kernel is sold to Europe, the USA, China, Japan and the Middle East. Selling NIS to China was considered 'theft' by processors and was minimal. Tropha Estate, part of the UK Jacoma Estates Limited, is a newer grower and processor with orchards in the north that are irrigated (unlike most of those in the rest of the country). They also assist smallholders to grow and process their crop. Overall, the industry directly employs over 10,000 people and is committed to sustainable development and social responsibility, with equal opportunities for female workers and for equal work.

Growth during the 2010s was rapid, with new plantings averaging 1,000 hectares per year from 2016 to 2020. By 2020, there were about 250,000 trees not yet bearing and 1,250,000 bearing trees producing 8,000 tonnes of NIS. With an average sound kernel recovery of 30 per cent, this made Malawi the seventh largest grower and producer in the world.[56] The International Macadamia Symposium in September 2023 listed 11,184 hectares, 42 per cent bearing, with more than 9,000 growers producing 11,400 tonnes of NIS in 2022.[57] Crop estimate for 2024 was 12,000 tonnes of NIS.[58]

A 'Review of Macadamia Production' report states that production could double from 2020 to 2025. Most marketing of raw kernel is through companies based in South Africa that represent large growers and processors in a number of African countries. The structure of the industry is wide, comprising corporate growers, smallholders, processors, aggregators who collect, handle, dry and deliver NIS, and organisations such as government and industry agencies.

The Malawi industry is generally sound and stable, producing quality kernel for the export market, and it will continue as an important smaller contributor to the global industry. Industry growth and confidence was driven by the very high global prices and demand during most of the decade from 2010, but the global marketing decline in 2021 and very soft market prices from 2022 are likely to result in much lower prices. The 2023 and 2024 crops each totalled about 12,000 tonnes. The industry has changed tremendously over the last decade and large corporate growers are likely to be able to weather the challenging times and successfully expand to grow, while smallholders may struggle to survive the periods of lower prices.

Mozambique

Until 1994, when Mozambique became a democracy and a member of the Commonwealth of Nations, there had been long-term civil wars, unrest and no confidence for corporations to invest there. Cashews had been a major industry, but production was badly impacted by the war.[59] The failed Komatipoort macadamia orchards in South Africa from the 1970s were an uninspiring example close to the border.

In neighbouring South Africa, the macadamia industry was expanding rapidly, facing increasing costs, had sound horticultural knowledge and was profitable with available funds. Mozambique was increasingly seen as being politically stable, with a government that encouraged development and export earnings. As well, it had a low risk of land claims, and land was plentiful, cheap, suitable, flat and with ample water. The right to use land is through the DUAT (Direito do Uso e Aproveitamento da Terra) system, essentially a forty-nine-year lease which can be doubled to ninety-nine years. Mozambique was seen as offering an opportunity mainly for established South African and Zimbabwean farmers. However, it was not until 2000 that farming land was declared free of land mines.

Howard Blight from Amorentia Nursery in South Africa initiated a large-scale macadamia and avocado orchard in Sussendenga, in the western highlands, which provided trees for early orchards. By 2006, several small plantings had been made totalling 20 hectares, but ambitious projections of 16,000 hectares of orchards by 2020 were being made.[60] There were many small and large plantings over the next fifteen years. Larger orchards were established rapidly, mainly in the provinces of Niassa, Manica and Maputo.

An international consortium, Valley of Macas, commenced in Serra Choa in 2006 and had an orchard of 70,000 trees by 2013. By 2016, around 500 hectares had been planted.[61] Macs in Moz built a modern factory in 2016, with a future capacity of 5,000 tonnes of NIS per annum. Beaumont is the main variety. Risk factors include cyclones, lack of infrastructure and endemic corruption. Advantages include support from South Africa, Malawi and Zimbabwe and access to a number of sea ports.

In the wilder north, with lions and buffaloes roaming the property, Graeme Stainbank from KwaZulu-Natal has led an orchard of 250 hectares on land with a ninety-nine-year lease.[62]

Production estimates are uncertain, due to some of the crop being processed in South Africa and Malawi. There were an estimated 600,000 non-bearing trees in 2021 and 1,200,000 bearing trees.[63]

In 2021–22, with the support of a French non-government organisation which has been providing biodiversity conservation programmes in Mozambique since 2013, the ACAMOZ project (focused on strengthening the cashew value chain) published a 43-page booklet on the Mozambique macadamia industry.[64] In summary, the booklet reports:

- The industry in 2022 consisted of forty-five macadamia estates, seven of which are producing commercially.

- Production is 1,500 to 2,500 tonnes of NIS per annum and likely to expand many times.

- There are few processing facilities in Mozambique, so NIS goes to South Africa or China through intermediaries, as the country has no certification to enable it to supply directly.

- Major growers are New Forest, Pamoja Impact, Niassa Macadamia, DD Farming, Murrimo Macadamias and Macs in Moz.

- Beaumont remains the main variety and, while infrastructure varies, sound and experienced management set high standards.

- The following recommendations were made to progress macadamias as an emerging industry, important to the country's economy:
  - clear government support, including research and technology transfer;
  - fostering of smallholders by the large estates partnering them;
  - encouraging processing within the country, and
  - using the Malawi industry as a model.

In 2020, the Mozambique Macadamia Association was formed to assist in co-ordination, lobbying and addressing industry restraints.

Part of Murrimo Macadamias, Zambezi Province, Mozambique, 585-hectare orchard with golf course.. Credit: Richard Hurly.

Mozambique has all the attributes to successfully produce quality macadamias at a low cost of production. Existing young orchards will significantly increase production and the country is set to become a global contributor.

Tanzania

In the 1960s, a German missionary in Lushoto near Tanga planted a few hundred seedling trees which came from Bobs Harries in Kenya. Some of these trees have survived and produced nuts, and they used to supply small quantities of NIS to the Kenya Nut Company in the early 1980s. In 1987, a Japanese company registered as Tanzanian Nutfields was allocated 20,000 hectares for the development of macadamia and other crops in Mahenge, in the south-west. The KNC provided technical support but the local government claimed large amounts in outstanding rates. The project was abandoned and the area became a game park.[65]

By the late 1990s, it was reported that 100,000 trees were planted in Tanzania, with early production of 50 tonnes of NIS.[66] It was reported in 2007 that 500 hectares of macadamias had been planted.

In 2005, an Indian business group from Kenya commenced planting macadamias in their coffee estates in Tanzania. Tanzania has been slower than other African countries to establish an industry. Much of the country is suitable, coffee is well established and increasingly both crops are being grown together. In 2017, the Jaquin Agricultural Company announced that it was entering macadamias in conjunction with the Lima Kwanza Limited group who would supply trees. In 2021, Farm for the Future[67] advised of large-scale orchards planned near Iringa. Another orchard is being established at Kilimanjaro.[68] The government's Tanzania Agricultural Research Institute (TARI) has commenced research and support for the developing industry.[69]

Zimbabwe

In the early 1960s, there were no more than fifty seedling trees planted as curiosities throughout the country, then known as Rhodesia. Three friends developed a passion for macadamias, combined to become pioneers and thankfully recorded their story. One, Arthur Lane, was an accountant, supported by his wife Anne, and the other, Ted Tonks, had a local engineering business.

In 1965, Arthur Lane bought seedling trees from

Tending to a young macadamia tree at Pamoja Farms, Tanzania. Credit: Eliza Powell.

Kilimanjaro forms the backdrop to Beaumont in flower at Pamoja Farms, Kilimanjaro, Tanzania. Credit: Eliza Powell.

Chris Reim in South Africa and scion wood from California. Arthur was the planner and administrator, while Anne developed the nursery and then the orchard at Umtalhi and Ted provided the enthusiasm and drive. They imported more seed nuts, grafted trees and received advice from Wells Miller at the California Macadamia Society, Len Hobson, Peter Allan, Andrew and Daphne Spurling from Malawi, and many more.[70] In 1967, they held a field day attended by fifty people and the highlight, which convinced many to enter the industry, was tasting roasted and salted macadamias.[71] By 1975, they had built a small factory but the Rhodesian Bush Wars dominated the country until the early 1980s. Eventually they were able to buy a Shaw Cracker and successfully establish a small industry. Their factory, PEC-MAC, was a co-operative and the name was short for Pecans and Macadamias. The Tree Nut Producers Association of Zimbabwe was formed and led by Ted Tonks and later by Anne Lane.

Ted passed away in the 1970s and Arthur in 1991. Anne had been a stabilising force and sold the business to Chris Barrow of Southdown Holdings at Naming'omba in Malawi. Macadamias in Zimbabwe had the reputation of being easy to grow, were productive and most profitable.

Southdown Estate, established in 1948, has had a long-term involvement in macadamias. It is now part of Ariston Holdings, which has three estates growing macadamias, is still planting and is the largest producer in the country. They purchased and operated PEC-MAC,[72] but it is not known if this factory still continues.

By 1998, there were forty growers with 138,000 trees, producing 396 tonnes of NIS and averaging 29 per cent kernel recovery.[73] From 2000, through Zimbabwe's Fast Track Land Reform Program, most farms had ownership transferred under the A1 model to smallholders or the A2 model to commercial farmers. For the next twenty years, NIS was sent to South Africa for processing. Many orchards were lost, production declined, land title and security were uncertain and, with associated political and economic instabili-ty, including violence, the industry declined.

While other African countries were expanding, the Zimbabwe industry was in limbo. Gradually, driven by increased demand, high prices and the ability to receive foreign credits, larger productive orchards were encouraged and from 2010 there was some confidence, the rejuvenation of existing orchards and new plantings. The Macadamia Empowerment Group Association (MEGA) was established to assist smallholders and, in 2021 and 2022, 300 new smallholders planted macadamias. A newer entrant is Avomac Farm, with 117 hectares of trees, which also supports and contracts new growers. Industry support groups are the Agricultural Marketing Authority (AMA) and Macadamia Association of Zimbabwe. Some markets in Europe are closed due to political concerns over land ownership.

In 2012, it was reported that there were one million trees on 5,224 hectares, most owned by five large estates, plus 18,751 trees owned by smallholders. Theft of nuts is a major problem. Most orchards are in the Chipinge district, with newer orchards in the districts of Manicaland and Mashonaland.

Gradually the government realised the importance of securing foreign exchange and there was confidence to resume plantings. There was no direct government support until, in 2020, the Marondera University co-ordinated most groups to develop economically self-sustaining projects largely for smallholders.[74]

While there have been expressions of interest in constructing a processing factory, it is believed that until at least 2022 all crop has been sent as NIS to South Africa or China. Philip Lee's report of October 2021 lists 600,000 trees under six years of age and 900,000[75] bearing. In April 2022, the World Macadamia Organisation forecast a 6,500-tonne NIS crop.[76] The industry is steadily recovering from the political disruptions, now has more trees planted than Hawaii and is starting to produce an increasingly significant amount for the global market.

Africa: industries in development[77]

Angola

Howard Blight, South African macadamia pioneer, visited the country in 2021 and advised that a small industry had commenced.

Democratic Republic of Congo

Following the interest and profitability in other African countries, small plantings were made by the Knoera Corporation in 2018.[78]

Egypt

Regarded as the founder of modern Egypt, Muhammad Ali Pasha in the mid-nineteenth century supported plant introductions, which resulted in macadamia trees being planted in the Botanic Gardens at the Ain Shams University and Moshtohor University. Today there are four specimen trees at the Mazhar Botanic Garden and one tree at the Orman Gardens. Apart from many individual trees, there is an orchard of several thousand at Nimos Farm, which has many types of horticultural crops.[79]

Eswatini

Officially known as the Kingdom of Eswatini (formerly Swaziland), this country has a small, emerging industry which has been developing since the early 2000s. A crop of 6 tonnes of NIS was reported in 2007. One group, the African Christian Church, has planted 14,000 trees and another group planted 6,000 trees in 2017.[80]

Mauritius

In 2018, the Government of Mauritius investigated the potential and economics of encouraging macadamia planting as an alternative to sugar-cane. By 2023, there have been no reports of commercial plantings and it is probable that the risk of cyclones is a deterrent.

Rwanda

An industry commenced with government support in 2016. The proposal was for one million trees to be planted over several years, with a target of 5,000 hectares by 2025.[81] Dorran Bungay reported in 2021 that ongoing planting had been occurring.

Uganda

Macadamias were introduced into the Bushenyi district in the 1960s, but no commercial plantings were reported until the early 2000s.[82] A large orchard with 100 hectares of macadamias supported by 400 smallholders was established near Kampala. In 2018, Besania Demonstration Farm at Mbarara in the south-west was growing and promoting macadamias.[83] Production in 2019 had increased to 201 tonnes of NIS and was expected to increase.[84]

The Government of Uganda in 2021 listed macadamias as a priority alleviation crop due to global demand and the potential for export earnings. The main grower is AMAFH Farms at Nambale, Mityana, 70 kilometres west of Kampala, established in 2003. In 2023, they have a reported 1,200 hectares of orchards and 1,000 farmers supplying nuts to their Nambale processing plant. Most product is exported, but they also sell nuts and oil under the Macarica brand.[85] Their nursery produces 3.6 million seedlings each year. Recommended varieties are the Kenyan selections Muranga, Kirinyaga, Embu and Kiambu, with 344 and 660 as cross-pollinators.[86]

In 2022, AMAFH Farms partnered with the National Agricultural Advisory Service to distribute 200,000 subsidised seedlings to promote macadamia growing.

Zambia

The first commercial orchard was of six hectares planted in 1996 by Priscilla Frost-d'Elbee. She is an innovative lady who processed and value-added her crop which she marketed under the brand 'Cilla's Completely Nuts'. Her example created interest and a small but expanding industry commenced in about 2007 where 200 hectares were reported as having been just planted by three growers led by Mark Winwood at Lusaka. Larger commercial orchards were planted in 2016 by South Africans and there are smallholder plantings. The government has used the smallholder macadamia industry in Malawi and Kenya as models.

The Amarium Limited company announced in 2015 that they proposed to plant 1,000 hectares near Serenje. NIS production in 2018 had rapidly increased to 210 tonnes and was expected to continue increasing as several commercial orchards were expanding.[87] A 2020 'Feasibility Study for the Cultivation of Macadamia in Zambia', produced by the Challenges Group Zambia Ltd, concluded that the Zambia macadamia industry is dominated by commercial farmers but that there is significant growth potential for both commercial and smallholders.[88] Low labour costs of $US 1.80 per day, a willingness from African farmers to sell their land and be employed, and government support to create employment and earn foreign exchange have together resulted in rapid expansion. In late 2022, a further 4,000 hectares were being planted. The effect of the global market decline from 2022 will be a factor in the future success of the industry.

Middle East and Europe

Iran

Iran had two hectares planted in 2002 and in 2007 further plantings were proposed.[89]

Israel

During the 1950s, macadamias were introduced and assessed. During the 1960s, several Hawaiian varieties were introduced and then some *M. tetraphylla* from California. By 1972, there were about ten acres in government and trial plots.[90] Macadamias were seen as a viable crop until a shortfall in water for irrigation in the 1990s forced the government to review priorities. As a result, this crop was no longer included in development plans.

Spain

The Banis family planted 2.5 hectares near Barcelona[91] and there are also small plantings in the Andalucia region of southern Spain.[92]

Chapter 17

ASIA AND THE PACIFIC

Asia: industry leaders – Asia: industries in development –
Pacific: industry leader – Pacific: industries in development

Asia: industry leaders

China

> Macadamias have brought prosperity to over 500,000
> farmers in the mountainous areas of Yunnan's border
> region.
>
> *(People's Daily Online,* 4 February 2024**)**

China has a rich, fascinating history of culture, evolution and revolution spanning 5,000 years, but it is
only in the last thirty years that the country has embraced the macadamia as a crop. Now, as a socialist
republic whose economy has become a capitalist
and people's powerhouse, China is destined to become a global leader both in the growing and the
consumption of macadamias.

Chinese immigrants who came to Australia seeking gold from the 1860s often became market gar-
deners, some at the Gympie goldfields. Although
there are no records, it seems likely that many who
returned home might have taken a few macadamias
with them. From the 1930s, conflict with Japan,
World War II and civil war were followed by massive
reforms in the Great Leap Forward and the Cultur-

al Revolution. All of this focused the country on survival, growth and sustainability. It is understandable
that the growing of macadamias would not have
rated highly. So although macadamias were introduced some 100 years ago, it has been only from
about 1995 that phenomenal industry growth has
occurred.

Early days

There are records of a macadamia tree having been
planted as a specimen in what was to become the
Taipei Botanical Gardens as early as 1910, and in
1923 the *Brisbane Courier* reported that macadamias had 'spread to China'.[1] In 1931, a mix of seed
nuts and plants was sent from Hawaii to China. The
South of the Five Ridges University at Guangzhou
experimented with macadamias but announced in
1951 that, while they grew satisfactorily, they were
not suited to being commercialised.[2]

China's proximity to Hawaii and known exchange of plants with Australia suggest that macadamias would have been quite well known and that at

least individual trees were being grown in a variety of locations, but it was not until 1979 that there was a record of grafted trees and seed nuts being sent. Both Australian and Hawaiian varieties together with 400 kilograms of seed nuts were introduced from the late 1970s to the early 1980s.[3] In 1980, there were reports of a small industry developing. Macadamia plantings were made on the island of Hainan, at Guangxi, but unsuitable environments and regular typhoons caused abandonment in Hainan and a move to the west.[4] In 1990, there were 200 hectares of orchards in the whole country, which included the Guangxi region.

In 1991, Duan Gouhui, Zong Hongxin and others planted 400 trees on Hongqi Mountain and more at Yongde County in Lincang City, as possibly the first commercial plantings. Following the abandonment of plantings closer to the coast, macadamias were assessed in the Yunnan Province, which has become the most widely planted region both in China and the world.

A market develops

China's macadamia industry was assisted by importing NIS and sometimes kernel from the early 1990s. Initially product was intended to be imported until orchards came into production, but this has continued and enormously increased as retail demand continues to be larger than the domestic crop can satisfy. In Australia, Brian Findlay and Winson Woo formed The Peninsular Group, exporting kernel to Malaysia and then entering the Chinese market in the early 1990s. Winson was from Hong Kong and Brian, while slowly developing his language skills, gradually established both an awareness and a market. They helped to create a premium image by acknowledging almonds and cashews as prized additions to a meal but suggesting that to really impress guests, macadamias were the ultimate nut. Chinese New Year was an opportunity to sell large quantities. The high rate of import tariffs for kernel and to a lesser extent for NIS was prohibitive, which for others later

led to evasion, some corruption, fines for breaches and even some jail terms. These tariffs were gradually phased down and from 2019 were eliminated.

Another Australian pioneer of macadamias into China was David Macrae from Pacific Plantations at Newrybar, in the Northern Rivers. Establishing a most attractive orchard followed by an innovative processing plant on the Northern Rivers site, Pacific Plantations commenced selling macadamias into China and other markets from the late 1980s. Both kernel and NIS were sold. This led to them entering into a joint venture processing business in China where David's marketing enterprise supported the early development of demand and image.

Pecans and walnuts-in-shell had been imported since the 1980s, mainly from the USA, which served as a model for macadamias. Over the next twenty-five years, most macadamia-growing countries supplied mainly NIS and from about 2006 onwards up to 25 per cent of world production was shipped to China as the market steadily expanded. The Kenya Government tried to prohibit exporting NIS, as it was crippling their local processors, but despite this NIS was still smuggled to China. In 2014, half of the South African crop was shipped to China. Initially, small entrepreneurs in Hong Kong and mainland China imported NIS and extracted the kernel in very basic factories. Some NIS was cracked and sorted by contracting to families at their home. Both retail companies and existing macadamia processors developed this market. Initially, NIS was dried and women cleverly cracked the shell using a hammer, which resulted in minimal kernel damage. Next, in the late 1990s, a hand-operated or wheel-operated sharp wedge machine was developed which again opened each nut, one by one. Then, in the early 2000s, an innovative automatic machine was invented to cut through about 70 per cent of the shell without damaging the kernel and a small hand tool was provided with each retail pack to twist open the shell and extract the kernel. This allowed brine and flavours to be soaked into the kernel before roasting. By selling the kernel in shell, the retail price ap-

Cracking macadamias in China using a hammer in the early 1990s. Credit: P. Zummo.

peared much lower. This technique proved most popular and was widely adopted both in China and other countries. Possibly half of roasted macadamia kernels now sold in China are in this form.

Importing through Hong Kong was gradually replaced by direct imports to mainland cities. Importation of NIS continues, and in 2023 it is even expanding despite China's own production increasing. Higher standards of quality control have been applied and new imaging technology is enabling NIS with defects to be removed.

The Chinese industry takes off

Ms Chen Yuxiu, based in Yunnan, has been the commercial pioneer and leader of the Chinese industry since 1992. With dedication, charm, passion, skills and leadership, she has been a continuing major presence in China's development and success. In 1992, as a young employee of Yunnan Grain and Oil Import and Export Corporation, she was tasked with purchasing 30 tonnes of seed nuts, a quantity not permitted by regulations. Her determination and eventual success led to an outstanding career, with her becoming Chair of the Yunnan Macadamia Society and in 2003 Manager of the Yunnan Macadamia Industry Development Co. Ltd.[5] She reported that John Wilkie was an ongoing consultant who provided much advice and enthusiasm, and Mr Cheng Hung Kay introduced macadamias in 1981 to the Research Institute at Zhangi-Jiang. John Wilkie Snr,

until his early death, continued to play a major role in advising the Chinese Government and researchers, and he was acknowledged in 2008 with a Gold Friendship Award in the Great Hall of the People.[6]

In order to help the local people who mainly lived in poverty in the mountainous areas, the Yunnan provincial government proposed developing small-scale trials and then promoting the growing of macadamias to earn foreign currency and create individual wealth.[7]

The Yunnan Tropical Institute led macadamia research and estimated that 4,600,000 hectares of land was potentially suitable. The first large commercial orchard was established in the extreme south of Yunnan Province, at Xishuangbanna (Banna), from 1995. The Tenth Five-Year Plan adopted by the Chinese Central Government and Yunnan Government in 2000 included macadamias. In 2002, officials from the Yunnan Government visited Australia to gather information to assist them in their planning.[8]

In 2000, the total crop was 6,700 tonnes of NIS rising to 18,300 tonnes in 2010. Two other provinces had smaller plantings. The Guizhou Province was assessed as having 240,000 hectares suitable for macadamias and, by 2018, 1,100 hectares had been planted.[9] By 2022, there were 253,000 hectares producing 32,000 tonnes of NIS in Yunnan Province and 46,000 hectares in Guangxi region producing 8,765 tonnes of NIS.

The second largest growing area is the Guangxi region, where macadamias were first introduced in 1974 and then small commercial plantings commenced in 1986. Trees have been planted on available flat, sloping and steep land. Through the China Tree Nut Association, the Guangxi Macadamia Growers Association was formed and in 2018 there were 100 growers with a total of 18,400 hectares producing 1,830 tonnes of NIS. Limited by the orchards being mainly small and widely dispersed, a lack of promotion and little local government support, the Guangxi region initially struggled but has now implemented a strategy to support the industry.[10]

Large-scale orchard, 2018. Credit: I. McConachie.

Typical orchard, China, 2018. Credit: I. McConachie.

Growth of the whole industry increased from about 2000 and after 2010 it was extremely rapid. This growth involved the production of nursery trees; orchard plantings; provision of horticultural advice and services; learning to harvest, dehusk and sort the crop; providing the infrastructure to store, dry and deliver NIS; processing NIS into kernel or sawing the shell; making value-added products; promoting these to the public; and distributing and marketing products to very different types of outlets. Despite employing consultants and undertaking research, there was a rapid learning curve and many problems and setbacks were experienced. Some growing environments proved unsuitable, as did many varieties. Understanding plant nutrition, horticulture and how to control pests and disease, as well as supporting the possibly up to 500,000 small-scale farmers were just some of the development challenges. Small-scale farmers who sometimes merged with others generally had larger plantings than smallholders in African countries. A major part of the government's goals was to encourage farmers to plant, and to provide them with advice and subsidise their establishment costs so that the family owners had an income until the orchards were producing. Most growers elected to receive only subsidised trees. The many setbacks were offset by the quantity and quality of research and extension services, and the work ethic, tenacity and positive attitude of all involved as they achieved their planting goals. Plant breeding resulted in new varieties being released.

In 2010, the Eleventh Five-Year Plan was adopted with a target of 15 million trees on 50,000 hectares with an estimated production of 150,000 tonnes of NIS by 2025. A national goal for China is for its people to be both healthy and rich. With a population of over one billion, and an advancing economy giving more people a disposable income, people were seeking healthy foods as well as snacks, which led to an increased demand for macadamias. In 2021, the targets of poverty alleviation had been addressed and macadamia growing was seen as helping people to increase their income and wellbeing.

Market development was initially fragmented, but it has become more national. Large corporations entered the retail market through e-commerce sales, such as those of Three Squirrels Inc. in 2012, where cartoon squirrels innovatively interacted with both the consumer and staff. The concept was from Meng culture (which translates as 'adorableness'), with cute squirrels causing consumers to relate to squirrels eating nuts. It was so successful that it took nine-and-a-half minutes after release to achieve sales of $A 20 million. The company had to pay compensation to people when orders could not be supplied. Online retailer Alibaba is another large producer and retail marketer of macadamias, mainly through e-marketing, and Cha Cha is another major brand. There was strong promotion of a wide range of innovative products with a high standard of packaging, and sales boomed. Sawn, roasted and flavoured NIS were the main products. Online retail sales accounted for 25 per cent of the market by 2015.

YMAC (Yunnan Macadamia Industry Development Company) is a whole-chain business engaged in breeding, growing, processing and marketing since 2013. In 2022, they launched Mum's Macadamias, which has been most successful. Other localised marketers are Songood Industries Co. Ltd, a large grower and processor in Lincang County, and Jiangcheng Sino-Aus Agriculture Co. Ltd (SAAT), another fully integrated grower, processor and marketer with a capacity of up to 10,000 tonnes of NIS. Brookfarm from Australia successfully market their retail range in China, including macadamia-based muesli.

There has also been an expansion of the international part of the industry, where orchards are purchased or developed in other countries and through business alliances. An example is the Chinese investor group Discovery Macadamias, growers in China who bought four orchards in Australia totalling 100,000 trees. Other business people followed and several of the large orchards established in Australia from 2010 had Chinese ownership. CL Macs, owned by David Ng and his family and managed by Peter Zummo, is a Hong Kong-owned business of over twenty years' standing that has, in Australia, progressively bought NIS, established new orchards and purchased existing orchards. It has also developed a large-scale, high-standard processing factory in Gympie, providing graded and dried high-quality NIS.

Many Australians have visited China to study the industry and provide advice. After a visit in 2011, Dr Russ Stephenson estimated that there were four million trees and that one single orchard of 100,000 trees was producing 800 tonnes of NIS.[11] In February 2014, Kevin Quinlan, Kim Jones and Kim Wilson undertook a study tour. They reported on a rapidly growing industry consisting of both large-scale orchards and many peasant farmers, with reasonably sound yields, high potential and good support. Many of the small plantings were on very steep, usually terraced slopes where all operations were completed by hand. Only growers were subsidised by the government and not processors. They estimated that there could be a 100,000-tonne crop averaging 30 per cent kernel recovery by 2020.[12]

The industry is regulated by the Chinese Government through the Specialised Committee for Nuts and Dried Fruits of China National Food Industry Association (CAN), based in Beijing. It is the central authority co-ordinating and regulating all nut and seed operations in conjunction with regional provinces. Regional associations were formed to represent their areas and work together on industry issues. In 2014, the Yunnan Macadamia Society was established, which has become the leading macadamia industry association in China, with goals of communication and development of the industry.

An example of the development of the industry can be seen in Lincang district, where in 2017 macadamias were being grown in 564 villages in eight counties, with orchards totalling 17,400 hectares projected to produce more than 12,000 tonnes of NIS, to be processed in nine factories. Often these orchards are on steep hillsides where machinery cannot operate, so all operations are carried out by hand. Soil erosion and control of pests and diseases will be difficult, although research is seeking biological controls.[13]

Macadamias grown on extremely steep slopes, China, 2018. Credit: I. McConachie.

China's retail demand forced the price of NIS to increasingly higher levels year by year, until 2021. Largely due to demand from China, NIS prices to growers in other countries tripled from 2000 to 2020, reaching never before realised levels. This stimulated interest in more and more plantings in China as well as the rest of the world.

In the early 2020s, most plantings and production were in the Yunnan Province, mainly on steep hillsides, with some large plantings on the flatter plains of Guangxi region and Guangdong Province. Smaller orchards and those planted on steep slopes will face greater challenges and soil erosion will become a problem. Other challenges include poor variety selections, variable and often poor management practices and variable quality. Advantag-

es were the amount of research being undertaken, government support and a proven record of perseverance in achieving goals.

To put macadamia growing in perspective, production in 2020/21 was 43,400 tonnes of NIS, compared to over I million tonnes of walnuts, 45,000 tonnes of almonds, 1,000 tonnes of pecans, and 500 tonnes of pistachios. However, existing macadamia plantings that are not yet bearing and new plantings will see production of NIS increase rapidly.[14] The Gain Report for 2022 forecast 50,000 tonnes of NIS or 10,000 tonnes of kernel from over 240,000 hectares of orchards.[15]

The global market downturn in 2023 that has resulted in NIS prices falling by up to 70 per cent from 2020 is unlikely to have a major effect in Chi-

175g　淮盐味壳果
Salted Nut in Shell

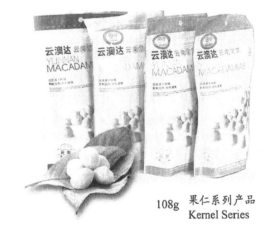

108g　果仁系列产品
Kernel Series

澳洲坚果系列产品
Series Products

礼盒系列产品
Gift Box Series

A range of premium retail packs, 2018. Credit: I. McConachie

Large-capacity, sound macadamia nursery, 2018. Credit: I. McConachie.

na. Both the Yunnan Government and Central Government have, in effect, made a commitment to growers by encouraging them to plant macadamias. It is probable that this will continue in some way by guaranteeing fair prices, particularly to formerly impoverished farmers.

The World Macadamia Organisation, a voluntary international body initially funded by eight countries, determined in 2022 that expansion of market demand should be focused on more promotion in China.

Philip Lee, a global authority from South Africa with a slightly different perspective, reported that in 2000 China had 200,000 trees and a decade later, 2.2 million trees. He observed that in 2010, the First Five-Year Plan to include macadamias aimed to plant 20 million trees by 2015, which was achieved. The next Five-Year Plan was to plant another 60 million trees, at the rate of 12 million per year, and it is claimed that this was also achieved. By 2022, he reported 84 million trees compared to the Republic of South Africa at 18 million and Australia at 11

million. An analysis of the Chinese industry listed that, in 2021, their trees over six years old averaged 1.4 kilograms of dry-in-shell (DIS), although this is likely to be misleading. The report concluded that yield projections are unlikely be realised, as up to 50 per cent of trees will not produce to commercial levels, and suggests that by 2030 the crop is expected to be in the range of 160,000 to 190,000 tonnes of DIS.[16]

Globally, no country has achieved their projections, due to many factors including crop variability between and within orchards, environmental adaptivity, climatic factors such as drought, frost, floods and high temperatures, together with losses due to theft, pests and diseases.

China with its resources has undertaken much research in plant breeding, horticultural practices, pests and diseases, quality and processing, which has led to many advances in machinery used to grade and process. The International Macadamia Symposium in September 2023 listed 309,000 hectares over five provinces producing 42,345 tonnes in

2022 and 85,400 tonnes in 2023.[17]

Cheng Lu, Deputy General Manager, CHK Trading, at the 2024 INC Congress estimated the 2024 crop at 69,000 tonnes of NIS at 3.5 per cent moisture content, where dry weather in Yunnan has reduced crop to 57,000 tonnes of NIS. The newer growing areas of Guangxi and Guangdong are expected to produce, respectively, 10,000 tonnes and 2,000 tonnes. Ongoing demand is mainly from younger consumers, with many new products sold largely through online retailers, which suggests market expansion if prices remain stable.[18]

China in 2023 is the largest grower in the world in terms of tree numbers, and before 2030 it should become the largest producer. The country may well go from being an importer of macadamias to being an exporter, and it is likely to continue to exert a strong influence on global markets and prices. China's research in many fields, its ability to set ambitious goals and largely achieve them, its persistence, commitment to reducing poverty and capacity to overcome obstacles, together with its potential domestic market, are likely to make this nation the dominant macadamia presence in the world. China's achievements in creating a massive new industry does its people much credit and they have progressed from a vision, to a plan, to establishment and progressively to being a mature producer of NIS. China's scale is likely to make it the dominating influence in the global industry.

芒市澳洲坚果加工厂
Mangshi Macadamia Processing Factory

Modern, high-standard processing factory, Mangshi Macadamia Processing Factory, 2018. Credit: Mangshi Yunnan

Myanmar

From 1982, a small number of seedling trees were planted in trials in Myanmar (Burma), partly to seek an alternative to growing opium poppies. Influenced by Thailand, Vietnam and China, a few smallholder growers and entrepreneurs demonstrated that macadamias were able to thrive and produce a crop but mainly in the north. Elevated districts in the Mandalay region and Shan State have satisfactory soils and the required subtropical environment, but the typical annual rainfall of 2.5 metres mostly falls from June to September, so irrigation is required to produce a reliable crop. Nearly all trees were seedlings and some were later top-worked with grafted varieties.

During the 1990s, small-scale farmers, larger-scale farmers and entrepreneurs continued to experiment with macadamias. In 2004, Kim Wilson from Australia undertook a feasibility study and by 2013 was providing nursery and orchard horticultural advice. The Karnsund family from Sweden planted a large commercial orchard in the early 2000s at Hsipaw, near Mandalay, that started producing in 2009. They estimated that there were then up to 3,400 hectares of orchards, most not yet bearing, and began forming an association of growers and negotiating to sell their NIS to a processor in China. In addition to the many small-scale farmers, a timber consortium was planning to plant 400 hectares.[19]

In 2017, a United States Agency for International Development (USAID) project provided the *Myanmar Macadamia Industry Guide,* a booklet based on Australian standards.[20] There were between seventy and one hundred commercial growers in 2020.[21] A government report of 2018 listed 200 hectares in the Shan region, 800 hectares in the Kachin region and 400 hectares in Mandalay.[22]

A major Chinese programme for small-scale farmers was the Lancang–Mekong Cooperation (LMC), a Belt and Road initiative established in 2016 with a special fund to support development in Cambodia, Laos, Myanmar, Vietnam and Thailand. Its goals were to help countries along the Lancang River and Mekong River with agriculture. In the case of macadamias in Myanmar, the goal was to improve production, quality and export capacity, and to reduce poverty of small-scale farmers. In 2022, with a budget of $US 384,000, LMC provided 118 farmers with a total of 31,000 trees and 206 small dehusking plants in the states of Shan and Kachin and the Mandalay region.

Typical larger orchard in Myanmar, 2011. Credit: Thomas Karnsund.

The Golden Triangle Macadamia Co. Ltd in Mandalay processes and markets three styles of roasted kernels.

Lawrie Raymond, an Australian processor in the 1990s, is now based in Myanmar and seeking to assist in the development of their industry. He advised in 2018 that development was slow but steady, although many seedling trees still had not been top-worked, and tree health and productivity was limited where there was no irrigation.[23]

By 2023, there were 5,000 hectares of macadamias planted but at early stages of bearing. There is a need to top-work seedling orchards and plant only grafted trees. Government assistance, LMC funding and the initiatives of large growers have all resulted in the establishment of a small, reasonably sound industry. While horticultural and quality standards are variable, Myanmar is likely to become a successful part of the South East Asian industry.[24]

Thailand

Thailand has had a colourful and distinctive odyssey with macadamias. The industry has not expanded there to the same extent as in other countries, but macadamias are highly regarded.

Thailand was introduced to macadamias from 1953 through a United Nations mission when plantings were made close to Bangkok, at the Fang Horticultural Experiment Station, and then later to the east and north of the country. These government-supported trials demonstrated that the subtropical north was suited to the cultivation of macadamias. The Fang Station continued to assess them through the 1960s, and from 1968 to 1988 scion wood was provided from Hawaii, with Australian horticulturist Tim Trochoulias providing sound advice through a joint project with the Australian Government. The Thai Royal family has played a major role is the development of the country's industry. The main growing areas were in the north, in the provinces of Chiang Mai, Chiang Rai and Phetchabun. In 1985, a selection of ten cultivars were planted in fifteen locations.

By 1992, there were reported to be 100 hectares planted by small-scale and larger growers, comprising 20,000 to 30,000 trees.[25] Terry Burnett from Australia installed a modern processing plant in 2005, and advised that there were about 6,000 hectares planted and that government research stations were providing extension services, supporting both small and moderate-sized orchards to expand steadily, and that NIS was supplied to five or six processing plants. By 2011, the Chiang Mai district had 40,000 trees, with processor-retailer Mountain Fresh marketing macadamias as part of its range of products.

During the next decade, plantings increased, more processors were established and NIS production expanded. Leesoaw Macadamia in Chiang Rai grows, processes, value adds, sells and is a tourist attraction, while Nannazz Fruits in Chiang Mai is a local outlet for macadamia sales.[26]

The Doi Tung Development Project is a heartwarming humanitarian initiative that has led the industry and added to the image of macadamias in Thailand. Northern hill tribes living in the so-called Golden Triangle were of several ethnic minority origins, including soldiers of Chiang Kai-shek who fled to Thailand at the end of China's civil war in 1949. The tribes or groups were living in extreme poverty in Thailand and facing ongoing threats to survival.[27]

The late Her Royal Highness Princess Srinagarindra, known as the Princess Mother (1900–1995), was dedicated to improving the lives of these hill tribes whose forests had been razed and who relied on income from opium poppies. Among many other initiatives in health and education, she established the Mae Fah Luang Foundation, which was the name that she herself had been given by the village people and meant 'Royal Mother from the Sky', as the only early access was by helicopter.[28] The Foundation runs the Doi Tung Development Project that covers twenty-nine villages in the Chiang Rai Province. The Foundation established a university and led reforestation projects, including the planting of macadamias. In the 1980s her daughter Queen Sirikit, known as the Queen Mother, became the macadamia patron. She funded and supported research, including sending many students to study in Australia and Hawaii, and took actions to promote a viable macadamia industry. In 2008, the trees that she supported produced 20 tonnes of NIS. The current modest but successful Thai industry would only be on a cottage scale without her vision and compassion for her people.[29] After her death in 1995, her daughter Royal Highness Princess Maha Chakri Sirindhorn assumed patronage of the Mae Fah Luang Foundation.

Local sales, tourism and exports to Europe, South Korea and China are the main outlets for the crop. Production figures are difficult to obtain, but the industry has grown considerably during the first two decades of the twenty-first century. Although plantings have continued and standards are generally high, it does not seem likely that Thailand will become a major global producer.

Vietnam

Possibly the oldest macadamia in Vietnam is a healthy tree planted in the 1960s and still growing in the Cadasa Resort gardens in Lam Dong Province.[30] Other macadamia trees were introduced to government experimental stations, botanic gardens and private gardens before or at that time, but World War II, the Indo-China War and the Vietnam War from 1955 to 1975 were not conducive to planting macadamia. It is likely that American servicemen brought with them retail packs of nuts from Hawaii, and the Vietnamese may have been aware of the small number of plantings by China in the 1970s and 1980s, just off the Vietnamese coast on the island of Hainan and in the southern provinces.

The first recorded planting of macadamias was in 1994, of twenty trees from unselected seed. They were planted by the Research Centre for Forest Tree Improvement at the Ba Vi Field Station west of Hanoi, at about 50 metres elevation. The trees survived and fruited but were growing in an unsuitable tropical environment. Professor Hoe Hoang visited Australia in 1998 for a study tour and in 2000 Deputy Prime Minister Nguyen Cong Tan came to Australia also, after which he recommended and championed the establishment of an industry in Vietnam. In 2002, the trees planted in 1994 were assessed following a visit to Australia to study the crop by Mr Nguyen Van Dang, the Vice Minister and Mr Nguyen Ngoc Binh, the Director General of Agriculture and Rural Development. They joked that while there were only twenty trees at that time, one day there would be 20 million, and this prophecy may well come true. From this visit, the Vietnamese and Australian governments, through the Australian Centre for International Agricultural Research (ACIAR), supported a programme to help with the sustainable development of a macadamia industry, primarily to assist small poorer farmers to replace opium poppies and to provide environmental and social benefits.[31]

A three-year project commenced in 2006, trial sites were established and quantities of seed nuts were imported as well as scion wood from a range of Australian, Chinese and Hawaiian cultivars. Small family farmers were selected for planting trials, workshops were held and horticultural information was obtained to commence and sustain a commercial industry in Vietnam. The project was initiated by the Ministry of Agriculture and Rural Development and led by Professor Hoe Hoang with Kim Wilson and Martin Novak, and later Kim Jones from Australia provided expert advice. From 2002 to 2012, they planted selected seed nuts and imported scion wood from China, Thailand, Hawaii and Australia as part of a sound feasibility study.[32]

Initially the goal was to encourage small family farmers to plant macadamias in conjunction with coffee and rubber. Elevations of 500 to 700 metres were recommended in the Northwest and Central Highlands regions. Overall, as would be expected in small-scale trials, yields and quality were variable. Most families or smallholders were skilled intuitive farmers who, once they believed in an economic future for macadamias, would become both persistent and enthusiastic. They were led by a range of government, private and university horticulturists, foresters and researchers. Gradually, a sound knowledge of soils, environment, varieties, yields, economics and tree culture was obtained and with this came the confidence that Vietnam could develop a viable industry, meeting both socioeconomic and long-term strategic government goals. No significant pests or diseases were encountered in the early trials.[33]

As an example, Pham Huu Tu, a farmer from the northern Thanh Hoa Province was given a document in 2006 suggesting he grow macadamias instead of his acacias. He ended up with 500 trees which produced 1.7 tonnes of NIS, earning his family $US 6,800 in 2014.

Interest in macadamia growing increased with larger-scale, better-funded farmers, investors from other countries and small family farmers. Orchards from less than one hectare to over 100 hectares were being established. So many people have contributed to the development of macadamias in Vietnam. Professor Hoe Hoang, Professor Nguyen Lan Hung, General Tran Dac Quang and Bui Xuan Trinh from the Government Offices were some of the advisors and supporters of this crop. A major factor was the involvement of the Lien Viet Post Bank, co-ordinated by their Vice Chairman Nguyen Duc Huong together with the State Bank of Vietnam and Him Lam J. Stock Company, all of whom liaised with various government offices, small and large potential growers, and were able to persuade the banks to commit large sums to small and large growers. $US 950,000 was reported as having been budgeted to finance macadamias from 2015 to 2020. Professor Hoe Hoang was an enthusiast, researcher, promoter, a leader in the formation of an industry association and remains as a guiding mentor.

By 2012, there were a number of government-sponsored projects in the Central Highlands and the Northwest regions, totalling over 50 hectares, and another 200 hectares planted on larger farms. Regular workshops had been held to inform and assist potential growers. From 2012 onwards, nurseries expanded, producing sound grafted varieties with known performance. A State Decree in 2013 pledged support for larger commercial orchards. By 2015, the Macca Corporation at Dien Bien was assisting smaller farmers; IDT International Company was also planting up to 400 hectares in the northwest of Dien Bien Province plus Vina Macca Corporation at Dak Lak in the Central Highlands. Macadamias were being established in the Kon Tum, Gia Lai, Dak Lak, Dalk Nong, Krong Nang and Lam Dong provinces, totalling over 1,500 hectares. Some domestic retail products were starting to become available, as well as cosmetic products made from macadamia oil.[34]

An example of an older orchard not typical of current plantings. Bui Quoc Hoan is on the left. Credit: B. Kaddatz

At the International Macadamia Conference in 2013, Kim Wilson sounded a note of caution with his observations that:

- there is limited land available;

- smallholders needed immediate income;

- soils in the Northern provinces are often poor but the Central Highlands have more suitable soils.

Possibly the interest of larger farmers and businesses, together with the enthusiasm and knowledge of many people, had driven development much faster than expected. The separate growth of an industry in China created a belief in Vietnam that there was a strong future demand for the purchase of both NIS and kernel. In addition, there was a potentially expanding retail demand within Vietnam. As with any new long-term crop, there are uncertainties due to the vagaries of nature, market cycles, variable horticultural standards, cli-mate change and problems in achieving sustainable yields of quality kernels. However, development of the industry is controlled or influenced by the central and local governments in the provinces and this has been a driving force not present in most emerging economies. As well, smaller family farms are generally larger than equivalent farms in many African countries, and funding and advice are available, so there is more confidence that yields and sound quality are obtainable. The high demand and prices from 2010 to 2020 were helpful, and the Australian Government, through ongoing consultancy services from Martin Novak, Kim Wilson and Kim Jones, continued to provide advice and support to the industry.

In response to the rapid expansion of the new industry, in April 2015 the Vietnamese Government proposed a cap on growth of the industry in order to prepare a Master Plan, to allow further assess-

Brice Kaddatz from Australia, inspecting a large nursery in Vietnam. Credit: Bui Quoc Hoan.

ment of environmental suitability, cultivars, knowledge of pests and diseases and to establish infrastructure. They considered the very rapid expansion as possibly leading to uncontrolled development, with a risk of many plantings not being viable. In summary, the cap would have allowed a maximum of 10,000 hectares to be planted until 2020, centred around suitable climates such as Dak Lak and Son La. Ten cultivars only were recommended; the cap encouraged conjoint planting with existing tea and coffee.[35] Due to industry pressure, these restrictions were withdrawn in 2016.

In 2016, the Vietnam Macadamia Association was formed to represent, co-ordinate and guide the new industry, which at that time consisted of about 1,000 farms, growing 1,400,000 trees on 10,000 hectares, of which at least 30 per cent were seedlings.[36]

At the Eighth International Macadamia Symposium, in China, in 2018, Professor Hoe Hoang strongly promoted the potential of the industry and

the availability of land, but he listed problems with seedling trees, poorly performing varieties, the need for industry infrastructure and varying productivity and quality.

There are three main processors, including the innovative Bui Quoc Hoan whose family company, Thuan Bao Khang Green Food, grows macadamias and has developed an innovative process to allow the shell to be opened and the kernel removed by hand. Initially supplementing the availability of NIS, he imported high-quality Australian product then sold this product in retail packs that were soon in high demand. He believes that future demand will be mainly for kernel. There were several other innovative products, including hair care, healing oil and skin rejuvenation.[37] Much of the production and probably kernel from small householder plantings that has reduced quality is sold in the domestic market, often as gifts, while premium kernel is sold to South Korea and Japan.

Vietnamese industry leaders meeting Australians Brice Kaddatz (left) and Jolyon Burnett (centre). Credit: unknown.

Production has increased from 262 tonnes of NIS in 2015 to 6,570 tonnes in 2020 and 8,500 tonnes in 2022. The Lien Viet Post Bank has continued to support mainly small householder growers by advancing loans until production is achieved. The Deputy Prime Minister, on behalf of the Ministry of Agriculture and Rural Development (MARD), released a detailed, optimistic plan to support a sustainable expansion to 130,000 hectares with the goal of producing up to 130,000 tonnes of NIS by 2030. Most of the plantings will be in the Central Highlands and some in the Northwest region. In 2020, there were sixty-five facilities to process the crop, with most being to remove the husk, sort, dry and deliver. These included full processors to the kernel stage, who have capacities between 10 and 1,000 tonnes per annum. Their MARD plan suggests that these should increase to 300 or 400 facilities, which would provide a service mainly to growers by co-ordinating deliveries, dehusking and drying, similar to Kenya.[38]

Philip Lee's 2021 report lists Vietnam as the eleventh largest macadamia-growing country in the world, with 700,000 trees under six years old and 800,000 bearing trees.[39] The global market decline of 2023 may reduce confidence for growth in new plantings, but existing early bearing and non-bearing trees have the capacity to at least double production.

Bui Quoc Hoan summarised the industry in 2022 by suggesting that although published statistics are often exaggerated, the Vietnam Macadamia Association forecast for 8,800 tonnes of NIS in 2023 and 20,000 tonnes in 2027 is possible, and that increasing amounts are being processed within the country with less being sold to China.[40] In May 2023, the Dien Bien local government approved investment to significantly increase the 350 hectares currently planted in the Northern province through thirteen separate macadamia projects.[41]

Being a new crop for Vietnam, the fledgling macadamia industry still faces the following challenges:

- understanding the environmental conditions suitable for macadamias to determine appropriate growing areas;
- determining the best cultivars for each growing area;
- suitable tree husbandry technique for mechanised and non-mechanised farms;
- economic viability of the various products and styles;
- meeting global quality standards;
- crop protection knowledge as different pests and diseases emerge;
- rationalising the current large number of crop handlers and processors;
- developing sound storage and market procedures;
- retaining confidence during the price collapse of 2022.

The crop in 2023 was 9,000 tonnes of NIS and a forecast of 10,000 tonnes was made in 2024.[42]

Given that the industry is largely controlled and supported by the government and its agencies, and finance is available, the proven adaptability, enthusiasm and skills of farmers combined with innovative processors and value-adders give assurance that Vietnam is set to become a major global producer.

Grafting or top-working a mature tree, to change to a more suitable variety. Credit: Bai Viet Duy Thanh.

Asia: industries in development

Indonesia

Macadamias were being assessed in 1982 by the Government Estate in East Java.[43] Plantings of commercial orchards were proposed in 2016, and orchards are reported at Solok in West Sumatra, Rembang in Central Java and Bondowoso in East Java.[44] Kim Wilson from Australia visited Indonesia and has supplied 10 tonnes of H2 seed nuts for nursery purposes but he is unaware if these have been grafted to produce commercial varieties.[45]

Japan

During the 1980s, small trial plantings were made at Okinawa and the southern part of Kyushu Island. Horticultural knowledge for macadamias was lacking and the trees did not perform well.[46] Some plants were established as Kyushu Island and Yonemoto sold 1,000 trees as ornamentals. Kim Wilson has supplied seed nuts.

Meiji Group have been long-term leaders in marketing macadamias in Japan. In 1976, they launched the first commercial chocolate macadamias in Japan. Credit: I. McConachie.

Laos

The Government of Lao People's Democratic Republic, like the governments of neighbouring Vietnam, Thailand and Myanmar, was seeking opportunities for farmers to alleviate poverty. Mr Lenbrook planted some macadamias in conjunction with coffee during the 1990s, at the Government Research Station, resulting in small experimental plantings of macadamias which provided information. Martin Novak, a horticultural consultant from Australia, first visited Laos and Vietnam in 2008.

The Australian Embassy in Vientiane has a small-scale grant programme called the Direct Aid Program (DAP), and a project has been developed to further assess potential to diversify coffee plantations by interplanting with macadamias.[47] Mike Askham, an Australian horticulturist living part-time in northern Laos in Itov, was passionate about assisting farmers with new crops. He brought seed nuts then scion wood from Australia and, on a small scale, set out to demonstrate their adaptability and how to grow them. He selected the Bolaven Plateau to plant small orchards. By 2008, a large corporation had planted an orchard at Paksong Tablelands at 1,200 metres elevation and also on the Bolaven Plateau.[48]

Plantings have expanded but information on numbers and production are not available.

Malaysia

Boo Yong Sea Estate has planted 12,000 trees on 400 acres in the State of Johor.[49]

Nepal

Macadamias were first introduced in the 1870s, but trial plantings did not commence until the 1970s. They were being considered as a commercial crop from about 2000, suitable for growing in the Middle Hills Districts, and were being reassessed in the high-demand period of 2017.[50] A positive feature is that they are not attractive to monkeys, which are a pest for many other crops. There is a small orchard in the Syangja District, with the nuts being processed in Kathmandu and sold locally.[51] Kim Wilson provided consultancy services in 2017 and inspected earlier trial plantings but there are no reports of expansion.

Philippines

There are reports of macadamias being introduced to the Philippines in the late nineteenth century and individual trees can be found growing at the Eden

Nature Park in Davao, Mindanao, and at residences. In 2017, the BPH – Baguio Experimental Station – had eight fifteen-year-old trees fruiting and being used for trials. There are reported plantings in the northern and central regions.[52]

In 2004 an Australian, Rowan Patterson, and colleague Kevin Heckleman, who both lived in the Philippines, enthusiastically set out to assess and establish an industry. They grew trees in Rowan's garden from Australian and Hawaiian varieties, which grew vigorously and cropped. They assessed small plantings at Negros, Panay and Mindanao Islands and liaised with the local Experimental Station. They imported a variety from California named Cannonball and determined that an elevation of 800 metres was required to induce flowering.[53]

Jorge Disuanco founded MacNut Philippines Inc. in 2010 and established a nursery, claiming to have 600,000 grafted seedlings, which he promoted to small and moderate-scale farmers.[54] There have been few reports on the success of these initiatives and it is believed that no significant industry has resulted.

Sri Lanka

Walter Hill provided macadamias in about 1870 but no records of their survival and assessment are available. An orchard of 2,000 trees and a nursery were established by 2002.[55] Kim Wilson in 2017 visited the country and led workshops, but he is not aware of increased plantings.[56]

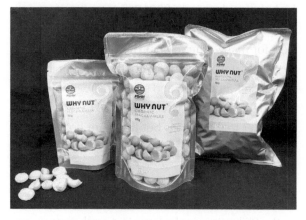

WhyNut product range sample. Credit: Macadamia Holdings NZ.

Pacific: industry leader

New Zealand

New Zealand growers have been a small but enthusiastic part of the global industry since the 1980s. Being a close neighbour of Australia, New Zealand sometimes shared botanists from the nineteenth century who were aware of native Australian flora and fauna. Several plants and animals were introduced, including possums which are a major and hated pest. The economy is largely dependent on exporting food crops, sheep and wool. In the North Island, fruit crops like avocado and kiwifruit, apples, dairy products, wine and meat are produced, and this is also where macadamias are grown. The South Island has a temperate climate and grows several of the above but there are no commercial macadamias.

Based on fossilised pollen and leaf records in New Zealand, in the Late Cretaceous period from 100 to 66 million years ago, there were many Proteaceae genera, including ancestors of the macadamia. Now only two Proteaceae species remain: *Knightia excelsa*, known as rewarewa, which has some morphological similarities to macadamia but no edible nuts, and the shrubby *Toronia toru* or toru.[57] A South American relative of the macadamia, *Gevuina avellana* or Chilean hazelnut, has been trial planted in New Zealand as suitable for growing in a temperate climate.

Macadamias were first introduced about 1875, possibly by Walter Hill from Brisbane, and a tree planted at Takapuna in north Auckland in 1877 was still healthy at least 130 years later. Small numbers of trees were planted in the 1920s and 1930s. Mr Jolly at Kerikeri planted the first orchard in 1932.[58] By 1986, there were 200 small cottage growers with a total of about 30,000 trees, of which only 1,000 were bearing.[59] New Zealand farmers are well recognised for their innovation, high standards, and ability to persevere to develop mainly export-oriented industries.

The climate in some districts in the North Island is barely subtropical, and where soils are adequate the macadamia trees are generally healthy. Disadvantages are persistent strong winds and frost, but the main problem is insufficient temperature or heat units to induce flowering in most regions and in most varieties. Until newer varieties were introduced, only the *M. tetraphylla* hybrid variety Beaumont flowered and fruited reliably, but it had disadvantages including that its nuts had to be hand-picked from the tree. Most Hawaiian and Australian varieties produced few flowers and nuts. It is possible that the Beaumont (HAES 695) variety in New Zealand is not identical to the same variety planted widely in South Africa.[60]

Despite these limitations there had been both enthusiasm and support, with the industries in Hawaii and Australia providing a vision to emulate. Intending growers, researchers and horticulturists often visited Australia and Hawaii to study the macadamia.

W. Fletcher from the Department of Agriculture and Fisheries in 1973 published the first guide to growing them commercially.[61] Dick Endt, a resourceful experimenter of a wide range of fruit trees,

formed the New Zealand Tree Nut Association in 1977 and his macadamias were closely studied by intending growers. Dr J. Taylor from the Department of Scientific and Industrial Research (DSIR) in 1980 unsuccessfully sought cold-tolerant germplasm from Australia and recommended caution in selecting land suited to the crop. Another publication on varieties and culture was published in 1981 by the Ministry of Agriculture and Fisheries and later revised.[62]

Orchards have been established in parts of the East Coast, Bay of Plenty, Waikato, Auckland, Coromandel and Northland. Tourism is an important industry with road-side stalls, tourist outlets and shops in cities. There are many attractive retail packs, and promotion of the New Zealand brand image accounts for many retail sales. Value-adding allows growers a way to expand their margin as most produce only small quantities of nuts. There are orchards ranging from a few hundred trees to what is considered an economic size of 1,500 trees or more.

John Brokx mowing at the WhyNut orchard, with his favourite Ferrari tractor. Credit: Macadamia Holdings NZ

Industry leaders from the 1980s were Ian Gordon and Neil Whitehead, who developed Macadamia Enterprises at South Head west of Auckland, selected local varieties and promoted the crop. Ian's son Nickoli continued his father's passion until 2020. Donald Boyes-Barnes had a nursery and provided advice to grower groups up to about 2000.[63]

Virginia Warren and husband Charles bought a steep, small property near Pukekohe, south of Auckland, in 1975, which they called 'Chineka'. They gradually planted 440 macadamias of several varieties and Virginia became a pioneer, marketer and leader of the industry. At 'Chineka' she built a very simple processing plant, value-added the kernels and sold them in a range of attractive retail packs. She united the industry and was founding President of the New Zealand Macadamia Society.[64] Like many others who marketed their own crop as retail products, her successful, innovative business was labour intensive.

The industry sought varieties suited to the environment and most orchards until recent years mainly had varieties such as Beaumont that had to be hand-picked from the tree. The industry, and to some extent the New Zealand Macadamia Society, is divided into generally older orchards having varieties described as 'pickers' and mainly newer orchards whose nuts are harvested from the ground and therefore described as 'droppers'. Growers with 'pickers' have a labour-intensive crop but believe that its quality is higher as there is no risk of deterioration from lying on the ground.

A long-term enthusiast and industry leader is Vanessa Hayes, who is changing and challenging the industry. Vanessa is energetic, passionate and strongly focused on the development of a viable industry. She with her partner Rod and children – Raelyn, Angelina and Walter – have been leading the industry over almost thirty years. A Maori woman descended from a Princess with land gifted from her Elders at Torere near Opotiki, on the eastern Bay of Plenty, she has imported a range of varieties, grafted, planted and assessed them, promoted them and now markets the end product. She has selected nine varieties, all 'droppers' which vary from early to late maturing. They have been selected based on yield, whole kernels and resistance to insect attack, as she requires them to be grown organically.[65] Part of her vision is to have macadamias planted to support Maori families on their land, and she is working through many obstacles to achieve this. In 2023, she has a 50,000-tree nursery, 1,500 mature orchard trees which were largely established as a variety trial block, and another 1,500 recently planted trees. She holds regular workshops and field days, and assists with horticultural and organic farming procedures. Vanessa regularly has test markets at Gisborne and

Vanessa Hayes and Rod Husband in Torere Nursery. Credit: Torere Macadamias, NZ.

purchases NIS which is processed under the Torere brand. She led the Society's 'New Zealand Industry 10-Year Plan, 2019 to 2029', whose three stages detailed goals for 1,000 hectares of organic orchards. A co-operative to process and market macadamias is part of the plan. All or almost all plantings since 2018 have been varieties developed by Vanessa.[66] The vision of the growers with 'dropper' varieties is that suitable varieties will provide an opportunity for the industry to expand and be viable.

While some older or 'picker' orchards have been abandoned, Vanessa believes that the industry will expand with proven 'dropper' varieties based on the research that she and others have undertaken.

As a contrast, at Cathedral Cove Macadamias north of Auckland, Brian and Sue Pilkington successfully grew, processed and marketed most attractive products. Now owned by Jillian and Doug Johnston, the enterprise remains as a successful, organic-certified orchard having varieties requiring hand-picking from the trees. Sue Vause and John Brokx, through their WhyNut brand, are also organic-certified grow-

ers and processors, who market a range of attractive retail packs.

The New Zealand Macadamia Society has represented the diverse goals of the industry, provided a meeting place, co-ordinated and fostered research and provided information and leadership. The global increase in NIS prices from 2010 stimulated plantings, but the downturn in demand and price from 2020 may weaken this.

Apart from a small number of what could be described as commercial orchards by global standards, the industry has been essentially cottage based. Most commercial orchards and processors have not been successful as yields are not economic, and they do not have the economies of scale to warrant new technologies and machinery. Simply, the growing environment is arguably marginal for macadamias, as trees grow satisfactorily but flowering is variable due to insufficient heat, resulting in low crops by global standards. In 2012, the country had about 12,000 trees with many bearing trees averaging only 2 kilograms of NIS per tree. The total

Th Kite gift range from Torere featured in TVNZ's *Country Calendar*, in one of the highest rating episodes of the decade. Credit: Torere Macadamias, NZ.

crop then was 180 tonnes of NIS, but was incorrectly projected to reach 450 tonnes in 2022. In 2021, there were eight small processors.[67] By 2023, the industry had grown only slightly, with a crop of about 300 tonnes of NIS.

In 1980, over 90 per cent of all trees were Beaumont but by 2020, this had reduced to 31 per cent. Now another 31 per cent are of the A series from Australia, and at least 30 per cent of New Zealand and specifically Torere selections. Pests such as green vegetable bug and guava moth are increasingly evident and will provide a challenge. In the global context, the industry is small but distinguished by the fact that many orchards are organic. Although this is a marketing benefit, it makes for increasing difficulty when the aim is to produce consistently sound, quality kernel. Low yields, the cost of labour and the cost to process and market make it difficult to be viable. The innovative approach by growers to add value and market their NIS and kernel remains. Macadamia sales through supermarkets and similar outlets are ongoing but often use imported kernel. The industry survives and to an extent prospers through high-priced, well-presented retail packs, largely through tourist and high-standard local markets as well as in supermarkets. However, macadamias are likely to remain a niche local product.

Pacific: industries in development

Fiji

Macadamia plants were introduced between 1880 and 1886, with a small orchard established in the late 1960s. Grafting, wind losses and tree maintenance were problems. A 1970 report advised of current plantings that totalled 60 hectares, and challenges including regular cyclones and poor yields.[68] Despite a visit to Australian orchards from the President in 1978, the industry did not develop and, in the 2020s, there is no commercial industry.

Papua New Guinea

Trial plantings were established in the highlands near Lae in the 1970s, but when they didn't crop in the second year they were abandoned.[69] Peter Shearer, who planted a large orchard on North Queensland's Atherton Tablelands some forty years ago, moved to New Guinea and is assessing suitability.[70]

Vanuatu

B. Spooner and T. Hannan in 1985 planted 5,000 macadamias in an investment project. Two rows were planted close together and with opposite limbs grafted together (inarching) to strengthen the trees against cyclones. Later they were neglected, still suffered from wind damage and were abandoned.[71]

Part 5

MAC FACTS
AND
THE FUTURE

Chapter 18

MAC FACTS

[A macadamia seed] is a baby plant in a box with its lunch.
(Fiona McMillan-Webster, 2022)[1]

This chapter is a mix of myths, trivia, anecdotes, possibly a few wise observations and some facts. The following, while in part repeating information scattered throughout this book may also provide a chuckle, new information or may surprise you.

The tree and the nut

- The *Macadamia genus* and species evolved and differentiated from other Proteaceae at least 60 million years ago (mya), its ancestors evolved about 28 mya and the four species that we know today, about 7 mya.[2] Most macadamias and closely related species have seeds that are inedible for humans, but two of the true macadamias, *M. integrifolia* and *M. tetraphylla*, have fruit or seeds that are outstandingly edible. Truly a gift from Mother Nature!

- We must remember that Mother Nature, whose bounty provided us with the macadamia, had a biological goal of evolving a plant whose seed could perpetuate itself. Nature did not consider the nut as a future food, but instead provided it with a husk that encased it until it fell on the ground, a hard protective shell that moisture, heat and enzymes would, in time, split, and the physiological processes to produce a new plant. This goal is very different from what humankind is seeking, so processing the nuts must stop the natural reproductive progression, to make the edible kernel stable and attractive.

- The macadamia tree is a subtropical rainforest evergreen, typically growing to ten metres or more. The commercial industry is almost entirely based on grafted or vegetatively propagated trees, so that their characteristics are known and consistent. Most commercial trees are vegetatively propagated by germinating a nut (seed), growing it as a seedling for up to twelve months and then grafting a scion from the desired parent variety. Usually, a limb from the donor known variety tree is cinctured or girdled for several weeks, to allow an accumulation of non-structural carbohydrate (energy) which sustains the graft until it forms a union with the seedling rootstock. Most varieties will commence to crop in about their fourth year. Most orchards reach maturity at ten to twelve years, depending on tree spacing. Early orchards were based on seedling trees and were very variable, had low kernel recovery and usually produced small crops. The two edible species were often grown together but had different sugar levels, which affected roasting. Of the nut-in-husk (NIH) on the tree, only about 17 per cent is edible kernel.

- In the 1960s, the *Fruit and Vegetable News* magazine asked for Australian examples of very large macadamia trees. Many were submitted, with the largest tree being situated below Mount Tamborine (Qld) and reported as 13 metres high and having a canopy diameter of 17.5 metres. The Tom Petrie Tree, planted at Murrumba at Petrie (Qld) in 1865, is even larger. In the rainforest, one tree in the Amamoor Valley (Qld) was measured by Dr Jodi Neal as 23 metres high.

- Macadamia trees produce proteoid roots which increase surface area and allow more effective nutrient uptake. Macadamia roots also have significant mycorrhizal fungi association.

- In Hawaii, many of the orchards on the Big Island are grown on pulverised aa lava fields, which are relics of early lava flows. Initially these were high in organic matter, had regular high annual rainfall and to some extent the macadamias could be considered to be growing hydroponically.

- Do koalas eat macadamia leaves? No, but they sometimes sleep in the trees where the limbs' crotch angles make a comfortable bed.

- Today it seems hard to understand that competent botanists for a hundred years recognised only one species, named *M. ternifolia*. In 1956, Lindsay Smith, the Queensland Government Botanist, set the record straight by separating and describing the then three known species. Leaf morphology had been the main criterion in describing the species until then, with nut size, bitterness, leaves to a node and other characteristics considered

A koala who regularly left eucalyptus trees to sleep in a macadamia. Credit: I. McConachie.

variations. In Greek or Latin, *terni-* from *ternifolia* means 'three', yet *M. tetraphylla,* which typically has four leaves to each node, was included in the category as well. In the last twenty years, genetic analysis has become the tool which finally determines the species, as well as also separating other closely related species, including North Queensland and overseas 'macadamias'.

- Many early reports stated that the true *M. ternifolia* was 'poisonous', apparently because Aboriginal people told settlers not to eat it. Almost certainly the Aborigines were signalling that it was extremely bitter. Research in 1995 listed the lethal dose of both *M. integrifolia* and *M. tetraphylla* as being 60 kilograms of kernel in one day, and about one kilogram for *M. ternifolia.* Yet the bitter taste makes it impossible to eat even one gram of *M. ternifolia.*[3]

- Did Australia and the rest of the growing world make a mistake with developing the industry based on the *M. integrifolia* species which includes some hybrids, rather than the *M. tetraphylla?* In the 1970s, consumer panels had a slight preference for the *M. tetraphylla* species, due to some selections being sweeter.[4] It was probably chance that Hawaii discarded *M. tetraphylla,* only because the nuts they received were considered inferior.

- In Australia, some of the settler population living in northern New South Wales and southeast Queensland from about 1880, and for the next sixty years, traditionally planted a macadamia and a mango tree in the backyard of their home. A Queensland Department of Agriculture and Stock report from the 1930s stated that there were at least 30,000 seedling macadamias at homes in Brisbane, as well as many more, further to the south and north.

- Botanically, the macadamia nut is not a nut but is a seed or drupe – as are walnuts, pecans and pine nuts. They are classified as dehiscent, as the seed splits open.[5] The rest of the world, apart from botanists, will continue to call it a nut.

- The macadamia is the only native food plant of Australia that has become an international food. The wild macadamia tree in the rainforest usually produced very few nuts, due to low levels of light, insect damage and competition for moisture and nutrients.

- In early days, removing the outer husk was basic. Sometimes a rock was used to tear the husk, then modified, hand-operated corn-shelling machines, later car or truck tyres and then steel scrolls. Nuts were dried in wooden frames with chicken-wire bases or spread on floors, particularly timber verandas, where there were air gaps. Children were sometimes encouraged, wearing suitable shoes, to walk on the nuts to break up the husk.

- On farm, the removal of the outer husk was difficult when production was small. An innovative approach was to jack up the rear wheel of the family car, fit a metal cage around the tyre and run the car gently in gear as the NIH was fed through. Another approach was to replace the metal blades of a motor mower with heavy plastic blades and repeatedly run the mower over the NIH. Neither of these techniques is recommended!

- Just imagine cracking between six and eight million nuts by hand, then roasting, salting and marketing them from a horse and cart. John Bucknell Waldron did this from the 1930s. He was based in Murwillumbah (NSW) and if he cracked and sorted four nuts per minute, just the cracking and sorting would have taken ten years of his working life.

- A newspaper reported that in the 1930s and 1940s most homes were built on timber stumps, so children collected the nuts and cracked them on indentations in the concrete floor of the downstairs laundry, using a hammer or with the aid of a vice.[6] This gave rise to the saying, 'You can tell a macadamia enthusiast by his (or her) satisfied smile and bruised fingers'.

- In Australia in the 1950s and 1960s, the Victa lawn motor mower was released, revolutionised

Girl with satisfied smile savouring a macadamia. Credit: I. McConachie, 2002.

the maintenance of household lawns and sold about 140,000 units a year. But the early base plates models had no lip, so when the mower ran over a macadamia, it fired the nut like a bullet. Following broken windows and injury to shins and pets, many backyard macadamias were removed.

- Norm Greber's response to the Queensland Department of Primary Industries (QDPI) horticulturists, when they told him that macadamias could not be grafted, is worth repeating. 'No one told me that', said Norm, as he showed them his successful technique.

- In Hawaii, two schoolboy horticultural students, Ralph Moltzau and Bill Storey, accidentally found the 'secret' of propagating or grafting macadamias some years prior to World War II. They were given a limb that had been partly broken for several weeks, which raised its carbohydrate levels. This, in time, resulted in limbs being cinctured or girdled prior to removal for grafting.

- DDT was developed during World War II to control mosquitoes that carried malaria. After the war, it became a widely used broad-spectrum insecticide. Rachel Carson's book *Silent Spring* (1962) reported massive environmental damage and DDT was subsequently withdrawn. However, before that point, it was seen as a 'miracle' chemical to control most insect problems in many crops, including macadamias. In the 1950s and 1960s, many farmers believed that if low rates of DDT use were effective, then much higher rates would be even more so. The chemical was massively overused. Possibly, with sound procedures, its benefits could have been achieved and its damage to the environment minimised?

- In the 1950s and 1960s, possibly the largest producer in Australia was a 1,000-tree seedling orchard owned by Bernard Mason at Gympie, producing about 10 tonnes of nut-in shell (NIS) a year. During World War II, Bernie had been allocated two Italian internees who, despite being controversial in the community and suspected of having an illegal radio and camera, saved Bernie's daughters when their truck caught fire. Geoff Garratt purchased the orchard in 1976 and the orchard continued until 2020.

- The spread of macadamias to many countries after World War II was enhanced by the belief that they were easy to grow and had little vulnerability to pests. Hawaii's benign climate with few extremes and few pests created the belief that macadamias were likely to be readily adaptable in many countries.

- Possibly an indication of macadamias being an emerging industry is the reluctance to remove trees or varieties that are not performing. In part, this is due to the time it takes for a replacement tree to reach full maturity.

- What makes macadamias special and probably the most highly prized of all nut kernels? There is no doubt that it has a delicate, tantalising flavour but what distinguishes it from all other nuts is its texture or 'crunch'. Peanuts and tree nuts are consumed with a moisture content usually at 4 to 5 per cent. Macadamias at that moisture lack character and should be consumed at 1.5 to 1.8 per cent moisture content.

- The market price of whole macadamia kernels is considerably higher than that of half-kernels. While handling, drying and cracking all contribute to the kernel splitting into its halves, macadamias are dicotyledons, having two seed leaves to each fruit. It took the use of medical X-ray machines in the 1970s to show that, even without any handling damage, the kernel within may consist of either half-kernels or whole kernels.

- The macadamia tree's cropping potential has been assessed as 4 to 5 tonnes of kernel per hectare, which is equivalent to 12 to 15 tonnes of NIS, yet productive orchards up to and in the 2020s are averaging about 25 per cent of that. The potential is debatably possible but it may take fifty years to achieve. Increasing productivity in the short term will largely depend on sound management of all aspects of tree culture and crop handling. In the longer term, there is believed to be potential in selecting rootstocks, and much greater potential in breeding using wild germplasm and selective crosses. Productivity potential is further anticipated in tree architecture and from maximising horizontal growing fruiting limbs, as well as from more efficient cross-pollination, use of productivity enhancing additives and in all aspects of tree culture and crop protection.

- Why a reasonably large tree produces so little edible kernel can be explained in part by the kernel's energy-dense composition, containing about 75 per cent oil. Two macadamia kernels contain similar energy to one orange fruit.

- There is considerable variability in the wild trees. Most shells are thick and the percentage of kernel recovery averages 25 per cent. Grafted commercial varieties once averaged about 35 per cent but this has now increased to over 40 per cent. Seedling trees also varied in the time to commence cropping from four to eight years, whereas modern varieties crop from three to six years. There may be a limit to the thinness of shells, due to their proneness to pre-germinate and be attacked by insects. As Norm Greber observed, 'It would seem that macadamias are so nutritious and tasty that no living thing passes a tree without having a nibble of some part of it'.[7]

- Rats (*Rattus rattus*) are a major pest to macadamia farmers. Light-sensitive cameras have shown that they can chew through the shell in eight seconds and use their claws to extract the kernel. Fortunately, after they have worked through a few shells, their teeth become blunt and the rats have to regrow or sharpen them. Feral pigs love macadamias also and an adult can eat up to 2 kilograms of NIS in a night. When caught, their stomachs contain only the kernel, so they must be able to spit out the shell. Both black and white cockatoos similarly love macadamias and their powerful beaks can crack the shells. They quickly learn which varieties have the thinnest shells and often pull off immature nuts for devilment.

- Research has indicated that macadamia flowers may have antimicrobial properties, that the inside of the shell has anti-fungicidal properties which protect the kernel as it is germinating, and that when the kernel starts to germinate it becomes bitter. The bitterness indicates the production of cyanogenic glycosides to deter attack from mice and other small creatures once germination commences.

The hard-working Executive Officer of the Macadamia Conservation Trust, Denise Bond, finding an *M. ternifolia*. Credit: I. McConachie.

- Nai Nai Bird, a Butchulla woman, wrote that lactating Aboriginal women would eat germinating macadamia kernels as the bitter component would stimulate milk production.[8]

- The industry focuses on oxidative rancidity where, due to exposure to oxygen over time, the oil is oxidised and staleness or rancidity occurs. This has a typical 'off' odour and flavour, which is measured by the level of PV (Peroxide Value). However, the industry ignores and has little understanding of hydrolytic rancidity where, due to moisture, temperature and enzymatic activity, there is a slight but measurable increase in free fatty acids. This has the result of flattening the 'freshness' of the kernel, and at a high level gives a slight soapy flavour. Hydrolytic rancidity occurs during storage of NIH, NIS and kernel when it is not fully dried and when cell damage to the kernel occurs.

- A mature macadamia tree produces between 20,000 and 40,000 flower racemes, each containing up to 200 florets. Taking an average of 500,000 flowers per tree, about 50,000 nutlets may be set. In late spring or early summer, over 90 per cent of these nutlets are shed as nature determines how many the tree might be able to carry to maturity. Usually, nutlets that are self-pollinated are shed first. A typical crop carried by a tree to maturity and harvest would be 2,000 nuts or 16 kilograms of NIS. So only about one flower in 200 survives to produce a mature nut. Nature is very generous, and being able to reduce nutlets lost could be a way to increase productivity, provided that the trees' supply of energy is sufficient. There is increasing evidence of the importance of cross-pollination and its ability to increase final yields.

- Maybe in the distant future there will be a smaller macadamia kernel that will be in demand due to more of them fitting in a packet of the same weight and their being more suited to confectionery, such as when chocolate coated? Experiments by Dr Cameron McConchie in cross-pollinating the commercial varieties with *M. jansenii* pollen produced smaller, attractive kernels with no trace of bitterness. The industry plant-breeding programme is also researching this.

- The understanding of and potential for macadamias to sequester carbon and reduce carbon dioxide emissions has expanded, and some growers and processors have begun to describe their operations as being 'carbon neutral'. Macadamias, in Australia, do not currently meet the criteria for carbon credits, but such credits could possibly be accrued from increases in soil carbon. This is a complex subject.

Processing and marketing

- An early goal for the industry was to produce macadamias with a shell so thin that they could be cracked in the mouth. This resulted in broken teeth and was not practical. This author's great-uncle was a dentist from the 1920s, when macadamias indirectly added to his income. He told of patients having attempted to crack macadamias with their teeth!

- Shells have a structural strength five times higher than other tree nuts and require considerable loads to fracture. Their strength in tension reaches or exceeds most ceramics. While macadamia shell structure varies, from the outer surface inwards their interlocking architecture offers a model for engineering such as bridge building.

- Processing results in the production of twice as much shell as kernel. In some processing plants, the shell is burnt for heat energy used to dry NIS and sometimes it is used to produce electric energy. Its calorific energy is almost as high as that of coal. It has been used as a ground cover, sometimes decoratively; as a mulch; in BBQs where its heat has often warped the plate and burnt the food being cooked; in making building boards; in making activated carbon and now as bio-char. It contains a small amount of oil but the extraction of this is uneconomic.

- An obstacle to a commercial industry was the difficulty in cracking or shelling to remove the kernel. Marketing NIS in retail packs was restricted by the difficulty and slowness of hand-cracking them. Particularly in Australia, the need to fully dry NIS before commercial cracking was not recognised. Drying, curing or dehydration causes the kernel to shrink and usually become loose inside the shell. A problem still not fully resolved is that the shoulder of the kernel sometimes adheres to the shell.

- Research has resulted in many concepts being trialled to crack the nuts. Inventors with lofty visions often refused to use the term 'cracker' but preferred 'sheller' or 'opener'. Cynical processors referred to their cracking machines as 'crunchers'.

- Many early crackers consisted of large rollers which required size grading and assumed that the NIS was round. Others used cam-operated plates against a steel base. Hacksaw blades sawing the shell and other innovations were tried as well. While various prototypes were built in the early 1970s, Paul Shaw from California developed a blade cracker where a flywheel with blades fitted sliced radially through the shell against blades in an angled chamber. It was simple and didn't require size grading.

- For fifty years, the Shaw-type blade cracker with minor improvements has been the main method of cracking the shell. Many new principles have been trialled and assessed, but all had certain disadvantages. In recent years, crackers have been developed in South Africa and China, claiming little kernel damage and a high percentage of whole kernel. In Vietnam, a single fixed blade cracker that is claimed to increase the percentage of whole kernels and reduce losses through kernel damage has been invented and is being manufactured in China.

- Kernel extraction in China developed differently from the rest of the world. In the 1990s when their domestic production was low, they imported most of their required nuts in shell and cracked them by hand using hammers, with which women became very skilled. Then came machines with a chisel tool that split the shells, one by one. Now there is a machine that cuts through most of the circumference of the shell, allowing salt and flavour to be introduced. These were roasted and marketed with a small tool to open them and have become a major retail product in China and other countries. The retail price appeared to be lower because of selling in shell (see next page).

- In terms of product range, apart from the traditional raw, roasted, chocolate-coated, confectionery, baking and food range of products, the

Sawn macadamia nuts. Macadamia shells sawn so they can be opened with a small hand tool. Credit: unknown.

versatility of the macadamia knows few bounds. In 2021, some of the new products marketed were a range of Macadamia Milks, Ice Creams, Aloha OatMac Powdered Milk, Mauna Loa Macadamia Milk Matcha Latte, Siggis's Probiotic Yoghurt, Kikkoman Chocolate Macadamia Milk, Tanglin Gin, Meiji Chocolate Koka Macadamia, Sally Hansen Hand Mask, and many others.[9] Further innovative products are Garam masala, a spice mix that makes them the perfect snack, and Kara Damia, a mousse of caramel milk and macadamias. A product that failed but deserved to be a success was a table spread based on macadamia oil. Marc Harrison from Husque in Brisbane invented a technique to produce beautiful display bowls using ground macadamia shells and an inert polymer.

- The former UK Queen Elizabeth II loved macadamias and her chef, Darren McGrady, often presented them on her breakfast table.[10]

- Up to the 1990s, it was accepted practice to roast macadamias in coconut oil as it imparted no flavour, but consumer concerns about its high level of saturated fatty acid led to its being replaced. Now most macadamias are dry roasted. Coconut oil is rarely if ever used.

- Low-grade or damaged kernel is often pressed to extract oil, usually in a cold process to minimise damage to the oil quality. The oil is filtered and sometimes refined and sold as a premium table oil.

- An accidental boost to publicity was 'Nutgate' or the 'Nut Rage Incident'. In December 2014 at New York Airport, a Vice President of Korean Air Lines, Heather Cho, was insulted when given a packet of macadamias rather than having them served on a plate. She ordered the plane to return to the terminal, required the Air Cabin Steward to beg forgiveness on bended knees and dismissed him. The resultant publicity went worldwide and one of the outcomes was that macadamia consumption in South Korea increased by 250 per cent. The Flight Attendant was reinstated and Ms Cho lost face.[11]

Composition and health benefits

- Macadamia kernels typically contain:[12]
- Oil 74%
- Moisture 1.5%
- Protein 9.2%
- Carbohydrate 7.9%
- Dietary Fibre 6.4%
- Trace Elements
- Vitamins
- Bioactivity
- Energy 3,080 kj per 100 g.

- It is hard to imagine when looking at a kernel that it is three-quarters oil. Some 81 per cent is monounsaturated, comprised of oleic acid (approximately 60 to 65 per cent of total), palmitoleic acid (14 per cent of total) and smaller amounts (1 to 5 per cent of total) of other fatty acids including linoleic acid. Research on the health impacts of (Omega 7: 14 to 20 per cent) continues, as there has been conflicting information available from current work. There is evidence to suggest that it enhances longevity and is an important dietary agent with multiple biological functions. Based on current research, palmitoleic fatty acid (Omega 7) contributes to significant health outcomes with no toxicity evidence.[13]

- Macadamia oil is comparable with olive oil in its salad and cooking uses. The oil has important value in cosmetics as well. Largely due to its palmitoleic fatty acid, it is compatible with the oil in human skin, nourishes it and is used in skin rejuvenation pharmaceuticals.

- Sugars make up more than half of the carbohydrate in kernels.

- Dietary fibre in macadamia kernels essentially comes from cell walls. It consists of insoluble cellulose and dietary or more soluble fibre, such as gums and lignins, all of which contributes to gut health.

- In terms of trace mineral elements, macadamias are rich in potassium, iron and magnesium; low in sodium; and contain zinc, copper and manganese.

- Vitamins include thiamin (Vitamin B1), precursors of Vitamin E, riboflavin, niacin, and folic acid, as well as bio nutrients (phytochemicals) including a range of natural antioxidants (see below).

- Macadamias contain a range of bioactive constituents including plant sterols, phospholipids, phenolic compounds and, in very small amounts, a wide range of constituents having antioxidant activity. With macadamias as with other nuts and foods, a synergistic effect is believed to apply, which essentially means that the overall combined effect on our health is greater than that of the individual constituents.

- Because of the macadamia's high oil (and thus energy) content, it had been assumed that it would contribute to weight gain. However, a number of studies have found that regular eaters have lower body mass index.[14] A recent study of obese patients showed that including macadamias in their diet did not lead to weight increase. In addition, macadamias are a key nutrient-dense

food within healthy dietary patterns, with an impressive range of health benefits.[15]

- There is no doubt that macadamias contribute to a healthy diet and can be classed as a Super Food or Functional Food. There have been many studies on the effect of macadamias on human health and longevity. Bio-markers of health and certain diseases show that consumption of macadamias may reduce inflammation, premature ageing, risks of cardio-vascular disease, some chronic diseases and also improve endothelial function.

- Macadamias contain no cholesterol. What distinguishes macadamia from other foods is its high level of palmitoleic fatty acid.

- The author, in attempting to highlight macadamias' health benefits in the late 1990s, presented a provocative but widely discussed poster: 'If I eat enough macadamias will I live forever?' Fortunately or unfortunately the answer was no, but do they improve your chance of living longer and healthier?

    - Palmitoleic fatty acid and the synergistic benefits of bioactive constituents suggest that longevity may be enhanced by macadamia consumption.[16]

    - Studies in Japan on persons over 100 years of age showed elevated blood serum levels of palmitoleic fatty acid.

    - A study by *PLoS Medicine* in 2022 reported that converting from a Western diet to an Optimal diet containing more nuts, legumes and grains could increase life expectancy by more than ten years.

- Macadamias very much fit into the requirements of the Mediterranean diet and also those of the Keto, Paleo or Vegan diets.

- Most tree nuts and peanuts are consumed where the kernel has a moisture content of 4 to 5 per cent. Macadamias are eaten at 1.5 per cent moisture content or a little higher so as to maximise the texture or crunch which characterises them.

Anecdotes

- Ambitious, controversial and maybe even fraudulent schemes are not new and they occurred with macadamias. Going back in time, Mutual Estates in 1933 and the Macadamia Production Pty Ltd in the period from 1944 to 1949 offered to provide land and trees and to maintain and harvest them, with 75 per cent of the income going to the investor, whose only expense was the cost of the trees. They claimed to have 3,199 acres of land available near the Brisbane suburbs of Archerfield, Wacol and Kingston and to have 500,000 seedlings. Four acres of land would be provided free. This offer led to several legal claims and the company was later having difficulty trying to sell its stock of trees. Only one known ongoing orchard may have resulted from the offer.

- An enterprising initiative was to market macadamia investment as 'The Money Tree'. Another on the label of Macadamia Honey jars was to advise that the product was 'extremely rare so only take one jar in order to allow others to enjoy it'. This, of course, resulted in the shelves being cleared. When Nutta Products in 1975 promoted and launched its 50-gram roasted and salted macadamias, storekeepers were each allocated a small quota, which created strong demand to secure more stock and added to their image.

- The AMS in its early days considered using the name 'Australian Nut' instead of 'Macadamia', which if adopted might have changed the marketing image of the product.

- To demonstrate its cosmetic affect, rub a little macadamia oil on one elbow before retiring each night and after a week compare your elbows in a mirror?

- One of the pioneers of the modern industry was Mr George Gray AM, an Ear, Nose and Throat Surgeon, who led investment in macadamia orchards for the medical profession from the 1980s. Apart from having a brilliant analytical mind, the delightfully eccentric George owned an impressive Rolls Royce and was known for parking illegally at the front entry to the Melbourne airport. No officials were brave enough to issue a parking fine, as this gentle-looking man might be someone of influence.

- Back in 1978, the proud young macadamia industry approached the Australia Post Office seeking to have a postage stamp issued featuring the truly Aussie icon. The reply was that they gave preference to native Australian plants, and it took thirty years of lobbying before they recognised the macadamia as an icon.

- Clive James, an iconic Australian writer, wrote about macadamias from England in 1976 when he said: 'The first person to import them into Britain will make a million pounds. The only way in the past to get to the kernel was with a large hammer since the nut came equipped with a casing of the same dimensions and consistency as ball-bearing ammunition. If your hammer swing was slightly angled the nut disappeared with the sound of a ricocheting bullet and you might see an old lady collapse in the street. Now, [he says,] the casing can be stripped off the kernels and presumably the casings are then sold as railway ballast or shrapnel.'[17]

- Professor Henry de Montmorency (c. 1880–1965) was a colourful character who claimed to be a descendent of Henri I de Montmorency (1534–1614), Marshal of France, who committed the unpardonable sin of allowing Roman Catholic items to fall into the hands of Protestants. In 1906, our Henry was charged as being a rogue and vagabond in seeking to obtain money fraudulently from a police sergeant and for claiming to cure ailments. He adopted the title of Professor Henry de Montmorency as a Phrenologist, Palmist, Astrologer and Crystal Reader and in 1907 established the Federal Herbal Institute in Lismore (NSW), later to become the North Coast Confectionery Works. He became a prominent person in Lismore and had, in 1926, an expanding business featuring macadamias, usually chocolate coated. He formed a public company during the Great Depression in 1931, which apparently failed.

His macadamia claim to fame was as one of the founders of the Australian Nut Association in 1932, and where he vigorously promoted the nut. One of his popular macadamia chocolates was named 'Nutty Cubes'. Later he lived at Pretty Gully near Tabulam, where he grew crops and searched for gold.

- The Australian novelist Eleanor Dark (1901–85), best known for her trilogy *The Timeless Land,* lived at Mill Hill Road in Montville, Queensland, in the 1950s. There she wrote *Lantana Lane*, a book of short stories about the local farmers. Now the road remains as a secluded residential lane, with her macadamia trees and those of her neighbour still standing. Her short story 'The Nuts that Were Ullaged' starts with neighbours believing that 'Bopples were not a bad little sideline' and drying them in wire beds in a wooden frame. Young Joy and local rats together find a hole in the wire at the corner of the shed and she uses the Bopple nuts as currency in a make-believe game. Surplus nuts are sold at the local store. The diminishing of the nut collection leads to Joy's discovery and contrition.

- John 'Doc' Buyers (1928–2006) was an American businessman sent to Hawaii to revive the C. Brewer agricultural businesses. Soon he acquired the business. 'Doc' was charismatic, a dreamer and doer who inspired, motivated and became a leader in the development of the Hawaiian macadamia industry. A larger-than-life, flamboyant man, he successfully intimidated this author when they had a chance to meet!

- E.R. (Ted or Teddy) Davenport was a legend and wonderful entertainer and raconteur, apparently living his life always wearing a bow-tie. He led

the CSR Macadamia Division and then became Administrative Manager of the AMS. Some of his sayings were: during World War II in England, 'Bugger Hitler'. Then later: 'Can we order the wine with a sense of urgency? – We probably won't get out of this world alive – Just a wee drop more.' When disciplined at Rotary, he would shout 'Outrageous!'

- In the 1960s, prominent Hollywood stars who invested in macadamia growing in Hawaii were Julie Andrews, James Stewart and Jim Nabors (best known for his signature character, Gomer Pyle).

- The Hon. Doug Anthony (1929–2020 CH, AC, Deputy Prime Minister of Australia) was one of Australia's most respected, competent and admired politicians. His father was a founder of the Australian Nut Association in 1932 and Doug maintained an active interest in the advances of the macadamia. His reminiscences of the early days of the Australian industry, visiting Hawaii, knowing Steve Angus and seeing the success were all satisfying memories to him.[18]

Of macadamias in general

- Macadamias are one of the very few new food plants that have become known across the world in the last 200 years if not the last 2,000 years.

- Global production in 1920 was less than 20 tonnes of NIS; in 1950 it was less than 400 tonnes, almost all in Hawaii; this had risen to 200,000 tonnes globally in 2020, and is projected to be close to 1,000,000 tonnes of NIS by 2035. The Australian crop increased by 1,000 times in the sixty years from 1960 onwards.

- Often described as the finest nut in the world, the macadamia kernel's wholesale price has averaged twice that of most other tree nuts. Maintaining that with increasing production will be an industry challenge.

- What makes macadamias special? Lynne Ziehlke from the AMS summarised that macadamias give a unique and distinct eating experience, as

Wild *M. integrifolia* in the Mary River Valley, smothered by cat's-claw creeper and madeira vine in 2018. Credit: I. McConachie.

The wild *M. integrifolia* in 2022, saved after removal of the creepers. Credit: I. McConachie.

they are: crunchy but soft and buttery, creamy yet chewy and have a balance of being both sweet and savoury. Apart from their distinctive gentle flavour, they more than other nuts have a crunchy texture.[19]

- Many wild macadamias are being threatened by noxious weeds and creepers. An example of one of the achievements of the Macadamia Conservation Trust is shown by the 'before and after' photos of a wild macadamia tree almost smothered by vines and its recovery several years later after the vines had been controlled.

- There is plenty of information about growing a macadamia tree at your home. In summary, they require a subtropical environment, protection from frost when young, plenty of mulch, moisture and patience. In more tropical areas, the absence of sufficient chilling is likely to prevent flowering. Plant grafted trees rather than seedlings. Dry the mature nuts in a way that they receive air circulation before cracking them.

- How long do macadamia trees live for? We know that in the rainforest, trees can remain dormant for very many years until there is light and favourable conditions. Currently through radiocarbon dating, and possibly enhancing annual growth rings, we may be able to age wild trees. It is possible that trees of at least several hundred years of age could be found. The planted Walter Hill Tree in Brisbane is over 160 years old and remains healthy. Tree age from a commercial perspective is based on economics. In Australia, we have orchards of more than sixty years still bearing commercially. Growers are usually reluctant to remove poor performing trees but with new precocious elite varieties, older trees should be replaced when an economic analysis is positive. A long-term growing perspective might be to replant, say, 5 per cent of the orchard each year.

- Jan Young lived near Mount Bauple and in about 2010 was getting a farm dam made when she found a number of what appeared to be fossilised macadamias. They created much interest, but Queensland Museum palaeontologists suspected they were concretions. Other palaeontologists are not so convinced.

- It still surprises some people that macadamias are native only to Australia and did not originate in Hawaii, although Hawaiians undertook their early domestication and promotion. For years in China, the characters for 'macadamia nut' were the same as those for 'Hawaiian' and 'the nut'.

Possibly a fossilised macadamia. Credit: B. Kaddatz.

- Macadamia orchards vary in size, from cottage or hobby farms to large corporate farms containing more than 100,000 trees. The owners of a 'traditional' orchard in developed economies usually have sufficient funds to purchase basic machinery and to adopt sound commercial practices. Possibly there are no more than 5,000 'traditional' macadamia farmers in the world. By contrast, farmers described as smallholders or small-scale farmers, mainly in Africa and South East Asia, total at least 400,000 and in those regions they may produce the majority of the macadamia crop. Smallholders can be described as having from one to ten hectares, on which they raise a range of animals and crops, a significant proportion of which is consumed by the family. Generally, they suffer from a lack of finance, often uncertain land tenure, inability to purchase fertiliser and crop-protection chemicals, and limited access to horticultural advice. Further, they are subject to large crop losses and are often exploited. The majority of the population in these countries lives on the land and produces food of various types. Governments in these countries are committed to alleviating poverty and, in this context, promoting macadamias is a worthy policy. In countries like Kenya, Malawi, Zimbabwe, South Africa, Vietnam and China, there is a balance between commercial macadamia orchards and smallholders. While macadamias are seen as being suited to cultivation by smallholders and having the potential to provide a substantial future income, there are still many challenges. Many organisations and government support policies have been increasingly formed to assist smallholders. In the past, almost all trees were seedlings and not grafted, the two edible species were planted together, cultural advice was unavailable, pests and diseases caused large losses, yields were often very low and crop handling procedures to maintain quality were lacking. Government support policies such as a guaranteed price irrespective of quality often resulted in low-quality kernel. While many actions to minimise these problems are being supported, these countries are faced with some low-quality nuts and suffer a reduced image. Influenced by the very high prices for NIS from 2012 to 2021, there have been many trees planted. As macadamia plantings become consolidated, there is an increasing risk of insect pests multiplying. In 2023, with a decline in prices, viable incomes for smallholders appeared unlikely to be achieved unless subsidised by government.

- Hawaii dominated global macadamia production and marketing until the 1990s, then Australia until the late 2010s, while South Africa is the largest producer in the early 2020s. It seems possible that by 2030 China will lead production, followed by South Africa, Kenya, Australia, Vietnam, Guatemala and other African countries.

Chapter 19

THE FUTURE

Where we were, where we are, where will we be?
– a crystal ball look at the future

Here for the love of macadamias.
(World Macadamia Organisation slogan, WMO Report, 2023)

An overview

As someone who was introduced to macadamias as a young lad in the 1940s and then was actively involved from the 1960s, I have observed and recorded the developing industry and its defining events, and have dwelt on its future. This is an attempt to document what may be the growth, issues, risks, challenges and opportunities that will shape the future, mainly in Australia but applicable to much of the world. In a rapidly growing and changing industry, there is the probability that when read in the future, this chapter will be seen as flawed and incomplete. When the writer has attended forums attempting to project the future, the diversity of opinions is huge.

The colonist discovery of the macadamia occurred less than 200 years ago and its commercial domestication occurred less than 100 years ago in Hawaii, seventy years ago in Australia, and just a few decades ago in the rest of the world. From a rainforest tree known only to Australia's Aboriginal people, macadamias have become a popular delicacy in most developed countries. Hawaii led the commercial industry from the 1930s to the latter years of the century, Australia then dominated and is now challenged by South Africa, China and others.

Macadamias are still a pioneer industry that is likely to double, and possibly triple, in global production by as early as 2030.

Table 10: Global growth, 1960–2030

Year	Global production (tonnes NIS)
1960	1,150t (1,120t Hawaii)
1970	3,800t (3,400t Hawaii)
1980	16,500t (14,000t Hawaii)
1990	33,400t
2000	75,000t
2010	110,000t
2020	230,000t
2030	700,000t (projected)

Source: Average from HMNA, AMS, INC Annual Reports, P. Lee, CEO's Report J. Burnett in *AMS News Bulletin, Winter* 2021, p. 3.

Where we were

The genesis of commercial plantings commenced about 1870 and, apart from Hawaii, there was no industry anywhere in the world until the 1960s. The rate of growth has been amazing and a testament to the fine qualities of the nut. From the 1970s, people in a number of countries were assessing the future of macadamias as a commercial crop. Hawaii had proven the potential, Australia had commenced a commercial industry, and South Africa, Kenya, Guatemala and Brazil all had a toe in the water. Hawaii provided the model for a viable industry, advancing tree horticulture, breeding better varieties, and developing technology for processing and marketing.

In 1981, AMS Patron Norm Greber wrote an article in the *AMS News Bulletin*, 'Nuts to the Knockers', pleading for confidence in the industry and emphasising the importance of breeding local varieties by the simple method of planting and selecting.

Both in Hawaii and Australia, there was ongoing research into understanding quality and its application in growing, crop handling, processing and packaging.

Kernel recoveries from wild trees direct from the rainforest vary widely, but average 25 per cent. From 1970 to 2000, commercial recoveries averaged 33 per cent in Australia and in 2022 new varieties averaged 40 per cent. A target of 50 per cent kernel recovery from thin-shelled varieties is achievable, but will require advances in preventing pre-germination, pest control and crop handling.

At the AMS Annual Conference in 2007, a panel and audience session entitled 'Hypotheticals' visualised the Australian industry in 2025.

Projections that have already been proved correct include:

- Intellectual property (IP) rights will apply to new varieties.

- Many small farms will have disappeared.

- There will be many corporate and large farms.

- There will be extreme pressure on land and water prices.

- It will be difficult to protect IP in China's growing industry.

- Trade barriers will have largely disappeared.

Predictions that largely did not eventuate were:

- An economic orchard will be 20,000 trees.

- NIS will be paid by its real value and will be batch processed.

- There will be only two processors.

- Macadamias will be genetically advanced.

- There will be single desk marketing.

- There will be consolidation at all levels.

- An appellation naming system will apply to origins.

- There will be a closed supply chain.

Although macadamias had always been less than 2 per cent of world tree nut production, by 2011, despite the industry's growth, this percentage had fallen even lower, due to increased planting of other tree nuts.

Profit to all sectors ultimately determines viability. Profit to the grower is dependent on three main factors: the cost to produce a kilogram of kernel, the number of kilograms produced, and the price received per kilogram. In retrospect, did the global industry take enough sound steps to develop and secure its future?

Why didn't Australia's macadamias take off sooner?

People within the industry have queried and sometimes been critical of Australia for missing the opportunity of developing an industry earlier and having a characteristic indifference to something uniquely Australian. Until Australia commenced its commercial industry from the early 1960s, Arthur Lowndes stated that there was only one orchard of 20 acres and no others larger than 3 acres and all were of low standards.[1]

Hawaii did not realise how fortunate they were in having initially an absence of pests and diseases, a benign climate that was favourable for macadamias and a territorial government that was most supportive in seeking new economic opportunities.

The reality in Australia was that with early plantings being seedling trees, endemic insect pests and diseases and low standards of tree horticulture, few if any orchards were commercially profitable. It took CSR to address this, seek commercial cultivars and learn to propagate these. Even CSR never achieved their projections. Up until the 1970s, government support in Australia was a fraction of that in Hawaii.

In retrospect, finding a rainforest plant and learning how to domesticate it would be expected to take many generations with funding, commitment, time, inevitable mistakes made and learned from, and goals slowly being achieved. Despite the very many steps that resulted in the industry of the 2020s, productivity is capable of much improvement and while trees are considered hardy, their ability to produce consistently high crops has a long way to go. Macadamias and their production are very much influenced by their growing environment, both annually and in the long term.

Where we are now

The focus in the early 2020s has been to expand demand, continue advances in cultural management, apply genetic research, meet legal and social obligations, breed new releases, ensure crop security, increase productivity and engage with all in the global industry.

With the COVID-19 pandemic dominating the world from early 2020 and large increases in product, the impact on the Australian and global macadamia industry was that inventory increased, which resulted in a major fall in prices, seriously affecting growers. Seasonal factors had reduced NIS production in Australia and South Africa, and the raw kernel market softened. Demand and retail sales declined during the pandemic and price to the growers, allowing for inflation, reached record low levels.

In May 2020, the AMS predicted that Australia's crop would be 46,900 tonnes, South Africa's 60,000 tonnes, Kenya's 33,000 tonnes and China's 29,000 tonnes. By 2030, it is possible that Australia,

from being the world's leading producer, in terms of production of NIS, may rank fourth. Global production in 2020 was estimated as 235,000 tonnes NIS, increasing to 420,000 tonnes by 2025 and to 670,000 tonnes by 2030.

From 2016, 3,000 hectares had been planted on the coastal flats south of Ballina (NSW) and 4,600 hectares in the Bundaberg (Qld) district. Projected plantings at the start of the new decade were at least 2,000 hectares per year and in the mid-2020s it looks as though plantings in the Maryborough (Qld) district might exceed 5,000 hectares.[2] In Maryborough, most of the arable soils and guaranteed irrigation water were used for growing sugar-cane until 2020, when the Wide Bay Sugar Factory announced its closure. Previously, some cane farms had been sold and macadamias planted, but the factory closure opened an opportunity to plant more macadamias. Macadamia Farm Management had planted 500 hectares by 2023 and Rural Funds Management, an agricultural investment group that had been growing macadamias in Bundaberg since 2007, purchased 5,400 hectares of the factory's sugar-cane farms for managed investment orchards, mainly macadamias.[3] It will take several years to assess price recovery to determine if the high rate of plantings of the late 2010s will continue.

The INC Congress in May 2024 estimated the Australian crop as 56,000 tonnes of NIS and global production as 339,000 tonnes – an increase of 7.5 per cent from the previous year.[4]

Political restrictions and trade barriers limited market opportunities in many countries. Due to persistent government lobbying, barriers are being reduced, eliminated or phased out in countries such as China, India, Japan and Korea.

Horticultural Innovation Australia has morphed into Hort Innovations (HI), a government-initiated, not-for-profit corporation, owned by horticulture growers. Its staff administer, manage and invest levy funds in research and development (R&D), marketing and cross-industry matters for thirty-seven different horticultural industries. Levy funds for R&D

receive matching funds from the Australian Government. The macadamia total budget for 2021 was $5,100,000, of which $1.3 million was matching funds. There is a long-term risk that matching funds for research could be reduced or withdrawn, and concern from some in the industry that HI may not be the most effective body to lead research and promotion. In 2021, only thirty macadamia growers were members of HI, illustrating the lack of grower engagement.

Table 11: Estimated world macadamia production, 2024*

Country of origin	Tonnes NIS at 3.5% moisture content
South Africa	92,000t
China	68,500t
Australia	56,000t
Kenya	46.000t
USA	15,000t
Guatemala	15,000t
Malawi	12,000t
Vietnam	10,000t
Brazil	6,500t
Colombia	1,100t
Other origins	17,100t
	—————
Total	**339,200t**

* Source: INC Congress 2024, May 2024.

The low prices to growers and the kernel market instability and volitivity in 2023 resulted in a stimulation of demand. The increased quantities of NIS exported from many countries to China also helped to reduce global inventories, and the raw kernel market by mid-2024 was struggling to secure enough product.

Where will we be?

In Australia, the rate of change in ownership, new plantings, new growing areas, cultural establishment and management was rapid in the early 2020s. As described, the softening of the market in 2021, the effects of COVID-19, the influence of China, and increasing production resulted in the boom prices disappearing and a new down cycle in industry growth. Global growth will create many challenges. In 2024, we can be bold and imagine a cautious glimpse into the future. The research, the skills of so many, even the mistakes made, the pioneers, growers, processors and marketers have combined to result in a table nut becoming available to more and more people. Many have considered the industry to be in early maturity but global expansions may show that our growth is only beginning.

Where will macadamias be grown in the future?

From 2000, the turn of the century, low labour costs and favourable exchange rates enabled many African countries and some in South East Asia and South America to increase their plantings. In the future, although these countries will face the problems of how to increase productivity and deal with pests and diseases, their large populations and advancing economies may provide internal markets. The large increase in small-scale growers in many countries will create challenges in supporting, co-ordinating, handling their crop and ensuring sound quality. In these countries, there may develop two types of growers, having different practices. Many countries may face political instability. Some countries will be leaders, some followers, some will struggle to produce sound quality.

Ongoing challenges

Some of these challenges are common to many agricultural products, while others are specific to macadamias.

Political, cultural and economic challenges

- unsupportive governments
- trade wars
- high interest rates
- global recession and instability
- taxation laws that discriminate against investments where income commences after a number of years
- government-supported prices to growers in some countries possibly creating an artificial and uncertain future market
- poorly funded and lower-cost producers dropping prices to clear stocks and so risking market instability and a possible lowering of quality standards. An example could be if a long-term price fall initiated by low-cost producers is used to market lesser-quality kernel
- less economically developed countries not having the resources or skills to expand demand, and so increasing the requirement to address sustainability goals.

Sustainability

This is a very broad topic but in the context of this chapter sustainable goals that must be addressed relate to farming practices that in any way affect environmental, social and economic outcomes. It is largely about achieving a balance between all aspects of farm operations, without compromising and damaging the long-term future. We have achieved advances in many areas, including food safety, residue testing and in quality standards.[5] International Quality Standards have been prepared but compliance will be difficult.

In Australia, growers have had to address increasing government controls over farm chemicals and their application. Urban encroachment on to agri-cultural land will continue. Many growers also must meet the requirements of the Great Barrier Reef Protection Measures, the goal of which is to reduce run-off of nutrients and sediments on to the reef.

The industry may consider that it has addressed food safety, and workplace health and safety, achieved a balance in minimising and responsibly using farm chemicals, sought to minimise and reverse erosion and to retain the biological health of soil. However, there will be ongoing and increasing expectations and legal obligations from government bodies, society and the consumer to reassess and improve these. Leoni Kojetin in 2022 defined future goals as reduction in waste, promoting diversity and atmospheric and soil carbon sequestration.[6]

Macadamia industry challenges

- Macadamia kernels are among the most costly to produce and a large, long-term fall in market prices could reduce viability.
- With at least four major price collapses in the last forty years and a resultant loss of confidence, cycles are likely to recur despite strong market development.
- The cycle is different in that the amount of kernel to be marketed in future years will increase massively.
- As countries reach even modest levels of production, developing markets with existing funding will be an immense challenge.
- The high cost of nursery trees will have an impact.
- Food safety issues could emerge, and any contamination with pathogens, toxins, chemical residues or heavy metals would result in a loss of consumer confidence.
- Crop handling will be refined and the NIS purchaser will need to assume more responsibility for drying, grading and storing.
- There is a need for less damaging crackers to be developed and an improvement in image technology to allow processing to become less labour dependent and more efficient.

- Segregation of varieties may become necessary to attract a premium.
- Kernel grades being customised for the desired end purpose should become the norm.
- Quality improvement programmes need to be ongoing.
- Variable and low quality must be addressed as the amount of lower-quality grades from some countries will increase.
- The impacts of chemicals, spray-drift and machinery noise need to be addressed.
- Maintaining sufficient processing capacity will not be easy.
- Encroaching urban and residential development, land prices, and 'right to farm' concerns may well collide.

With current knowledge, macadamia trees are relatively easy to grow, but to consistently produce high levels of premium kernel is much harder. The industry must focus more on the end goal of producing sound kernel. In terms of kilograms of marketable kernel per hectare, increases in productivity have the potential to secure the industry or individual growers. Few macadamia orchards in the world average more than one tonne of kernel per hectare, and this has applied for the last sixty years.

Drone capable of monitoring tree health, pests and disease, and leaf nutrition symptoms. Credit: AMS.

Most growers have not learnt to manage trees in terms of their canopy and architecture, and are reluctant to replace poorly performing varieties because of the time required for the new trees to reach maturity. Managing decline of tree health and productivity in older orchards is an ongoing challenge that may be solved with tree replacement and horticultural management. There is a need to better understand replacing both the tree's energy requirements and energy lost with the crop.

The edible kernel averages only about 16 per cent of the nut-in-husk that falls from the tree and 35 per cent of the NIS. This energy-rich kernel, which contains about 75 per cent oil, requires considerable resources or inputs for the tree to manufacture kernel.

In 1998, John Chapman and Dr Russ Stephenson, researchers from the Queensland Department of Primary Industries (QDPI), challenged the industry by stating that the long-term productivity potential of macadamias was 12 to 15 tonnes of NIS per hectare. Russ Stephenson and John Wilkie Jnr later confirmed this as a realistic, very long-term potential yield, but expressed it as four to five tonnes of kernel per hectare. This prediction was conservatively based on actual increases in the productivity of other long-established crops such as citrus, pome fruits and avocados.[7]

To achieve this potential, the effect of rootstocks was estimated to provide only modest increases in productivity. Larger increases were possible from plant breeding, and all aspects of culture, including tree architecture, dwarf or smaller trees, pollination, tree and soil health and chemical manipulation.

Theoretical current maximum kernel recovery from varieties is generally accepted to be what is achieved from Australian and Hawaiian trials and orchards. This varies due to the environment and cultural standards, but in many countries is much lower. If Australian orchards are averaging 35 per cent total kernel recovery, many countries are comparably in the 15 to 25 per cent range. This can be due to a high level of unsound kernel, but is also related to the growing environment as well as management. Environmental factors are partly understood

Bees, both native (left) and European (right), are necessary for pollination of macadamia flowers. Credit: at right, C. Webster.

but countries or orchards that are achieving lower levels are at a significant disadvantage. Whole kernels have always sold at higher prices than half-kernels, so varieties, horticultural, handling practices and processing must focus on achieving higher output of whole undamaged kernel.

Other future growing and processing issues include the increasing cost of land, energy, water, farm chemicals and labour. Further issues are government regulations, climate change, tree spacing, developing smaller-canopy trees, pollination, early fruiting limbs, learning from other crops, reducing losses at all levels, tree replacement with more productive varieties, pest and disease and canopy management.

New varieties are likely to be more compact, precocious, selected for different environments, and have higher yields and kernel recovery. As larger crops are produced, tree management will be more demanding to replace the energy from the crop.

The technology to crack macadamias has barely advanced in the last fifty years. Current crackers are mostly modified blade and flywheels that tend to damage the kernel. Reducing kernel damage and cracking loss, minimising kernel that adheres to shell, and optimising the percentage of whole kernel, all have to be addressed. Discussed in the previous chapter 'Mac Facts', there are developments in cracker design that more effectively allow the kernel to be removed with less damage and loss, resulting in more whole undamaged and attractive kernels.

Climate change

Climate change is a long-term trend and is different from natural weather variations that have been observed over centuries. There is overwhelming evidence that it is occurring and its impacts will compound and become increasingly evident. In Australia, there will be the certainty of more extremes in weather events, higher overall temperatures and days over 40°C, more droughts, more floods, storms and hail and wind events. Possibly because of their rainforest origin, the macadamia species have poor tolerance to weather variation, so tree health, longevity and crop performance will be affected. The failure of most countries' production to reach projections over the last twenty years has in part been due to adverse weather conditions. As an example, from 1910 to 1920, extreme heat days in Australia averaged one per year. This rose to five days in 1970–80, and in 2010–20 it was eighteen days.[8]

CSR Limited in Australia in the 1970s studied macadamias in growth chambers at their sugar research laboratories in Brisbane. They found that growth ceased above 37°C, and plant death commenced above 40°C.[9]

Climate change will increasingly challenge and affect the industry with rising temperatures, reduced chilling (necessary to initiate flowering), more weather extremes and increased fire risk. Dr Michael Powell in 2010 predicted that in Bundaberg by 2050, rising temperatures will be too hot and possibly not provide sufficient chilling for macadamias to be viable. Some growing districts will no longer be reliable, and strategies will need to be developed to achieve long-term sustainability.

Apart from long-term global warming, where research into expanding the climatic adaptability of macadamias is required, the availability of suitable economically viable land where irrigation can be provided may be limited or require major capital expenditure. Probably in all countries and certainly in Australia, South Africa and China, plantings have been made in marginal sites where climate, soils, slopes, rainfall and water availability will restrict sustainability. In Australia, population growth – with people preferring to live close to the Pacific Ocean – and rising land prices are reducing possible planting areas. However, apart from concerns about long-term warming, both the Bundaberg and Maryborough farming areas can sustain large new plantings. Other areas such as Mackay and to its south could be considered, but cyclonic winds will be a risk. Further expansion in New South Wales, south of the Big Scrub and along the coastal flats, is probable. The large, near-coastal plantings south of Ballina sustained considerable losses due to flooding, drainage and possibly acid subsoils in 2023. There are hundreds of thousands of hectares inland further to the west from Coffs Harbour to Gladstone which will be assessed in the future. These are large areas, mainly along the eastern sides of the Great Dividing Range, which are potentially available although water, high summer temperatures and frost render many of them marginal. Brave but wise farmers will adapt to these areas.

In the past, it was considered necessary to establish wind protection despite the fact that this reduced the plantable area, caused competition for moisture and nutrients, and the windbreaks were often ineffective. Many of the orchards established in Australia in the last ten years risk major damage and tree losses from cyclonic winds.

The industry needs to develop horticultural and varietal adaptation to counter climate change and marginal environments.

Opportunities

Optimising productivity is a major factor in profitability. It is a priority which, despite being addressed, is not leading to significant increases. In Australia and overseas, sound tree husbandry is resulting in earlier yields, but fully mature orchards often decline in productivity. Crowding and light interception, declining soil fertility and tree health are factors. Although the vagaries of nature are part of the problem, much is within growers' control. A major advance is addressing soil health, which in the past was largely about inorganic nutrients, whereas now the importance of biological health is being addressed. A range of soil additives are being used in many higher-performing orchards. Increasing soil carbon levels is also encouraging sound biological processes.

Carbon sequestration

Apart from the benefits in raising carbon levels in soils, the importance of sequestration of carbon dioxide from the atmosphere was recognised as being increasingly important to protect the environment from global warming. Macadamias are carbon friendly. The trees absorb carbon dioxide and so does their orchard soil and its cover. In July 2022, Australia joined many other countries and legislated to reduce greenhouse gas emissions. The goal is to reduce such emissions sufficiently to become carbon neutral by 2050 and to create a carbon sink whereby more carbon gases are absorbed than are emitted. Since macadamia trees are significant absorbers of carbon, overall tree numbers can contribute to a more sustainable global climate.

Ongoing research directed to making the Australian macadamia industry 'carbon neutral' resulted in initiatives such as planting carbon-absorbing plants throughout the orchard. The *Bundaberg Today* reported that Hinkler Park macadamia orchard was now carbon positive. Their 3,000-hectare macadamia farms had achieved total greenhouse gas reduction and removal of 17.670 tonnes of carbon dioxide equivalent between 2020 and 2021 across its macadamia production systems through carbon sequestration and cutting energy and fertiliser use.[10]

Plant breeding and genetic potential

Breeding of more productive varieties is likely to lead to the greatest advances in productivity. Genetic research will advance and accelerate plant breeding outcomes. The available and largely untapped gene pool in Australia contains wild germplasm of unknown but possibly major potential. Highly productive varieties will require correspondingly high standards of management.

As an example, Dr Cameron McConchie at CSIRO succeeded in about 2000 in crossing *M. integrifolia* with *M. jansenii,* which resulted in a smaller, attractive kernel with typical flavour and texture as well as high-kernel recovery, enhanced fatty acids, possibly requiring less chilling to initiate flowering and possibly able to tolerate higher temperatures. With global warming, the ability to grow in higher temperatures may become a major criterion of breeding. Future hybrid crosses could result in a 'new', smaller macadamia to expand retail products and be ideal for chocolate coating.

Intellectual property

As the industry matures, Plant Breeder Rights or other forms of protection such as for innovative new retail products will become increasingly recognised and honoured. As is the case for other crops, productivity and product range will expand.

Marketing

Macadamias, because of their characteristic flavour and texture and also because of their rarity, have sold, mostly in wholesale form, at higher prices than other tree nuts. Macadamias have a high image as a special nut with superb eating qualities and versatility. However, marketing in the past has arguably been, at times, unsophisticated, and much trading could be considered commodity based. Despite initiatives to better create demand, the market will at times be controlled by traders. As production increases and a large number of countries become suppliers, maintaining and enhancing this image will be a challenge. While the industry believes that macadamias are the premium nut, brokers and kernel users may consider them interchangeable and their price a limiting factor.

India, with a population of 1.4 billion and more people becoming middle-class income earners, has the potential to become an important market. With a very small domestic almond production in India, imports of almond have expanded 400 per cent since 2010 and now amount to 200,000 tonnes per year, which suggests a similar potential for macadamias.

At a June 2021 AMS industry forum, broad goals were defined as communication, and market and industry development. These were summarised as New Order Proposed Services and included advocacy, social responsibilities, training, non-levy funded R&D, international leadership and collaboration and regional support.

Product versatility

Macadamias have always been versatile, having many different end uses. There has been much ongoing innovation with new products and this will continue. The AMS in December 2020 listed the 'Top 20 of 2020' – the most inspiring products of the year – which included macadamia-flavoured herbal teas. Examples not covered in the preceding chapter are Toyo Nuts for Beauty, macadamia latte and milks, activated carbons made from macadamias, vegan nut burgers, non-dairy yoghurt, dry gin, honey, skin care products and granola. There is increasing development of macadamia-based 'milks'. Themes intended for our new Gen Z include the culture of wellness, food rituals, cultural discovery and feel-good foods that are healthy and delicious.

Health benefits

Research will reveal more about the health benefits of macadamias, including greater understanding and promotion of palmitoleic fatty acid which comprises from 16 to 20 per cent of macadamia oil, a component whose level distinguishes macadamias from all other foods. The previous chapter 'Mac Facts' provides more information on macadamias, human nutrition and their place in a healthy diet.

Quality

Ongoing quality improvement will lead to a more uniform product, with marketing grades suited for their end purpose. Appearance, shelf-life, uniformity, marketing grades suited for specific end purposes, rewarding and penalising growers for the value of their NIS, understanding how to protect quality in the orchard, and supply chain monitoring – all these factors need to be addressed. There is the potential in many ways to even better meet consumers' expectations and provide a more attractive end product with longer shelf-life.

Mechanisation

Mechanisation and robotics will assist the more developed countries and help high labour cost countries to be cost efficient. The potential of robotics in reducing labour costs in Australia should be progressed.

An Australian vision

Australia will face high costs to produce kernel and possibly unfavourable exchange rates. Issues that could affect future growth include:

- climatic variations reducing the annual crop
- the fact that many orchards, mainly in the Northern Rivers districts, are thirty or more years old, their soil health has deteriorated, while land values and subdivisional pressures are increasing, and tree canopies are too large, resulting in reduced light and crops. With the fall in NIS prices from 2021, many orchards have been cleared, subdivided and will now have different land uses.

In May 2019, the AMS presented its 'Strategic Blueprint 2030', which is possibly the most perceptive current projection of the Australian industry. It stressed the importance of core values, innovation, leadership, collaboration, creating ongoing demand and responsibility. These are all characteristics of mature industries. Some of its goals are:

- a fair gross margin to the grower of 50 per cent
- planting at least 1,000 hectares per year in Australia
- increasing productivity to average 6 tonnes NIS per hectare
- national production of 200,000 tonnes NIS per year.

The writer's 2024 observation suggests that none of these, apart from 1,000 hectares of planting per year, is on target to be achieved.

The AMS has led and represented macadamia growers for fifty years and is recognised as having the vision and resources to continue to play a major role in global matters. Now with South Africa and China producing much more than Australia, the Australian industry cannot be expected to contribute the funds and leadership of the past. An ongoing challenge for the AMS is to achieve the right balance between supporting the growth of a global macadamia industry (encompassing demand, standards, reliable supply chains, etc.) and enhancing the prospects for Australian growers to be able to sell their nuts to a global market. What is sure is that ongoing co-operation between the various sectors of the Australian industry is critical to address this challenge and obtain premium kernel prices for growers.

A global vision

Macadamias are different from other nuts with their flavour and crisp texture. Demand and their retail price are largely based on their image as a special, although becoming less rare, exotic nut. The image of macadamias as The Premium Tree Nut must be maintained. Loss of this image through oversupply, poor quality or adverse publicity could relegate macadamias to just another tree nut. With increasing availability, brokers and major buyers understandably would like to drive prices of bulk kernel down. However, supply of low quality due to age or low standards could damage the image of the whole industry. Macadamias are a natural product and subject to contamination with pathogens, toxins such as aflatoxin, chemical residues and other health risks. Responsible processors have sound procedures in place and pasteurisation of kernel is widely adopted.

In February 2021, the World Macadamia Organisation was formed and is led by Jillian Laing from New Zealand. It aims to represent macadamia-growing countries with the goal of developing a greater understanding and appreciation of macadamias while stimulating growth, creating demand, and providing the long-term foundations for a strong, sustainable market. As part of this, the AMS in 2021 invited Australian growers to contribute a voluntary marketing levy of 3 cents per kilogram of NIS to develop demand and markets. Other countries contributed as well. Formation of the World Macadamia Organisation recognises the need for major growing countries and industry participants to collaborate to address growth issues and work towards becoming a mature, sustainable industry, able to control its future.

Australia has led and funded much of the research and market development over the last thirty years, joined in recent years by South Africa. China is undertaking research in many fields but access to outcomes is limited. Now the global industry faces a less viable future unless major ongoing demand is created. It is believed that the Californian almond industry has a marketing budget of $US 80 million per year whereas the WMO budget is approximately $US 1 million. In 2023, seven member countries are providing its funding. They are: Australia, South Africa, Kenya, Guatemala, Vietnam, Zimbabwe and Brazil. All producing countries have an obligation to support this essential initiative. The marketing crisis of 2022–23 resulted in growers being largely unable to support the WMO and this has been taken up by processors and handlers and some allocation of levy promotion funds in Australia plus similar initiatives from other member countries.

The global industry is fragmented in at least a marketing sense, due to internal competition, the necessity for each industry to remain sound, being widespread and having variable quality standards. Macadamias have high capital investment requirements and generally only attain a positive cash flow from year six or later. Some orchards will never be viable due to environment, funding and management factors. Macadamias have been 'flavour of the month' with agricultural investors and farmers, but a downward price cycle will test this. Many of the plantings made in the last ten years were based on increasingly higher prices to growers and an assumption that these prices would be maintained.

Established macadamia-growing countries generally have a sound infrastructure in place or are working towards this. The market base must be expanded, as 75 per cent of all consumption is in only five countries: Australia, USA, Japan, China and Europe.

The global industry is increasingly considering and implementing plans for a sustainable future. Only time will tell whether projections of global production for 700,000 tonnes of NIS in 2030 is sound, but production is certain to increase and the even longer term is impossible to imagine.

From the early 2010s, there was a massive increase in global plantings, partly driven by demand and prices from China. The resultant increase in production from the early 2020s emphasised the need for more effective planning for the future. The COVID-19 pandemic, rapidly increasing global production and global events reduced marketing and market development. These events will more likely result in setbacks than viable growth. Some in the industry may acknowledge that the 2021 market cycle could have been avoided or minimised. The big risk is that if supply exceeds demand on a more or less ongoing basis, the future will be tested troubled in terms of many growers' viability.

Other tree nuts have corporations or investment companies planting very large orchards as part of their industry growth and this is now happening with macadamias and is likely to continue. This has many positive benefits but some risk of dominance or companies operating independently of the overall industry.

Wild Macadamia Conservation: protecting wild macadamia genetic resources

Macadamias are endemic to Australia and thus the responsibility for maintaining genetic diversity of the crop's wild relatives falls to Australia. It is in the interests of the global industry to support Australia in its efforts to conserve the remaining wild macadamias. In 2024, the Macadamia Conservation Trust, formed and supported by the AMS as its Trustee for nearly twenty years, became Wild Macadamia Conservation, an independent not-for-profit environmental organisation. Bravely steering the group through this transition is Paul Chapman, a grower from northern New South Wales with a long involvement in both macadamia industry R&D and habitat restoration. Wild Macadamia Conservation will continue to work with AMS and other partners to protect wild macadamias and the habitats that sustain them.

Conclusion

For sixty years, the industry was a mainly confident but excitable juvenile, for thirty years it has been in adolescence and now it needs to mature and take sound, strong control of its destiny. It is becoming mature, in the sense that it is now a major tree crop industry, globally distributed and providing quality macadamias in a range of forms to much of the world. While we currently have some vision of the macadamia industry of 2030, it is certain that unforeseen and unknown factors will intrude. Beyond 2030 is unknown, but it should consist of ongoing growth. How quickly will the global industry mature? How many false steps or unexpected situations will arise? It will require people with vision, having the industry's best interests at heart, with funds and a will to invest in the future rather than for today and tomorrow.

Dr William Storey, in the very early industry days of 1965, stated: 'Macadamias represent the work of many people' and he also said: 'The success of the industry and appreciation for the qualities of the macadamia did not just happen. It happened because people were making it happen.'[11]

There have been many unsung heroes: researchers, their assistants, pioneers, growers, processors and marketers; many who risked and often lost; almost all were enthusiasts and believed in this great industry. A little of its story has been told. Most is still in the future.

The sky is the limit. Credit: I. McConachie.

Notes

1: Macadamia Country

1   M. Dettmann and D. Jarzen, 'The Early History of the Proteaceae in Australia: The Pollen Record', *Australian Systematic Botany*, 11 (4) 1998, p. 429.

2   Don Seccombe, Channel 7 Brisbane news-reader, personal communication with author, 1980.

3   Robert Hill, 'Origins of the Southeastern Australian Vegetation', *Philosophical Transactions of the Royal Society*, 29 October 2004, p. 1537; https://royalsocietypublishing.org/doi/10.1098/rstb.2004.1526.

4   'Physical Earth', *National Geographic Magazine*, May 1998.

5   *Wild Plants of Greater Brisbane*, 2000, p. 3.

6   Dettmann and Jarzen, 'The Early History of the Proteaceae in Australia', p. 429.

7   M. Pole, 'The Proteaceae Record in New Zealand', Australian Systematic Botany, 11 (4) 1998, pp. 343–72.

8   M. Dettmann and H. Clifford, 'Fossil Fruit of the Macadamieae', *Nature,* Queensland Museum, 2010.

9   P. Barker and E. Thomas, 'Origins, Significance, and Paleoclimatic Influences of the Antarctic Circumpolar Current', *Earth Review*, June 2004, pp. 43–162.

10 A. Mast et al., 'A Smaller Macadamia from a More Vagile Tribe', *American Journal of Botany*, 95 (7) 2008, p. 859.

11 Ibid.

12 'Burleigh Heads National Park', Queensland Government Environmental Protection Agency, Queensland Parks and Wildlife Service, undated.

13 Mast et al., 'A Smaller Macadamia', p. 859.

14 Thuy Mai, 'Genomic-Assisted Exploitation of Wild Germplasm for Improvement of Macadamia', PhD thesis, University of Queensland, 2020.

15 N.R. Grebe, 'Macadamia Country', *California Macadamia Society Yearbook*, 1963, p. 35.

16 M. Powell and L Gould, *Macadamia Species Recovery Plan 2020–2026*, Macadamia Conservation Trust.

17 I. McConachie, personal observations, 1979–2021.

18 C.J. Nock et al., 'Wild Origins of Macadamia Domestication Identified Through Intraspecific Chloroplast Genome Sequencing', *Frontiers in Plant Science*, 10 (334) 2019; doi: 10.3389/fpls.2019.00334.

2: Botany: Meet the Macadamias

1   A. Mast et al., 'A Smaller Macadamia from a More Vagile Tribe', American Journal of Botany, 95 (7) 2008, p. 843.

2   Yuval Noah Harari, Sapiens: *A Brief History of Humankind*, 2015.

3   C.J. Nock et al., 'Genetic Diversity and Population Structure of Wild and Domesticated

4   Mast et al., 'A Smaller Macadamia'.

5   L.A.S. Johnson, *'Macadamia tetraphylla'*, *Proceedings of the Linnean Society of New South Wales*, 79 (1–2) 1954.

6   J.H. Maiden and E. Betche, 'On a New Species of Macadamia together with Notes on Two Plants New to the Colony', *Proceedings of the Linnean Society of New South Wales,* 21, 1897, pp. 624–5.

7   L.A.S. Johnson, *'Macadamia ternifolia F. Muell* and a Related New Species', *Proceedings of the Linnean Society of New South Wales*, 79 (1–2) 1954, pp. 15–18.

8   W.D. Francis, *Australian Rain-Forest Trees*, 2nd edn, 1951.

9   Johnson, *'M. tetraphylla'*.

10 L. Smith, *Proceedings of the Royal Society of Queensland*, 67 (5) July 1956.

11 A. Shapcott and M. Powell, 'Demographic Structure, Genetic Diversity and Habitat Distribution of the Endangered, Australian Rainforest Tree *Macadamia jansenii* Help Facilitate an Introduction Program', *Australian Journal of Botany*, 59, 2011, pp. 215–25.

12 A. Shapcott, 'Threatened Plant Translocation Case Study: *Macadamia jansenii* (Bulburin nut), Proteaceae', Australasian Plant Conservation, 28 (1) June–August 2019.

13 Dr C. McConchie (CSIRO), 'What Does *Macadamia jansenii* Represent?', Restricted Report to Australian Macadamia Society, 21 August 2012.

3: Aboriginal People: The First to Enjoy Macadamias

1   Philip A. Clarke, Aboriginal Plant Collectors, 2008, pp. 74–5.

2   W.J. Ellwood, J.B. Campbell and G.J. Susino, 'Agricultural Hunter-Gatherers: Food-Getting, Domestication and Farming in Pre-Colonial Australia', in Jan Michal Burdukiewicz (ed.), *Understanding the Past: Papers Offered to Stefan K. Kozlowski*, 2009, pp. 115–22.

3   *Burnum Burnum's Aboriginal Australia: A Traveller's Guide,* edited by David Stewart, 1988.

4   *Bunya Mountains Aboriginal Aspirations and Caring for Country Plan*, Bunya Mountains Elders Council and Burnett Mary Regional Group, 2010.

5   Z.D.S. Skyring, 'Hunting with the Wide Bay Blacks', Article 11 in Gympie Jubilee 1867–1917, supplement to *Gympie Times*, 16 October 1917.

6   PhD research that cannot be confirmed as the DNA test used is now obsolete.

7 Maurizio Rossetto et al., 'From Songlines to Genomes: Prehistoric Assisted Migration of a Rain Forest Tree by Australian Aboriginal People', 2017, online https://pubmed.ncbi.nlm.nih.gov/29117184 and also https://doi.org/10.1371/journal.pone.018666; Maurizio Rossetto, interview with author, August 2019.

8 Love Dispain and Woo, 'Recent Excavations on North Stradbroke Island', *Point Lookout Surf Life Saving Club Excavation Report*, 2018.

9 Eleanor Crosby, letter to author, 9 September 2019.

10 R. Leis, letter about Walter Petrie, 9 August 1977.

11 Senator Ron Boswell, interview with author, 1996.

12 Sue Gallagher, CREEC, interview with author, 1998.

13 Germaine Greer, *White Beech: The Rainforest Years*, 2014, p. 277.

14 Michael Aird, Curator Aboriginal Studies Studies, Queensland Museum, interview with author, September 1998

15 Dr Rhys Jones, Department of Prehistory, Australian National University, interview with author, 1998.

16 Online: https://www.wettropics.gov.au/fire.

17 Dr Lennox Davidson, interview with author, 1979.

18 Dr Joanne Blanchfield, Aboriginal Medicines, University of Queensland, interview with author, 1 November 2013.

19 Nai Nai Bird, *Guerum Woolana: Wild Food*, 1990, p. 17

20 'Coolibah', *Queenslander,* 11 August 1917, p. 2.

21 F.J. Watson, *Vocabularies of Four Representative Tribes of South Eastern Queensland*, supplement to Journal of the Royal Geographical of Australasia (Queensland), 1944, p. 92.

22 Walter Petrie in *Brisbane Courier*, 30 August 1932, p. 15.

23 *Sydney Morning Herald*, 5 November 1867, p. 5.

24 *Southern Moreton Bay Islands Heritage Trail,* Redland City Council, 2008.

25 Ian Fox, Cultural Heritage Consultant, interview with author, 9 September 2019.

26 J. Maldon Thompson, Department of Public Lands, Timber Regulations, Brisbane, 18 May 1870.

27 Penny Watsford, *Forest Bountiful*, 2010, p. 61.

28 *Rockhampton Bulletin and Central Queensland Advertiser,* 27 October 1868, p. 2.

29 Ray Kerkhove, *Aboriginal Campsites of Greater Brisbane,* 2015, p. 71.

30 Marcel Aurousseau, *The Letters of F.W. Ludwig Leichhardt*, 1968.

31 *Maryborough Chronicle,* Wide Bay and Burnett Advertiser, 18 March 1876, p. 2.

32 Aunty Theresa Williams, personal communication with author, September 2023

33 A.E. Harch, letter, 21 August 1977

34 J.G. Steele, *Aboriginal Pathways: Southeast Queensland and the Richmond River,* 1984, p. 229.

35 Ibid.

36 Printed with permission of Uncle Glen Miller, son of Aunty Olga Miller

37 Colleen Wall, personal communication with author, 2023. Printed with permission

38 *Bunya Mountains Aboriginal Aspirations and Caring for Country Plan*, 2010, p. 30

4: European Discovery: Explorers and Botanists

1 *Queenslander, 17 April 1875*

2 Charles Fraser's Journal entry of 26 July 1828, cited in J.G. Steele, *The Explorers of the Moreton Bay District, 1770–1830*, 1972, p. 230.

3 Charles Fraser's Journal entry of 26 July 1828, cited in J.G. Steele, *The Explorers of the Moreton Bay District, 1770–1830*, 1972, p. 230.

4 Clunie's Report, given in Steele, The Explorers of the Moreton Bay District, p. 359. Charles Bateson, Patrick Logan: Tyrant of Brisbane Town, 1966, p. 167.

5 W.G. McMinn, *Allan Cunningham: Botanist and Explorer*, 1970, p. 93.

6 W.G. McMinn, personal communication, 1990

7 McMinn, *Allan Cunningham*, p. 102.

8 Dr Rob Lamond, personal communication with author.

9 Germaine Greer, *White Beech*, 2014, p. 266.

10 Maurizio Rossetto et al., 'From Songlines to Genomes: Prehistoric Assisted Migration of a Rain Forest Tree by Australian Aboriginal People', 2017, online https://pubmed.ncbi.nlm.nih.gov/29117184 and also https://doi.org/10.1371/journal.pone.018666.

11 Joanna Egan, Black Bean *Castanospernum australe* in 'Illustrated: Australian Plant Seeds', Australian Geographic, 25 August 2017; online https://www.australiangeographic.com.au/topics/science-environment/2017/08/illustrated-australian-plant-seeds/.

12 Uncle Wayne Fossey, personal communication with author on Gumburra or Gumbar nuts, 19 September 2019.

13 Jim Low, 'Ludwig Leichhardt: Above the Ordinary?', https//:www.simplyaustralia.net, accessed 10 May 2006.

14 T. Archer, *Recollections of a Rambling Life* (1897), facsimile edn, 1988, p.

15 Dan O'Donnell, 'Ludwig Leichhardt and the Sunshine Coast', *Queensland Agricultural Journal,* September–October 1989, p. 246.

16 Colin Roderick, Leichhardt: *The Dauntless Explorer*, 1988, p. 207.

17 C.L. Gross and P.H. Weston, 'A Review of *Macadamia*

F. Mueller', National Herbarium of New South Wales, 1992.

18 L.S. Smith, Government Botanist. Note attached to botanical drawing of species, 4 November 1955.

19 D.A. Herbert, 'Bidwill, John Carne 1815–1853', *Australian Dictionary of Biography*, Vol. 1, 1966, pp. 98–9; online https://adb.anu.edu.au/biography/bidwill-john-carne-1778/text1997, accessed 8 March 2024

20 Tom Ryan, 'John Carne Bidwill', unpublished manuscript, 2010, p. 4

21 Stuart Read, 'Bidwill of Wide Bay', *Queensland Review*, 19 (1) 2012, p. 75.

22 L.A.S. Johnson, *'Macadamia ternifolia F. Muell* and a Related New Species', *Proceedings of the Linnean Society of New South Wales*, 79 (1–2) 1954, p. 15.

23 K.J. Frawley, 'Forest and Land Management in North-East Queensland: 1859–1960', PhD thesis, Dept of Geography, University of Queensland, 1983.

24 Clive Vievers, *Walter Hill*, Bulletin 371, Historical Society North Queensland Inc., August 1991.

25 Ross McKinnon, 'Hill, Walter (1819–1904)', *Australian Dictionary of Biography*, Supplementary Vol., 2005; online https://adb.anu.edu.au/biography/hill-walter-12981/text23461, accessed 8 March 2024.

26 C.J. Nock et al., 'Genetic Diversity and Population Structure of Wild and Domesticated Macadamia', Final Report MC18004, Horticulture Australia Ltd, 2021.

27 Herbert J. Rumsey, *Australian Nuts and Nut Growing in Australia,* 1927, p. 27.

28 Walter Hill, Letter to the Editor, *Brisbane Courier*, 5 March 1867.

29 Gordon Smith, *Walter Hill of Brisbane's Botanic Gardens,* 2008, p. 60 29

30 Extract on Walter Hill's distribution of macadamias to ships from presentations made by Guides at Brisbane Botanic Gardens, 2008.

31 J. Dahler, C. McConchie and C. Turnbull, 'Quantification of Cyanogenic Glycosides in Seedlings of Three Macadamia [Proteaceae] Species', *Australian Journal of Botany*, 43 (6) 1995, pp. 619–28.

32 R.W. Home, 'Ferdinand von Mueller, Botanist of Honour', *Australian Heritage,* Winter 2006, pp. 66–71.

33 *Transactions of the Philosophical Institute of Victoria,* January–December 1857, Vol. II, pp. 62, 72

34 Ibid., p. 72

35 Dr D.M. Churchill, personal communication with author, 1977

36 George Bentham assisted by Ferdinand Mueller, *Flora Australiensis,* Vol. 2, 1864

37 Royal Botanic Gardens Victoria, Von Mueller Correspondence Project, https://www.rbg.vic.gov.au/science/library/von-mueller-correspondence-project/

38 Deidre Morris, 'Mueller, Sir Ferdinand Jakob Heinrich von (1825–1896), *Australian Dictionary of Biography*, Vol. 5, 1974; online https://adb.anu.edu.au/biography/mueller-sir-ferdinand-jakob-heinrich-von-4266/text6893, accessed 8 March 2024.

39 K.F. Russell, 'Macadam, John (1827–1865)', *Australian Dictionary of Biography*, Vol. 5, 1974; online https://adb.anu.edu.au/biography/macadam-john-4054/text6453, accessed 8 March 2024

40 J. Williamson, *Football's Forgotten Tour: The Story of the British Australian Rules Venture of 1888*, 2003, pp. 12–1

5: Industry Pioneers, 1860–1900

1 *Sydney Morning Herald*, 5 November 1867.

2 Dr Ray Kerkhove, personal communication with author, 2020, on reports of German settlers holding Bauple Nut Festivals in the late nineteenth century.

3 Alice Graham, personal communication with author, 1977.

4 Qld Parliament, 'Report on the Brisbane Botanical Gardens', 6 May 1861, p. 2.

5 Agricultural', *Brisbane Courier*, 20 April 1867, p. 2.

6 Ibid.

7 Penny Watsford, *Forest Bountiful*, 2010, p. 28.

8 Harold W. Crawford, 'Brisbane's Two Botanic Gardens', *Australian Horticulture*, March 1983, p. 91.

9 Toowoomba Botanic Gardens', *Brisbane Courier*, 15 May 1897, p. 7.

10 Peter Osbourne, 'The Queensland Acclimatisation Society', *Royal Historical Society of Queensland Journal,* November 2008, pp. 337–47.

11 A.G. Lowndes, 'Two Historic Macadamia Trees in Australia', *California Macadamia Society Yearbook*, 1966, pp. 29–32.

12 *The Gardener's Chronicle and Agricultural Gazette,* 3 September 1870, pp. 1181, 1231 & 1869.

13 Zachariah Skyring, 'Hunting with the Wide Bay Blacks', *Gympie Jubilee 1867–1917,* pp. 64–7.

14 J. Maldon Thompson, Department of Public Lands, 18 May 1870

15 *Gazette*: Queensland Land Department, 'Timber Regulations', 1 March 1887.

16 *Capricornian* (Rockhampton, Qld: 1875–1929), Saturday 25 March 1876, p. 202.

17 Walter Hill, 'Report on Botanic Gardens', *Australasian*, 30 September 1876, p. 29.

18 'The Field Naturalists at Sankey's Scrub', *Brisbane Courier,* 27 March 1888.

19 F. Turner, *The Cultivation of the 'Australian Nut'*, 1893.

20 'The Queensland Nut', *Brisbane Courier*, 4 April 1893.

21 *Sydney Morning Herald*, 3 August 1911.

22 J.H. Maiden, *The Useful Native Plants of Australia*, 1889, p. 40.

23 Jan Petrie Hall, personal communication with author, 1990.

24 'Nut Growing Experiments', *Queenslander*, 8 October 1931, p. 13

25 Dr Joseph Bancroft, 'The Australian Nut', *Australian Town and Country Journal*, Saturday 6 May 1899, p. 29.

26 Tewantin Correspondent, 'Restdown Orangery and Apiary', *Gympie Times*, 15 August 1895.

27 Shirley Lahey, *The Laheys: Pioneer Settlers and Sawmillers*, 2003, p. 125.

28 Lowndes, 'Two Historic Macadamia Trees in Australia'.

29 C.J. Nock et al., 'Wild Origins of Macadamia Domestication Identified Through Intraspecific Chloroplast Genome Sequencing', *Frontiers in Plant Science*, 10 (334) 2019; doi: 10.3389/fpls.2019.00334.

30 A. Collins, letter in *Queensland Fruit and Vegetable News*, 20 September 1965, p. 618.

31 A. Collins, in *Queensland Fruit and Vegetable News*, 23 December 1965.

32 Richmond River Historical Society, Research Report, 1979.

33 Hewitt, Thomas George (1841–1915)', *Obituaries Australia*; online: https://oa.anu.edu.au/obituary/hewitt-thomas-george-14158.

6: Early Commercial Years to 1940

1 E.H.F. Swain, *The Timber and Forest Products of Queensland*, 1928, p. 472.

2 'Nuts for the King', *Cumberland Argus and Fruitgrowers Advocate*, Saturday 8 November 1902, p. 8.

3 *Richmond River Historical Society Bulletin*, 125 (March) 1988, pp. 15–17 and R*ichmond River Historical Society Bulletin*, 126 (June) 1988, pp. 12–13, being extracts from *Tweed Daily* (Tweed Heads, NSW), 14, 15 & 18 July 1932.

4 *Northern Star* (Lismore, NSW), 2 September 1988.

5 Tewantin Correspondent, '"Restdown" Orangery and Apiary', *Gympie Times*, 15 August 1895.

6 Records of T.G. and N. Hewitt, 1977. The Hewitts recorded, often without references, reports and notes from 1890 to 1940.

7 Sue Gallagher, interview with author at CREEC, Burpengary, May 2006.

8 Geoff Clare, Queensland Forestry, field trip with author, 1982.

9 Marjorie Puregger, personal communication with author, 1968.

10 Penny Watsford, *Forest Bountiful*, 2010, p. 97.

11 Hazel Lon, Gympie Regional Forum, 29 October 2021.

12 *Gympie Times*, 22 October 1907, p. 3.

13 Richmond Rivers Historical Society, research by staff, June 1979.

14 Jeanne Middleton, interview with author, 9 September 2006

15 R. Johansen, *Tales of Our Times*, 1997, p. 68.

16 Reports from N.C. Hewitt to various newspapers, 1927.

17 *Walkabout*, 1 May 1950, p. 17.

18 *Queensland Agricultural Journal*, June 1923, p. 539

19 P. Rickard, letter to author, 4 November 2002.

20 Shirley Lahey, *The Laheys: Pioneer Settlers and Sawmillers*, 2003, p. 61.

21 Godfrey McCullough, letter to author, May 1978

22 'The Queensland Nut', *Queensland Agricultural Journal*, July–December 1939, p. 165.

23 *Daily Mail* (Brisbane), 4 February 1933, p. 9.

24 Catherine Chambers, 'The H2 Hinde Tree', 7 September 2009, Environment and Resources Queensland Government.

25 Ron Johansen, 'John Waldron: A Cracking Pioneer', *Tales of Our* Times, 2003.

26 N.C. Hewitt, 'Nut Culture', *Daily Mail* (Brisbane), 30 November 1926.

27 Johansen, 'John Waldron: A Cracking Pioneer', p. 26.

28 Herbert J. Rumsey, *Australian Nuts and Nut Growing in Australia*, 1927.

29 Foreman Crawford, *Duck Creek Mountain now Alstonville*, 1983.

30 L. Fleming, 'My Great Uncle Norman Rae Greber OAM and His Wife Kathleen Houston', *The Gympie Researcher* [Newsletter of Gympie Family History Society Inc.], September 2017. See also Ian McConachie, 'Greber, Norman Rae (Norm) (1902–1993)', *Australian Dictionary of Biography*, Vol. 19, 2021; online https://adb.anu.edu.au/biography/greber-norman-rae-norm-25514/text33859, accessed 12 March 2024.

31 *Queenslander* (Brisbane), 8 October 1931, p. 27.

32 J. Hall and W. Petrie, 'Rollo Petrie Reminiscences', 2004.

33 Queenslander (Brisbane), 15 September 1923, p. 3.

34 'Another Industry for the Foreigner', *Daily Examiner* (Grafton, NSW), 11 September 1925, p. 4.

35 'Queensland Nut – Neglected in Australia – an American Delicacy', *Tweed Daily* (Tweed Heads, NSW), 5 November 1928, p. 5.

36 *Daily Mail* (Brisbane), 30 November 1926.

37 E. Cheel and F.R. Morrison, *The Cultivation and Exploitation of the Australian Nut*, 1935, pp. 23, 50 & 105.

38 *Grafton Argus and Clarence River General Advertiser*, 26 August 1910, p. 8.

39 David Harrison, Letter to Editor, Northern Star (Lismore, NSW), 2 September 1988

40 *Richmond River Historical Society Bulletin*, 125 (March) 1988, pp. 15–17, *Richmond River Historical Society Bulletin,* 126 (June) 1988, pp. 12–13, being extracts from *Tweed Daily* (Tweed Heads, NSW), 14, 15 & 18 July 1932.

41 A. Steven, Australian Nut Association, letters to Commonwealth of Australia, Commerce Dept, re quality.

42 N.C. Hewitt, Report to Australian Nut Association, April 1933.

7: The Aftermath of World War II to 1970

1 Macadamia Production Pty Ltd, 'Prospectus', 1950.

2 *Brisbane Telegraph*, 8 July 1953. p. 9.

3 'Macadamia Nut Growers' Conference', *Queensland Fruit and Vegetable News,* 16 January 1958, p. 66.

4 Bernie Mason, interview with author, 1976

5 Barry Mason and family, interview with author, July 2022.

6 A. Ross, 'The Macadamia Nut', *Queensland Agricultural Journal*, 1952, pp. 21–32.

7 Queensland Department of Agriculture and Stock, 1954 Survey.

8 Catherine Chambers, 'The H2 Hinde Tree', Queensland Department of Environment and Resource Management, 2009.

9 *California Avocado Society 1959 Yearbook*, 43, pp. 67–71

8: CSR Enters and Leaves, 1960–1986

1 Arthur Lowndes, 'Memoir', typescript, 1989, p. 43.

2 Lincoln Doggrell, personal communication, c. 1988.

3 Kerry Brown, *Our Sweetest Icon: Sunshine Plantation's Big Pineapple 1971–2011*, 2011, pp. 101–2.

4 Dr Lennox Davidson, personal communication, 1979.

5 Arthur Lowndes, *South Pacific Enterprise: The Colonial Sugar Refining Company Limited*, 1956.

6 Arthur Lowndes, *The World Pattern: A Geography for Secondary Schools,* 1947.

7 Arthur Lowndes, 'Some Observations from Australia', *California Macadamia Society Yearbook*, 1964, p. 29.

8 Lowndes, 'Memoir', p.44.

9 'News Letter', Colonial Sugar Refining Company, April 1970.

10 Arthur Lowndes, letter to author, 1978.

11 *Nambour Chronicle* (Qld), 25 September 1974.

12 John Simpson, Report, 16 May 1974.

13 'Australian Macadamias: The Inside Story', CSR Ltd 1980.

14 'Australia's Own Nut to Become Big Export', *Panorama* [magazine], April 1980, pp. 4, 5.

9: The Australian Macadamia Society

Apart from the sources cited below, information in this chapter has been taken from *AMS News Bulletins* from 1974 to 2023.

1 *Queensland Fruit and Vegetable News,* 28 February 1974.

2 Minutes of Formation Meeting of Macadamia Nut Society, Beerwah, 7 March 1974.

3 Formation of South of Brisbane Committee advice to AMS, Ross Wilson, 2 March 1976.

4 Australian Macadamia Society Constitution, undated but c. 1976.

5 *AMS News Bulletin,* Vol. 1 (1), July 1974.

6 Profile: 'Australian Macadamia Society Limited: History of AMS', undated.

7 Ibid.

8 Sunshine Plantation Pty Ltd Report to AMS, 3 October 1984.

9 E. Davenport, 'AMS Review, 1997', *AMS News Bulletin,* 27 June 1997.

10 AMS Letter to Growers, seeking support for marketing trial, 13 December 1984.

11 I. McConachie, 'Twenty-Five Years Old', *AMS News Bulletin*, 1999.

10: Australia, 1970–1980

1 R. Benson and J. Chaseling, *Investment in a Macadamia Nut Orchard,* Misc. Bulletin 17, NSW Department of Agriculture, 1972.

2 Cloutier and Brosgarth (Lismore), 'Investment in Macadamia Production', June 1971.

3 'Delicious Aroma Started Industry', *Sunday Mail* (Brisbane), 7 January 1979.

4 Tom Hoult, 'Moving in on Macadamias Nuts', *Australian* (newspaper), 26 October 1972, p. 2.

5 A. Goodhall. 'Cashing in on Nuts', *Weekend Australian,* 18 February 1978.

6 New South Wales and Queensland Departments' reports, 1980.

7 R. Misfeld, letter to author, 30 July 1977.

8 *Propagation of Macadamia,* NSW Department of Agriculture Bulletin H225, 1972.

9 Hon. V. Sullivan MLA, 'Two New Macadamia Varieties for the Industry', Queensland Department of Primary Industries, 10 February 1979.

10 *Nambour Chronicle,* 10 November 1977.

11 'Trees on a Trellis', *Queensland Country Life*, 31 March 1977, p. 29.

12 I. McConachie, 'Macadamias in Australia', *California Macadamia Society Yearbook,* 1975, pp. 45–6.

13 D. Ironside, 'Macadamia Money Saver', *Nambour Chronicle,* 9 August 1979.

14 AMS Style Brochure, 1980.

15 I. McConachie, 'Nutta Products Qld Pty Ltd', letter to industry, 18 November 1974.

16 'A Prophet Unknown', *National Farmers,* 22 July 1993, p. 2.

17 Ash and Bill Viola, 'Personal Story', manuscript, c. 2008.

18 Bob Lake, 'Nut Market Is Limitless', *Nambour Chronicle,* 14 September 1978.

19 See Sunshine Coast Council website: https:// heritage.sunshinecoast.qld.gov.au/Stories/ celebrations-and-achievements/royal-visits.

20 C. Ogimori, 'Macadamia Specialist Predicts a Big Future for Local Growers', *Northern Star* (Lismore), 30 March 1977.

21 R. Brown, letter to AMS re Australia Post, 30 May 1978.

22 Shirley Stackhouse, 'Nuts: A Sensible Crop', *Woman's Day,* 13 June 1977.

23 Clive James, 'Cracking Nut for Culture', *Observer Review,* 27 June 1976.

11: Australia, 1980–1990

1 'CSR Grows, Processes, and Markets Macadamia Nuts', CSR brochure, January 1980.

2 Suncoast Gold Macadamias report, 31 August 1990.

3 AMS, 'Notice to Growers', 1980.

4 'Firm Expands Macadamia Holdings to North Queensland', *Queensland Country Life*, 7 March 1981.

5 *Queensland Country Life*, 7 March 1981. Consolidated

6 'Big Plan for Nut Orchard', *Rockhampton Morning Bulletin,* 29 August 1988.

7 'New Era for Macadamias', *Sunshine Coast Daily*, 25 September 1986

8 Macadamia Farm Management, investment brochure, 1988.

9 '18,000 Trees Take Hold at Bundaberg', *Queensland Country Life,* 22 September 1988, p. 27.

10 Macadamia Processing Company, official opening invitation, 22 May 2005.

11 'Tax Plan in a Nutshell No Longer Looking Like a Cracking Idea', *Australian*, 30 June 1983.

12 B. Burnett, 'Irrigation', The Land, *Sunshine Coast Daily,* 1 December 1981.

13 'Economic Potential of Selected Horticultural Crops: Macadamias', Bureau of Agricultural Economics, July 1984.

14 B.D. Spooner, interview with author, 8 September 2005.

15 'Macadamia Society Seeking Information on Macadamia Habitat', *Coffs Harbour Advocate,* 20 December 1989.

16 Peter Brown, 'A Nut with a Future', *Australian Horticulture.* November 1984.

17 Ross Loebel, 'Macadamia Production Expected to Quadruple', New South Wales Department of Agriculture, 17 September 1985.

18 R.A. Stephenson, 'The Macadamia: From Novelty Crop to New Industry', *Agricultural Science.* November 1990, pp. 38–43.

19 K. Corcoran, 'Macadamia Nut Industry Faces Turmoil', *Northern Star* (Lismore), June 1990.

12: Australia, 1990–2000

Apart from the sources cited below, information in this chapter has been taken from *AMS News Bulletins* from 1990 to 2000.

1 Len Green, from 'Aussie Nuts', *Bondi Bonbons: A Selection from Joys of a Jongleur,* 1993. The full verse was also published in the *AMS News Bulletin,* May 1994, with Dr Green's permission, and subsequently in the *California Macadamia Society Yearbook.*

2 ABC News Report, 13 July 1990.

3 I. McConachie, report to Queensland Department of Primary Industries (QDPI), 22 March 1990.

4 I. McConachie, report to AMS, February 1998
Vale: Stanley B. Fenner', *AMS News Bulletin,* March 1999, p. 7.

5 R. Mason and I. McConachie, 'A Hard Nut to Crack', *Food Australia,* October 1994

6 E. Davenport, 'Vale: K.J. Ainsbury', *AMS News Bulletin,* March 1999, p. 6.

7 Jeff Blakland, 'Renaissance Man Broke Ground', *Australian* (newspaper), 29 April 1994, p. 17.

8 Neil Jones, 'Ingenious Farmer with Eye for Ideas', *Australian,* 10 January 1994, p. 13.

9 AMS, 'Season 1990: Quality Initiatives'.

10 'Australian Macadamia Industry Coordination Scheme', *AMS News Bulletin,* July 1983.

11 AMS, 'Macadamia Industry Outline', 1994.

12 Australian Macadamias, 'The Peninsular Group: Company Profile', 1995.

13 'The Growers' Independent Macadamia Processor, Stahmann Farms Inc.', 1 June 1994.

14 Macadamia Processing Co., 'Macadamia Magic', tourist brochure, c. 1993.

15 Macadamia Processing Co., 'Newsletter to Shareholders', March 1994.

16 'Macadamia Market Expansion', *Northern Star* (Lismore), 5 August 1998.

17 'Agrimac International', *Northern Star,* 21 April 1995,

pp. 4–6.

18 'Macadamia Firm Plans Expansion', Crestnut Products, press release, 1998.

19 'Big Future in a Nut Shell', Leisuretimes, *Gympie Times*, 18 March 1997, pp. 26–35.

20 'Big Future in a Nut Shell', Leisuretimes, *Gympie Times*, 18 March 1997, pp. 26–35.

21 E. Stephens, 'Australia Proud to Lead the World', *Food News*, 7 April 1996, p. 7.

22 AMS Ltd, 'The Macadamia Story', brochure promoting the industry, c. 1998.

23 Kim Jones, 'Industry Statistics: Trees and Production', 2000.

13: Australia, 2000–2010

Apart from the sources cited below, information in this chapter has been taken from *AMS News Bulletins* from 2000 to 2010.

1 *Top 300 Jared Diamond Quotes*, No. 20; online; https://quotefancy.com/jared-diamond-quotes.

2 E.G. Glock (Germany), 'Year 2000: Australian Macadamia Nut Crop Report', 5 August 2000.

3 J. Twentyman, *AMS News Bulletin*, May 2000, pp. 30–1.

4 Macarthur Agribusiness Report, 2002.

5 General Manager's Report, *AMS Annual Report*, 2007, p. 3.

6 I. McConachie, Report to AMS, October 2009.

7 Australia Post, letter to AMS, December 2011.

8 'Response to Cyclone Larry', *AMS News Bulletin*, September 2007, pp. 56–7.

9 Macadamia Industry Report 2001/2002, p. 7.

10 'Macadamia Industry Brief', *AMS News Bulletin*, January 2003, pp. 26–7.

11 Jennifer Wilkinson, 'Macadamias Grown in Australia but Processed Offshore', *Australian Nutgrowers*, September–November 2002.

12 'Gympie Switched on to Nutshell Power', *Gympie Times*, 18 October 2001.

13 'Jack Merton Gowen OAM', *AMS News Bulletin*, November 2004.

14 Stephanie Jackson, 'Going Nuts over Bush Tucker', *Blues Country*, January–February 2003.

15 MEA [Macadamia Exports Australia]: Buyers of Macadamia Kernel and NIS', Company profile, 24 December 2004.

16 A. Gorman, 'Ice-cream of the Crop: Anthony and Teena Mammino', *Courier Mail*, 15 December 2013.

17 C. Hardner, 'Squirrelling the Macadamia', *Ecos*, 107, April–June 2001, p. 5.

18 B. Spooner, interview with author, 8 September 2005.

19 'Strategic Plan. Data. Core Industry and AMS KPI', HAL and AMS, 2010.

20 K.J. Wilson, 'The Past, Present and Future of the Australian Macadamia Industry', *AMS News Bulletin*, 2008.

14: Australia, 2010–2023

Apart from the sources cited below, information in this chapter has been taken from *AMS News Bulletins* from 2010

1 Geoffrey Blainey, *The Story of Australia's People*, Vol. 2, 2015, p. 380.

2 'Australian Macadamias Strategic Plan 2009–2014', *AMS News Bulletin*, November 2008, p. 17.

3 Horticulture Australia Limited, 'Strategic Investment Plan 2014–2019'.

4 Paul Tollis, letter to author, 21 November 2015.

5 L. Kiernan, 'HAIG Divests Australia's Largest Macadamia Portfolio to Belgian Sugar Producer Finasucre in $60M Deal', *Global AgInvesting*, 22 October 2019.

6 'New Merger in Macadamia Industry', *Australian Tree Crop*, 12 March 2020.

7 'Floodplain Orchard Development', AMS Factsheet, 2018.

8 'Boombera Park: Arrow Funds Management Update', 8 March 2022.

9 B. Fitzgerald, 'West Australia Numbers Bloom', *ABC Rural Report*, 23 September 2014.

10 *AMS News Bulletin*, Winter 2017, p. 59.

11 *AMS Yearbook*, 2017.

12 Nelson Yap, 'Zadro Family Hoping Investors Will Go Nuts over Macadamia Offering', *Australian Property Journal*, 27 July 2022.

13 NAB Small Business Report, Costi Farms, 7 September 2022.

14 L. White, 'Change in Ownership for Large Macadamia Farms', *Tweed Daily* (Tweed Head, NSW), 11 May 2016.

15 C. Shepherd and J. McKichnie, 'Mapping the Growth of Macadamia Orchards', Horticulture Innovations, *AMS News Bulletin*, Winter 2021.

16 Larry Schlesinger, 'Canadian Strike Record Deal for Bundaberg Macadamia Orchards', *Financial Review*, 30 January 2023.

17 J. Price, 'In-Shell Demand Surge from China Sees Market Dynamics Shift', *AMS News Bulletin*, Summer 2023, pp. 4–5.

18 Queensland Government media statement, 'Paradise Dam Progresses with $116 Million Big Build Spend', 21 June 2023.

19 Tallis Miles. 'Flagship Weekend', *Australian AgJournal*, May 2024, pp. 20–5.

20 INC Congress 2024, Roundtable Recap, 17 May 2024.

21 'World Macadamia Organisation Needs Australian Support', AMS News, 18 November 2022.

22 'Interim Funding Model for WMO', AMS News Bulletin, Summer 2023, pp. 24–5.

23 Australian Nut Industry Council, DAFAT report, 10 February 2022.

24 J. Price, AMS News Bulletin, Spring 2023.

25 L. Kojetin, 'Symposium Roundup', AMS News Bulletin, Summer 2023, pp. 56–8.

15: The Americas

1 Sandra Wagner-Wright, History of the Macadamia Nut Industry in Hawaii 1881–1981, 1995, p. 16.

2 Ibid, p. 8.

3 Willis T. Pope, The Macadamia Nut Industry in Hawaii, 1929, p. 4.

4 Wagner-Wright, History of the Macadamia Nut Industry in Hawai'i, pp. 4–5.

5 Ibid, p. 6.

6 Ibid, p. 10.

7 Ibid, p. 14.

8 Ibid.

9 Ibid, p. 18.

10 John S. Pritchett, 'What Is the Bishop Estate?', 1999; online: https://www.pritchettcartoons.com/bet_essay.htm.

11 Wagner-Wright, History of the Macadamia Industry in Hawaii, pp. 18–24.

12 Ibid, pp. 20–2.

13 Craig Hardner, Mark Nikum, Jacqueline Batley and Ainnatul Adawiyah binti Ahmad Termizi, 'Reconstructing the Domestication Pathway of Macadamia from Australia via Hawaii and California', American Society for Horticultural Science (ASHS) Annual Conference paper, 2013; Craig Hardner, 'Macadamia Domestication in Hawai'i', Genetic Resources and Crop Evolution, 63 (8) 2015, pp. 1411–30.

14 Hardner et al., 'Reconstructing the Domestication Pathway of Macadamia'.

15 David Reitow, 'A Hard Nut to Crack: Macadamia in Hawaii', HortScience, 47 (10) October 2012, p. 2.

16 State Department of Agriculture, Paper 16, undated, p. 17.

17 US Macadamia Nuts: Economic and Competitive Conditions Affecting the US Industry, USITC Publication, 1998.

18 Alan Yamaguchi, 'International Research', AMS News Bulletin, 2007, pp. 67–70.

19 Apart from resources cited above, a further detailed report of the Hawaiian industry is available from G.T. Shigeura and H. Ooka, Macadamia Nuts in Hawaii: History and Production, 1984. The California Macadamia Society Yearbooks from 1954 to 2010 also provide much information.

20 William Kemper, 'Distribution of the Macadamia in California', California Macadamia Society Yearbook, 1959, p. 25.

21 California Macadamia Society Yearbook, 1955, p. 1.

22 Paul Shaw, California Macadamia Society Yearbook, 1971, p. 60.

23 Nambour Chronicle [newspaper], 14 September 1978, p. 38.

24 California Macadamia Society Yearbook, 1974, p. 9.

25 Tom Cooper, California Macadamia Society Yearbook, 1983, p. 28.

26 Sonia Rios, Gary Bender and Ben Faber, 'Macadamia Production in California: A Hidden Gem of an Industry?', Topics in Subtropics [blog], University of California, Agriculture and Natural Resources, March 2020; online: https://ucanr.edu/blogs/blogcore/postdetail.cfm?postnum=39358.

27 El Cultivo de la Macadamia en Costa Rica, Centro Agricola de Turrialba, 1984, p. 1.

28 Robert C. Axtell, 'Some Observations of Macadamia Trees in Central America', California Macadamia Society Yearbook, 1967, p. 36.

29 Orlando Rojas, report to author, 3 September 1993.

30 A. Volio, 'Macadamias in Costa Rica Hawaiian Grown', [Australian] Tree Crop [magazine], Summer 1996, p. 2.

31 La Nación [Costa Rica newspaper], 24 September 1982, p. 34.

32 United States Department of Agriculture (USDA), 'Macadamia Nuts: Economic and Competitive Conditions Affecting the US Industry', report, 1998, Section 4, p. 10.

33 S.T. Nakamoto, 'Macadamias in Costa Rica, 1992', Tree Crops Journal, Hawaii, 1 (3) 1993, p. 7.

34 C. Nottebohm, report to author, 3 February 2004.

35 Philip Lee, Macadamia 101: A Guide to Macadamia Production, SAMAC, October 2021, Table 2.1.

36 INC Congress 2024, Roundtable Recap.

37 Axtell, 'Some Observations of Macadamia Trees in Central America', p. 3.

38 E.R. Davenport, CSR Ltd, Report to Macadamia Division, 1978.

39 Valhalla Experimental Station Report, 1 February 2013.

40 Agropecuaria Patzulin SA and Agropecuaria Guatemala CA, company profile, 2005.

41 T. Nottebohm Jnr, interview with author, 22 August 2006.

42 USDA, Farm Reports, 1991–2015.

43 International Nut Council, Global Statistical Review, 2020.

44 L. Kojetin, 'Symposium Roundup: IMS 23', AMS News Bulletin, Summer 2023.

THE MACADAMIA: AUSTRALIA'S GIFT TO THE WORLD

45 A. Yamaguchi, Mauna Loa Macadamias Report, 10 March 2005.

46 T. Nottebohm, INS Roundtable Recap, May 2024,

47 'Mexico', *California Macadamia Society Yearbook*, 1960, p. 43.

48 E. Trask, 'Macadamias in Mexico', *California Macadamia Society Yearbook*, 1960, p. 43.

49 C. Nottebohm, report to author, 6 September 2005.

50 *California Macadamia Society Yearbook*, 1960, p. 36.

51 Gabriel Rivas Ross, Finca Kassandra report to author, 2006.

52 Foundation For His Ministry (FFHM), Oaxaca Mission Baja, website: https://www.ffhm.org/our-missions (accessed 26 August 2015).

53 Ross, Finca Kassandra report to author, 2006.

54 Gabriel Rivas Ross, Finca Kassandra brochure, 2015.

55 Nottebohm, INC Congress 2024.

56 David Anderson, report to author, 14 September 2000.

57 Lee, *Macadamia 101*.

58 Australian Macadamia Society, Conference Report, October 2016.

59 R.A. Hamilton, 'Macadamia Nut Development and Prospects in Brazil', Report to University of Hawaii, c. 1978.

60 Pedro Luis Blais de Toledo Piza and Isabel Piza, presentation at Eighth International Macadamia Symposium, China, 2018. Pedro Piza is an industry historian who has confirmed the author's records in Brazil and other South American countries.

61 Mr Takitani, verbal report to author, 1974.

62 Hamilton, 'Macadamia Nut Development and Prospects in Brazil'.

63 João Dierberger, *California Macadamia Nut News*, 2 (2) 2001, p. 3.

64 USDA, Gain Report, 15 September 2006.

65 Philip Lee, 'Report on a Visit to Brazil', *AMS News Bulletin*, January 2002.

66 Ibid.

67 Piza, presentation at Eighth International Macadamia Symposium, 2018.

68 Naomi Michall, *Food Navigator Newsletter*, 20 March 2019.

69 'Global Production Growth 2022 and Beyond', *AMS News Bulletin*, Autumn 2017.

70 Lee, *Macadamia 101*, Table 2.1.

71 L Kojetin, 'Macadamia Production in Brazil', *AMS News Bulletin*, Winter 2024, pp. 14–17.

72 Edgar Dorronsoro, report to author, 2006.

73 Heinz Gattringer, 'Report on the Macadamia Industry in Colombia', 11 April 1998.

74 Lee, *Macadamia 101*.

75 Donald Brainard, report to author, 1995.

76 Matthias Tapernoux, 'Macadamias in Ecuador', presentation at the International Macadamia Symposium, Brazil, October 2006.

77 Nuez de Macadamias, La Concordia, 'Profile Sheet', 2006.

78 Tapernoux, presentation at International Macadamia Symposium, 2006.

79 Matthias Tapernoux, report to author, 2 July 2022.

80 Clarence Johnson, 'Macadamia Growing in Paraguay', *California Macadamia Society Yearbook*, 1969, pp. 18–21.

81 C. Johnson, 'A Report from Paraguay', *California Macadamia Society Yearbook*, 1972, p. 57.

82 C. Johnson, 'Excerpts from a Letter', *California Macadamia Society Yearbook*, 1968, p. 71.

83 P. Piza, 'Paraguay', report to author, 2016.

84 Dorothy Burt, in *Proceedings*, Second International Macadamia Symposium, 2003, pp. 14–15.

85 *Proceedings*, Eighth International Macadamia Symposium, China, 2018.

86 Lee, *Macadamia 101*, p. 27.

87 W.B. Storey, 'The Macadamia in Venezuela', California Macadamia Society Yearbook, 1971, pp. 62–9.

88 Elizabeth E. Roebling, 'Macadamia Tree Offer Lifeline to Small Growers', *IPS News*, 27 June 2011.

89 *AMS News Bulletin*, 2017, p. 55.

90 C.H. Deichman, 'Macadamias in Jamaica', *California Macadamia Society Yearbook, 1963*, p. 38.

91 C.H. Deichman, 'Second Report from Sherwood Forest', *California Macadamia Society Yearbook*, 1965, pp. 34–5.

92 W.B. Storey, 'The Macadamia in Trinidad', *California Macadamia Society Yearbook*, 1967, p. 4.

16: Africa and Beyond

1 Professor Peter Allan, 'Highlights of 30 Years with Macadamias', *SAMAC Yearbook*, 1996, pp. 56–63.

2 Mark Penter, 'Progress in New Cultivar Trials with the South African Macadamia Industry', SAMAC, *In a Nutshell*, July–September 2016, p. 24.

3 Vibetnum, 'The Flower Garden', *Western Mail* (Perth, Australia), 8 August 1913, p. 14.

4 Penter, 'Progress in New Cultivar Trials'.

5 C.A. Schroeder, 'Recent Developments in the Macadamia Industry of South Africa', *California Macadamia Society Yearbook*, 1975, p. 77.

6 Chris Reim, 'Developments in South Africa', *Proceedings* HMNA, 1976, pp. 116–19.

7 C. Reim, 'Early History of Macadamias in South Africa', SAMAC Mini Symposium, 1991, pp. 1–2.

8 Macadamia and Pecan Nut Symposium, Politsi,

272

South Africa, 1978.

9   Macadamia Mini Symposium, SAMAC, 1991.

10 C.A. Schroeder, 'The Macadamia in Malawi and South Africa', *California Macadamia Society Yearbook*, 1970, pp. 20–5.

11 Macadamia Mini Symposium, SAMAC, 1991.

12 Allan, 'Highlights of 30 Years with Macadamias'.

13 Mark Penter, 'Beaumont Drive to be Paved by Len Hobson', *SubTrop Quarterly Journal*, April–June 2015, pp. 56–8.

14 P. Lee, 'South Africa and Africa', Proceedings Second International Macadamia Symposium, 2003.

15 'Macadamia Finance Investment' [brochure]. Distributed throughout the 1970s.

16 'Lowveld Macadamias Industries Limited Prospectus', August 1976.

17 R. Schormann, 'History and Statistics of the Southern Lowveld', *Mini Macadamia Symposium Proceedings*, 1991, pp. 4–5.

18 R.F. Hurly, 'A Theoretical Analysis of the 1993 SAMAC Tree Census', SAMAC Yearbook, 1994, pp. 11–16.

19 Derek Donkin, 'By the Numbers', SubTrop Marketing Symposium, November 2016.

20 Philip Lee, report to author, 18 August 1996.

21 'Australian Growers Study Tour Report', 2015.

22 Jill Whyte, 'What's Happening in South Africa', *Proceedings AMS Conference*, 2005.

23 P. Lee, 'South Africa and Africa', *Proceedings Second International Macadamia Symposium*, 2003.

24 Myles Osbourne, AusMac Conference Presentation, Australia, November 2022.

25 Philip Lee, *Macadamia 101: A Guide to Macadamia Production*, SAMAC, October 2021, pp. 36–9.

26 L. Godard, J.-B. Roelens, M. Béringuier and F. Montfort, 'Report on the Mozambican Macadamia Value Chain', ACAMOZ, 2021–22, p. 13; online: https://iam.gov.mz/index.php/download/report-on-mozambican-macadamia-value-chain-acamoz-nitidae/.

27 Suncoast Gold Macadamias, 'Report to Shareholders', October 2023.

28 INC Congress 2024, Roundtable Recap, May 2024.

29 Richmond River Historical Society (Lismore, NSW), record of NIS sent to Kenya in 1926.

30 Mr Mashiro Shiota, Kenya Nut Company, report to author, 2006.

31 E.R. Davenport, 'History and Structure of Kenyan Macadamia Industry', CSR Ltd, 28 June 1983.

32 Y. Sato and J. Waithaka, 'Kenya Nut Company Report', 1998, pp. 1–12.

33 Ibid.

34 Agricultural Attaché Report, Nairobi, 14 March 1994.

35 'Protests over Kenya's Government Ban on Exports of Nut in Shell Macadamias', SAMAC, In a Nut Shell, May–June 2010, p. 7.

36 International Macadamia Symposium Report, China, September 2018.

37 Ten Senses Africa, 'The TSA Story'; online: https://tensensesafrica.com/the-tsa-story/.

38 NutPAK, 'The Building Blocks of a Sustainable Macadamia Industry in Kenya', presentation, September 2012.

39 Camellia PLC, website November 2022; see https://www.camellia.plc.uk/.

40 L. Kojetin, 'Symposium Roundup: IMS23', *AMS News Bulletin*, Summer 2023, p. 56.

41 INC Congress 2024, Roundtable Recap, May 2024.

42 Osbourne, AusMac Conference Presentation, Australia, 2022.

43 IMC Symposium, September 2023.

44 Lindi Bohya, 'Malawi Macadamias', *The Macadamia Magazine*, 2020.

45 A. and D. Spurling, 'Macadamias in Malawi', *California Macadamia Society Yearbook*, 1971, p. 71.

46 Edward Tonks, 'Macadamias Revisited', *California Macadamia Society Yearbook*, 1972, p. 45.

47 *Commonwealth Development Corporation: Partners in Development, Finance Plus Management*, London, The Corporation, 1976, and website, 2021. (Since 2021, the CDC has been renamed as British International Investment.)

48 E.J. Zuza, K. Maseyk, S. Bhagwat, A. Emmott, W. Rawes and Y.N. Araya, 'Review of Macadamia Production in Malawi', *Agriculture*, November 2021.

49 Daphne Spurling, 'The Infant Macadamia Industry in Malawi', *California Macadamia Society Yearbook*, 1973, p. 86.

50 Zuza et al., 'Review of Macadamia Production in Malawi', p. 52.

51 Russ Stephenson, 'Some Impressions of the Macadamia Industry in South Africa and Malawi', November 1989.

52 Bohya. 'Malawi Macadamias'.

53 The Emmott Family; see https://www.nutcellars.com/about-us (accessed 24 May 2023).

54 *AMS Conference Proceedings*, 2012.

55 Ibid.

56 Lee, *Macadamia 101*.

57 Kojetin, 'Symposium Roundup: IMS23', p. 56.

58 INC Congress, Roundtable Recap, 17 May 2024.

59 J. Hanlon, 'Power Without Responsibility: The World Bank and Mozambican Cashew Nuts', *Review of African Political Economy*, 83, 2000, pp. 29–45.

60 Jaco Botha, MozMac Ltd, Report 4 January 2007.

61 Ibid.

62 Colleen Dardaqan, 'Mozambique Venture', *The Macadamia Magazine,* February 2020.

63 Lee, *Macadamia* 101, p. 27.

64 Ibid.

65 Southern African Development Community, 'Sugar and Macadamias', SADC **Sugar Digest**, 2019, p. 14.

66 Ibid.

67 Ibid.

68 See online: https://www.pamoja-impact.com/.

69 Farm for the Future. See website: https://ffftanzania.com/.

70 Edward Tonks, 'A Short Report on the International Macadamia and Pecan Nut Symposium Held in Dunmow [UK], 1971', *California Macadamia Society Yearbook,* 1971, p. 59.

71 *Business Weekly,* Zimbabwe, 23 February 2022.

72 Rocky Stone, 'Zimbabwe Macadamias', SAMAC, *In a Nutshell,* 1999, pp. 17–19.

73 I. Mharidzo, J. Mwandifura, L. Chikazhe, J. Manyeruke and N. Mashayakure, 'An Insight into the Macadamia Nuts Industry of Zimbabwe: Its History, Current State and Constraints', *IJRISS [International Journal of Research and Innovation in Social Science],* February 2022, pp. 686–94; online: https://www.rsisinternational.org/journals/ijriss/Digital-Library/volume-6-issue-2/686-694.pdf.

74 *The Macadamia Magazine,* 2020.

75 Lee, *Macadamia* 101, pp. 25–6.

76 World Macadamia Organisation, April 2022.

77 These reports cover the development of macadamias in many African countries: *The Macadamia Magazine,* 2020, p. 15; Australian Macadamia Conference Report, October 2016; Africa Press, 13 April 2021.

78 Knoera Corporation, 'Production Statistics, Republic of Congo', January 2019.

79 Therese Labib, personal communication to author, Mazhar Botanic Garden report, 26 June 2023.

80 Southern African Development Community, 'Sugar and Macadamias'.

81 Ibid.

82 'Macadamia Trees Growing in Uganda', *Afrinspire* Report, 15 November 2019; online: https://www.afrinspire.org.uk/post/macadamia-trees-growing-in-uganda.

83 Ibid.

84 Selina Wamachi, 'Uganda Macadamia Nut Market Insight', report.

85 See online: https://www.monitor.co.ug/uganda/magazines/farming/drumming-the-beat-of-macadamia-4181198.

86 Joshua Kato. online: https://www.newvision.co.ug/category/news/ugandans-called-upon-to-embrace-macadamia-gro-135925.

87 *AMS News Bulletin,* November 2009, p. 43.

88 Challenges Group Zambia Ltd, 'Executive Summary' in 'Feasibility Study for the Cultivation of Macadamias in Zambia', 2020; online: https://prospero.co.zm/app/uploads/2021/11/MacadamiaFeasibilityStudyPublic_2021.pdf.

89 Mohammed Taqi Minbashi Moeini, 2 October 2007.

90 Reuben Ohad, 'A Progress Report on Macadamia Culture in Israel', *California Macadamia Society Yearbook,* 1972, p. 51.

91 *California Macadamia Society Yearbook,* 2006, p. 77.

92 E. Ortega et al., 'Characteristics and Lipid Profiles of Macadamia Nuts',w February 2014, pp. 33–9.

17: Asia and the Pacific

1 Macadamia Nut Basic Information, China, *Wikipedia* (accessed 15 March 2013).

2 Ibid.

3 Craig Hardner, 'A Review of Genetic Resources', March 2007.

4 P. Ito, N. Subang and M. Yun, 'Macadamia Acreage and Production in the People's Republic of China', *HMNA Proceedings,* 3 May 1996.

5 Ms Chen Yuxiu, 'Macadamia Production in China, 18 September 2012', in published *Proceedings* of the Eighth International Macadamia Symposium, China.

6 Ms Chen Yuxiu, information booklet, Eighth International Macadamia Symposium, China, 2018.

7 Ms Chen Yuxiu, report to author, 29 September 2023.

8 Ms Chen, information booklet, 2018.

9 Presentation, Eighth International Macadamia Symposium, China, 2018.

10 Guangxi Macadamia Nut Association, 10 October 2018.

11 Dr Russ Stephenson, 'Impressions on Macadamias in China', *AMS News Bulletin,* 2012, pp. 32–5.

12 K. Jones, K. Quinlan and K. Wilson, 'Study Tour of the China Macadamia Industry, February 2014', HAL Report, MC12702.

13 'Beautiful Lincang – Fragrant Nut', Eighth International Macadamia Symposium, China, 2018.

14 *Pacific Nut Producer Magazine,* Report, 10 February 2020; see also website

15 Adam Branson, in *Tree Nuts Annual,* Tree Nuts Annual_Beijing_China People's Republic of_CH2022_0105, USDA Foreign Agricultural Service, 2022.

16 Philip Lee, 'A Final Word on China', in Philip Lee, *Macadamia 101: A Guide to Macadamia Production,* SAMAC, 2022 edn, sections 2.1–2.4.

17 'Suncoast Gold Macadamias Shareholders Report', October 2023.

18 INC Congress 2024, Round table Recap, 17 May 2024.

19 Thomas Karnsund, personal communication.

20 See also online: https://www.researchgate.net/publication/322581487_Myanmar_Macadamia_Industry_Guide_to_Orchard_Best_Management_Practices.

21 'LMC Development of Macadamia Industry for Small Farmers in Lancang–Mekong Region', China Embassy, 2011.

22 K.J. Wilson, report to author, 19 July 2013.

23 Lawrie Raymond, report to author, 28 February 2018.

24 Myanmar News Agency, 13 January 2023.

25 Supamatee, Ito and Jalahan, 'Hawaiian Macadamia Cultivars in Thailand', *HMNA Proceedings,* 1992, p. 107.

26 'Over the Hills and Far Away, *The Nation* (newspaper, Thailand), 20 February 2018.

27 Warangkana Srichamnong, 'Macadamia in Thailand', report to author, 7 February 2010.

28 See website: https://www.maefahluang.org/en/about/ (accessed 13 June 2023).

29 See website: doitung.com/en/ (accessed 13 June 2023).

30 Professor Hoe Hoang, report to author, November 2016.

31 Ibid.

32 'Results of Genetic Testing and Vegetative Propagation of Macadamia in Vietnam', Vietnamese Academy of Forest Sciences Report, c. 2015.

33 'Adaptability and Nut Yield of Macadamias in North West Vietnam', World Agroforestry Centre and Australian Government, c. 2012.

34 Hoang, report to author, November 2016.

35 Vietnamet [online newspaper, Vietnam], 28 April 2015; see website: https://vietnamnet.vn/.

36 Hoang, report to author, November 2016.

37 Libra Handcare, May 2015.

38 *Vietnam News* [newspaper], 5 April 2022; see website: https://vietnamnews.vn/.

39 Philip Lee, *Macadamia 101: A Guide to Macadamia Production,* SAMAC, October 2021, p. 13.

40 Bui Quoc Hoan, report to author, 5 September 2022

41 See online: http://dienbien.gov.vn/en-us/Pages/2023-5-31/The-delegation-of-the-two-northern-provinces-of-Lakq1y2r.aspx (accessed 15 June 2023).

42 INC Congress 2024, Roundtable Recap, May 2024.

43 Ir Koestono, *California Macadamia Society Yearbook,* 1982, p. 131.

44 See online: https://revitalization.org/article/indonesia-turns-to-macadamia-nuts-to-reforest-barren-lands-restore-biodiversity-and-revitalize-rural-economies/ (accessed 16 June 2023).

45 K.J. Wilson, personal communication with author, 19 July 2023.

46 John Yonemoto, *California Macadamia Society Yearbook,* 1991, p. 23.

47 'Macadamias for Lao', [Australian] Direct Aid Program (DAP), 2014–21.

48 Mike Askham, interview with author, March 2008.

49 International Macadamia Symposium report, 20 December 2015.

50 A. Barrueto, J. Merz and T. Hammer, 'A Review of the Suitability of Macadamias for Growth in Nepal', University of Bern, Switzerland, undated.

51 See online: https://mahaguthi.com.np/product/macadamia-nuts (accessed 15 June 2023).

52 Joshua Cooper, 'Macadamia Trees and Their Popularity in the Philippines', Healing Plant Foods, 1 January 2023.

53 Rowan Patterson, interview with author, April 2004.

54 Lovely Carillo, *AMS News Bulletin,* November 2010, p. 55.

55 International Macadamia Symposium report, 2002.

56 Wilson, personal communication with author, 19 July 2023.

57 Dr Mike Poole, 'The Proteaceae Record in New Zealand', *Australian Systemic Botany,* 11 (4), November 1998.

58 Bernard Coleman, 'The History of Macadamias in New Zealand', Masters thesis, Massey University, 2003.

59 Ian McConachie, report to AMS, undated.

60 Ian McConachie, assessments of NIS at Suncoast Gold Macadamias, 2001.

61 W. Fletcher, *Growing Macadamia Nuts in New Zealand,* (NZ) Department of Agriculture and Fisheries, 1973.

62 S.N. Dawes, *Macadamia Varieties and Culture,* (NZ) Ministry of Agriculture and Fisheries, 1981.

63 Ian McConachie, 'Macadamias in New Zealand', *AMS News Bulletin,* 13 (2) July 1986.

64 Virginia Warren, 'Macadamias in New Zealand', 'Chineka Macadamias' flyer, undated.

65 Rod Husband and Vanessa Hayes, 'Macadamias in New Zealand', AMS Conference presentation. 2012.

66 Vanessa Hayes, report to author, 23 June 2023.

67 Lee, *Macadamia 101,* 2021, p. 27.

68 M. Iqbal and R. Hampton, 'Macadamia Nut Production in Fiji', *Fiji Agricultural Journal,* 1977, 39 (2) 1977, pp. 97–103.

69 B.D. Spooner, report to author, 8 September 2009.

70 K.J. Wilson, report to author, 19 July 2023.

71 Spooner, report to author, 8 September 2009.

18: Mac Facts

1   Fiona McMillan-Webster, *The Age of Seeds,* 2022.

2   A. Mast et al., 'A Smaller Macadamia from a More Vagile Tribe', *American Journal of Botany,* 95 (7) 2008, pp. 843–70.

3   J. Dahler, C. McConchie and N. Turnbull, 'Quantification of Cyanogenic Glycosides in Seedlings of Three Macadamia (Proteaceae) Species', *Australian Journal of Botany,* 43, 1995, pp. 618–28.

4   Queensland Department of Primary Industries, Sandy Trout Food Preservation Laboratory taste panels. R.E. Leverington, 1975.

5   ASU School of Life Sciences, University of Arizona, USA.

6   'Your Brisbane – Past and Present', *Courier Mail* (newspaper), 20 May 2009.

7   *California Macadamia Society Yearbook,* 1979, p. 77.

8   Nai Nai Bird, *Guerum Woolana Wild Foods,* 1990, p. 17.

9   *AMS News Bulletin,* Autumn 2022, pp. 59–61.

10 'Macadamia Nuts from A to Z; 26 Things to Know', *Fine Dining Lovers,* 22 March 2018; online: https://www.finedininglovers.com/article/macadamia-nuts-z-26-things-know.

11 'Flight Attendant Kicked Off Korean Air Flight Alleges Cover Up, *New York Times*, 18 December 2014.

12 '2009 Nutrient Composition of Raw, Unsalted Tree Nuts', *Nuts for Life*, Horticulture Australia Ltd, 2011.

13 Nancy Morse, 'A Toxicity/ Safety Assessment of Dietary Palmitoleic Fatty Acid (POA)', 2016.

14 Joan Sabate, 'The Macadamia Nut Effects on Adiposity and Cardiovascular Risk Factors', *Journal of Nutritional Science,* May 2023.

15 'Nuts for Life Report', *AMS News Bulletin,* Winter 2024, pp. 38–9.

16 W. Hu, M. Fitzgerald, B. Topp, M. Alam and T.J. O'Hare, 'A Review of Biological Functions, Health Benefits, and Possible de Novo Biosynthetic Pathway of Palmitoleic Acid in Macadamia Nuts', *Journal of Functional Foods,* 2019, pp. 1–12; online: https://doi.org/10.1016/j.jff.2019.103520.

17 Clive James, 'Clive James Down Under Cracking Nuts for Culture', *Observer Review,* 27 June 1976, p. 17.

18 Author interview with Doug Anthony, 2014.

19 Lynne Ziehlke, AusMac Conference, 2016.

19: The Future

1   Arthur Lowndes, *California Macadamia Society Yearbook,* 1964, p. 29.

2   L. Schlesinger, 'Ryral Funds Nuts Out Macadamia Earnings', *Financial Review,* 24 August 2023.

3   AMS News Report, 17 July 2021.

4   INC Congress 2024, Roundtable Recap, 17 May.

5   Leoni Kojetin, Ausmac Conference, Gold Coast, November 2022.

6   Leoni Kojetin, AMS Conference, 9 November 2022.

7   J. Chapman and R. Stephenson, 'Yield Potential of Macadamias: A Report and Updates from 1998 to 2022'.

8   'Australia's Changing Climate', CSIRO and Bureau of Meteorology', 2021.

9   Advice from John Simpson, 1988.

10 *Bundaberg Today,* 2 February 2022.

11 Dr William B. Storey, *California Macadamia Society Yearbook,* 1966, p. 9.

Bibliography

1) Government Publications (by date)

Australian Federal Government

'Economic Potential of Selected Horticultural Crops: Macadamias', Bureau of Agricultural Economics, July 1984.

New South Wales Government

F. Turner, *The Cultivation of the 'Australian Nut'*, Dept of Agriculture (NSW), 1893.

R. Benson and J. Chaseling, *Investment in a Macadamia Nut Orchard*, Misc. Bulletin 17, NSW Department of Agriculture, 1972.

Propagation of Macadamia, NSW Department of Agriculture Bulletin H225, 1972.

Ross Loebel, 'Macadamia Production Expected to Quadruple', NSW Department of Agriculture, 17 September 1985.

Queensland Government

J. Maldon Thompson, Department of Public Lands, *Timber Regulations*, Brisbane, 18 May 1870.

E.H.F. Swain, *The Timber and Forest Products of Queensland*, Brisbane, Qld: Queensland Government Printer, 1928.

Hon. V. Sullivan (MLA), 'Two New Macadamia Varieties for the Industry', QDPI, 10 February 1979.

'Economic Potential of Selected Horticultural Crops: Macadamias', Bureau of Agricultural Economics, July 1984.

I. McConachie, report to QDPI, 22 March 1990.

C. Horsburg, *Macadamia Nuts: A Marketing Analysis*, QDPI, 1996.

Catherine Chambers, 'The H2 Hinde Tree', Environment and Resources Queensland Government, 7 September 2009.

Media statement, 'Paradise Dam Progresses with $116 Million Big Build Spend', 21 June 2023.

'Burleigh Heads National Park', Queensland Government Environmental Protection Agency, Queensland Parks and Wildlife Service, undated.

2) Select Industry Documents/ Publications (by date)

Australian Macadamia Society (AMS)

Australian Macadamia Society News Bulletins, 1974 to 2024.

I. McConachie, 'Nutta Products Qld Pty Ltd', letter to industry, 18 November 1974

Constitution, undated but c. 1976.

R. Brown, letter to AMS re Australia Post, 30 May 1978.

AMS Style Brochure, 1980.

'Australian Macadamia Industry Coordination Scheme', *AMS News Bulletin*, July 1983.

Sunshine Plantation Pty Ltd Report to AMS, 3 October 1984.

AMS Letter to Growers, seeking support for marketing trial, 13 December 1984.

Ian McConachie, 'Macadamias in New Zealand', *AMS News Bulletin*, 13 (2) July 1986.

AMS, 'Season 1990: Quality Initiatives'.

I. McConachie, report to AMS, February 1998.

'The Macadamia Story', AMS brochure promoting the industry, c. 1998.

E. Davenport, 'AMS Review, 1997', *AMS News Bulletin*, 27 June 1997.

———— 'Vale: K.J. Ainsbury', *AMS News Bulletin*, March 1999, p. 6.

'Vale: Stanley B. Fenner', *AMS News Bulletin*, March 1999, p. 7.

I. McConachie, 'Twenty-Five Years Old', *AMS News Bulletin*, 1999.

J. Twentyman, *AMS News Bulletin*, May 2000, pp. 30–1.

'Macadamia Industry Brief', *AMS News Bulletin*, January 2003, pp. 26–7.

'Jack Merton Gowen OAM', *AMS News Bulletin*, November 2004.

Jill Whyte, 'What's Happening in South Africa', *Proceedings AMS Conference*, 2005.

General Manager's Report, *AMS Annual Report*, 2007, p. 3.

'Response to Cyclone Larry', *AMS News Bulletin*, September 2007, pp. 56–7.

Alan Yamaguchi, 'International Research', *AMS News Bulletin*, 2007, pp. 67–70.

K.J. Wilson, 'The Past, Present and Future of the Australian Macadamia Industry', *AMS News Bulletin*, 2008.

'Australian Macadamias Strategic Plan 2009–2014', *AMS News Bulletin*, November 2008, p. 17.

I. McConachie, Report to AMS, October 2009.

'Strategic Plan. Data. Core Industry and AMS KPI', HAL and AMS, 2010.

Australia Post, letter to AMS, December 2011.

Dr C. McConchie (CSIRO), 'What Does *Macadamia jansenii* Represent?', Restricted Report to AMS, 21 August 2012.

Dr Russ Stephenson, 'Impressions on Macadamias in China', *AMS News Bulletin*, 2012, pp. 32–5.

'Global Production Growth 2022 and Beyond', *AMS News Bulletin*, Autumn 2017.

'Floodplain Orchard Development', AMS Factsheet, 2018.

C. Shepherd and J. McKichnie, 'Mapping the Growth of Macadamia Orchards', Horticulture Innovations, *AMS News Bulletin*, Winter 2021.

'World Macadamia Organisation Needs Australian Support', AMS News, 18 November 2022.

L. Kojetin, 'Symposium Roundup: IMS 23', *AMS News Bulletin*, Summer 2023, pp. 56–8.

J. Price, 'In-Shell Demand Surge from China Sees Market Dynamics Shift', *AMS News Bulletin*, Summer 2023, pp. 4–5.

'Interim Funding Model for WMO', *AMS News Bulletin*, Summer 2023, pp. 24–5.

L Kojetin, 'Macadamia Production in Brazil', *AMS News Bulletin*, Winter 2024, pp. 14–17

California Macadamia Society (USA)

California Macadamia Society Yearbooks, 1955 to 2007.

William Kemper, 'Distribution of the Macadamia in California', *California Macadamia Society Yearbook (Yearbook)* 1959, p. 25.

E. Trask, 'Macadamias in Mexico', *Yearbook*, 1960.

C.H. Deichman, 'Macadamias in Jamaica', *Yearbook*, 1963, p. 38.

N.R. Grebe, 'Macadamia Country', *Yearbook*, 1963.

Arthur G. Lowndes, 'Some Observations from Australia', *Yearbook*, 1964, p. 29.

C.H. Deichman, 'Second Report from Sherwood Forest', *Yearbook*, 1965, pp. 34–5.

Arthur G. Lowndes, 'Two Historic Macadamia Trees in Australia', *Yearbook*, 1966, pp. 29–32.

Robert C. Axell, 'Some Observations of Macadamia Trees in Central America', *Yearbook*, 1967, p. 36.

W.B. Storey, 'The Macadamia in Trinidad', *Yearbook*, 1967, p. 43

Clarence Johnson, 'Excerpts from a Letter', *Yearbook*, 1968, p. 71.

———— 'Macadamia Growing in Paraguay', *Yearbook*, 1969, pp. 18–21.

C.A. Schroeder, 'The Macadamia in Malawi and South Africa', *Yearbook*, 1970, pp. 20–5.

Edward Tonks, 'A Short Report on the International Macadamia and Pecan Nut Symposium Held in Dunmow [UK], 1971', *Yearbook*, 1971, p. 59.

W.B. Storey, 'The Macadamia in Venezuela', *Yearbook*, 1971, pp. 62–9.

A. and D. Spurling, 'Macadamias in Malawi', *Yearbook*, 1971, p. 71.

Edward Tonks, 'Macadamias Revisited', *Yearbook*, 1972, p. 45.

Reuben Ohad, 'A Progress Report on Macadamia Culture in Israel', *Yearbook*, 1972, p. 51.

Clarence Johnson, 'A Report from Paraguay', *Yearbook*, 1972, p. 57.

Daphne Spurling, 'The Infant Macadamia Industry in Malawi', *Yearbook*, 1973, p. 86.

Ian McConachie, 'Macadamias in Australia', *Yearbook*, 1975, pp. 45–6.

C.A. Schroeder, 'Recent Developments in the Macadamia Industry of South Africa', *Yearbook*, 1975, p. 77.

CSR Ltd

'News Letter', Colonial Sugar Refining Company, April 1970.

E.R. Davenport, CSR Ltd, Report to Macadamia Division, 1978.

'CSR Grows, Processes, and Markets Macadamia Nuts' [brochure], January 1980.

'Australian Macadamias: The Inside Story' [booklet], 1980.

Hawaiian industry

Hawaii Macadamia Producers Association Proceedings, 1961 to 1982.

Hawaii Macadamia Nut Association Proceedings and *Yearbooks*, 1983 to 1999.

Proceedings of the First International Macadamia Research Conference Hawaii, 1992.

Horticulture Australia Limited

'2009 Nutrient Composition of Raw, Unsalted Tree Nuts', *Nuts for Life*, 2011.

'Strategic Investment Plan 2014–2019'.

International Nut Council (INC)
Global Statistical Review, 2020.

International Trade Commission (ITC/ USITC)

US Macadamia Nuts: Economic and Competitive Conditions Affecting the US Industry, 1998.

Macadamia Conservation Trust (Australia)

M. Powell and L. Gould, *Macadamia Species Recovery Plan 2020–2026*.

SAMAC (South/Southern Africa Macadamia Industry)

SAMAC, *In a Nut Shell*, 1993 to 2009.

C. Reim, 'Early History of Macadamias in South Africa', SAMAC Mini Symposium, 1991, pp. 1–2.

R.F. Hurly, 'A Theoretical Analysis of the 1993 SAMAC Tree Census', *SAMAC Yearbook (Yearbook)*, 1994, pp. 11–16.

Professor Peter Allan, 'Highlights of 30 Years with Macadamias', *Yearbook*, 1996, pp. 56–63.

Rocky Stone, 'Zimbabwe Macadamias', SAMAC, *In a Nutshell*, 1999, pp. 17–19.

'Protests over Kenya's Government Ban on Exports of Nut in Shell Macadamias', SAMAC, *In a Nut Shell*, May–June 2010, p. 7.

Mark Penter, 'Progress in New Cultivar Trials with the South African Macadamia Industry', SAMAC, *In a Nutshell*, July–September 2016, p. 24.

Philip Lee, *Macadamia 101: A Guide to Macadamia Production*, SAMAC, October 2021 (also update 2022).

United States Department of Agriculture (USDA)

Farm Reports, 1991–2015.

'Macadamia Nuts: Economic and Competitive Conditions Affecting the US Industry', report, 1998.

Gain Report, 15 September 2006.

3) Select Newspaper & General Interest Periodical articles (by date)

Australasian (Melbourne, Vic.)

Walter Hill, 'Report on Botanic Gardens', 30 September 1876.

Australian (incl. Weekend Australian; national)

Tom Hoult, 'Moving in on Macadamias Nuts', 26 October 1972, p. 2.

A. Goodhall. 'Cashing in on Nuts', 18 February 1978.

'Tax Plan in a Nutshell No Longer Looking Like a Cracking Idea', 30 June 1983.

Neil Jones, 'Ingenious Farmer with Eye for Ideas', 10 January 1994, p. 13.

Jeff Blakland, 'Renaissance Man Broke Ground', 29 April 1994, p. 17.

Australian Town and Country Journal (Sydney, NSW)

'The Australian Nut', Saturday 6 May 1899, p. 29.

Brisbane Courier (Qld)

Walter Hill, Letter to the Editor, 5 March 1867.

'Agricultural', 20 April 1867, p. 2.

'The Field Naturalists at Sankey's Scrub', 27 March 1888.

'The Queensland Nut', 4 April 1893.

'Toowoomba Botanic Gardens', 15 May 1897, p. 7.

Courier Mail (Brisbane)

'Your Brisbane – Past and Present', 20 May 2009.

A. Gorman, 'Ice-cream of the Crop: Anthony and Teena Mammino', 15 December 2013.

Coffs Harbour Advocate (NSW)

'Macadamia Society Seeking Information on Macadamia Habitat', 20 December 1989.

Cumberland Argus and Fruitgrowers Advocate (NSW)

'Macadamia Society Seeking Information on Macadamia Habitat', 20 December 1989.

'Nuts for the King', Saturday 8 November 1902, p. 8.

Daily Examiner (Grafton, NSW)

'Another Industry for the Foreigner', 11 September 1925, p. 4.

Daily Mail (Brisbane, Qld)

N.C. Hewitt, 'Nut Culture', 30 November 1926.

Financial Review (Australia)

Larry Schlesinger, 'Canadian Strike Record Deal for Bundaberg Macadamia Orchards', 30 January 2023.

Food Australia

R. Mason and I. McConachie, 'A Hard Nut to Crack', *Food Australia*, October 1994.

Gympie Times (Qld)

Tewantin Correspondent, '"Restdown" Orangery and Apiary', 15 August 1895.

'Big Future in a Nut Shell', Leisuretimes, 18 March 1997, pp. 26–35.

'Gympie Switched on to Nutshell Power', 18 October 2001.

Nambour Chronicle (Qld)

Bob Lake, 'Nut Market Is Limitless', 14 September 1978.

D. Ironside, 'Macadamia Money Saver', 9 August 1979.

New York Times (USA)

'Flight Attendant Kicked Off Korean Air Flight Alleges Cover Up, 18 December 2014.

Northern Star (Lismore, NSW)

C. Ogimori, C., 'Macadamia Specialist Predicts a Big Future for Local Growers', 30 March 1977.

'Big Scrub to Grow Again', 23 May 1983.

K. Corcoran, 'Macadamia Nut Industry Faces Turmoil', June 1990.

'Agrimac International', 21 April 1995, pp. 4–6.

'Macadamia Market Expansion', 5 August 1998.

Observer Review (UK)

Clive James, 'Cracking Nut for Culture', 27 June 1976.

Panorama (inflight magazine of Ansett Airlines)

'Australia's Own Nut to Become Big Export', April 1980, pp. 4, 5.

Queensland Agricultural Journal

'The Queensland Nut', July–December 1939, p. 165.

Queensland Country Life

'Trees on a Trellis', 31 March 1977, p. 29.

'Firm Expands Macadamia Holdings to North Queensland', 7 March 1981.

'18,000 Trees Take Hold at Bundaberg', 22 September 1988, p. 27.

Queensland Fruit and Vegetable News

'Macadamia Nut Growers' Conference', 16 January 1958.

A. Collins, letter, 20 September 1965, p. 618.

Queenslander (Brisbane)

'Coolibah', 11 August 1917, p. 2.

'Nut Growing Experiments', 8 October 1931, p. 13.

Rockhampton Morning Bulletin (Qld)

'Big Plan for Nut Orchard', 29 August 1988.

Sunday Mail (Brisbane, Qld)

'Delicious Aroma Started Industry', 7 January 1979.

Sunshine Coast Daily (Qld)

B. Burnett, 'Irrigation', The Land, 1 December 1981.

'New Era for Macadamias', 25 September 1986.

Tweed Daily (Tweed Head, NSW)

L. White, 'Change in Ownership for Large Macadamia Farms', 11 May 2016.

Woman's Day (magazine)

Shirley Stackhouse, 'Nuts: A Sensible Crop', 13 June 1977.

4) Published articles with by-lines, also chapters and books (by name)

Professor Peter Allan, 'Highlights of 30 Years with Macadamias', *SAMAC Yearbook*, 1996, pp. 56–63.

Archer, T., *Recollections of a Rambling Life* (1897), facsimile edn, Bowen Hills, Qld: Boolarong, 1988.

Aurousseau, Marcel, *The Letters of F.W. Ludwig Leichhardt*, Cambridge University Press, 1968.

Axell, Robert C., 'Some Observations of Macadamia Trees in Central America', *California Macadamia Society Yearbook*, 1967.

Barker, P. and E. Thomas, 'Origins, Significance, and

Paleoclimatic Influences of the Antarctic Circumpolar Current', *Earth Review*, June 2004, pp. 43–162.

Bateson, Charles, *Patrick Logan: Tyrant of Brisbane Town*, Sydney: Ure Smith, 1966.

Benson, R. and J. Chaseling, *Investment in a Macadamia Nut Orchard*, Misc. Bulletin 17, NSW Department of Agriculture, 1972.

Bentham, George, assisted by Ferdinand Mueller, *Flora Australiensis*, Vol. 2, London: Lovell Reeve & Co., 1864.

Big Scrub Rainforest Landcare Group, 'Rainforest Rescue', 2021.

Bird, Nai Nai, *Guerum Woolana: Wild Food*, self-published, 1990.

Blainey, Geoffrey, *The Story of Australia's People*, Vol. 2, Melbourne: Viking, 2015.

Blakland, Jeff, 'Renaissance Man Broke Ground', *Australian* (newspaper), 29 April 1994, p. 17.

Bohya, Lindi, 'Malawi Macadamias', *The Macadamia Magazine*, 2020.

Branson, Adam, in *Tree Nuts Annual*, Tree Nuts Annual_Beijing_China People's Republic of_CH2022_0105, USDA Foreign Agricultural Service, 2022.

Brown, Kerry, *Our Sweetest Icon: Sunshine Plantation's Big Pineapple 1971–2011*, Qld: Love of Books, 2011.

Brown, Peter, 'A Nut with a Future', *Australian Horticulture*, November 1984.

Bunya Mountains Aboriginal Aspirations and Caring for Country Plan, Bunya Mountains Elders Council and Burnett Mary Regional Group, 2010.

'Burleigh Heads National Park', Queensland Government Environmental Protection Agency, Queensland Parks and Wildlife Service, undated.

Burnett, B., 'Irrigation', The Land, *Sunshine Coast Daily*, 1 December 1981.

Burnum Burnum's Aboriginal Australia: A Traveller's Guide, edited by David Stewart, Sydney: Angus & Robertson, 1988.

Burt, Dorothy, in *Proceedings*, Second International Macadamia Symposium, 2003, pp. 14–15.

Chambers, Catherine, 'The H2 Hinde Tree', 7 September 2009, Environment and Resources Queensland Government.

Cheel, E. and F.R. Morrison, *The Cultivation and Exploitation of the Australian Nut*, Sydney: Government Printer, 1935.

(Ms) Chen Yuxiu, 'Macadamia Production in China, 18 September 2012', in published *Proceedings* of the Eighth International Macadamia Symposium, China.

Clarke, Philip A., *Aboriginal Plant Collectors*, Kenthurst, NSW: Rosenberg Press, 2008.

Collins, A., letter in *Queensland Fruit and Vegetable News*,

20 September 1965, p. 618.

'Coolibah', *Queenslander* (Brisbane), 11 August 1917, p. 2.

Cooper, Joshua, 'Macadamia Trees and Their Popularity in the Philippines', *Healing Plant Foods*, 1 January 2023.

Corcoran, K., 'Macadamia Nut Industry Faces Turmoil', *Northern Star* (Lismore), June 1990.

Crawford, Foreman, *Duck Creek Mountain now Alstonville*, self-published, 1983.

Crawford, Harold W., 'Brisbane's Two Botanic Gardens', *Australian Horticulture*, March 1983.

Dahler, J., C. McConchie and C. Turnbull, 'Quantification of Cyanogenic Glycosides in Seedlings of Three Macadamia [Proteaceae] Species', *Australian Journal of Botany*, 43 (6) 1995, pp. 619–28.

Dardaqan, Colleen, 'Mozambique Venture', *The Macadamia Magazine*, February 2020.

Davenport, E. 'AMS Review, 1997', *AMS News Bulletin*, 27 June 1997.

———— 'Vale: K.J. Ainsbury', *AMS News Bulletin*, March 1999, p. 6.

Dawes, S.N., *Macadamia Varieties and Culture*, (NZ) Ministry of Agriculture and Fisheries, 1981.

Deichman, C.H., 'Macadamias in Jamaica', *California Macadamia Society Yearbook*, 1963.

———— 'Second Report from Sherwood Forest', *California Macadamia Society Yearbook*, 1965.

Dettmann, M. and H. Clifford, 'Fossil Fruit of the Macadamieae', *Nature*, Queensland Museum, 2010.

Dettmann, M. and D. Jarzen, 'The Early History of the Proteaceae in Australia: The Pollen Record', *Australian Systematic Botany*, 11 (4) 1998; online https://www.publish.csiro.au/sb/sb97022.

Egan, Joanna, Black Bean *Castanospernum australe* in 'Illustrated: Australian Plant Seeds', *Australian Geographic*, 25 August 2017; online https://www.australiangeographic.com.au/topics/science-environment/2017/08/illustrated-australian-plant-seeds/.

El Cultivo de la Macadamia en Costa Rica, Centro Agricola de Turrialba, 1984.

Ellwood, W.J., J.B. Campbell and G.J. Susino, 'Agricultural Hunter-Gatherers: Food-Getting, Domestication and Farming in Pre-Colonial Australia', in Jan Michal Burdukiewicz (ed.), *Understanding the Past: Papers Offered to Stefan K. Kozlowski*, University of Warsaw, 2009, pp. 115–22.

Fitzgerald, B., 'West Australia Numbers Bloom', *ABC Rural Report*, 23 September 2014.

Fleming, L., 'My Great Uncle Norman Rae Greber OAM and His Wife Kathleen Houston', *The Gympie Researcher* [Newsletter of Gympie Family History Society Inc.], September 2017.

Fletcher, W., *Growing Macadamia Nuts in New Zealand*, (NZ) Department of Agriculture and Fisheries, 1973.

Francis, W.D., *Australian Rain-Forest Trees*, 2nd edn, Canberra: AGPS, 1951.

Godard, L., J.-B. Roelens, M. Béringuier and F. Montfort, 'Report on the Mozambican Macadamia Value Chain', ACAMOZ, 2021–22, p. 13; online: https://iam.gov.mz/index.php/download/report-on-mozambican-macadamia-value-chain-acamoz-nitidae/.

Goodhall. A. 'Cashing in on Nuts', *Weekend Australian*, 18 February 1978.

Gorman, A., 'Ice-cream of the Crop: Anthony and Teena Mammino', *Courier Mail*, 15 December 2013.

Grebe, N.R., 'Macadamia Country', *California Macadamia Society Yearbook*, 1963.

Green, Len, 'Aussie Nuts', from *Bondi Bonbons: A Selection from Joys of a Jongleur*, Bondi, NSW: Six Ways X-Rays, 1993.

Greer, Germaine, *White Beech: The Rainforest Years*, Bloomsbury Publishing, 2014.

Gross, C.L. and P.H. Weston, 'A Review of *Macadamia F. Mueller*', National Herbarium of New South Wales, 1992.

Hall, J. and W. Petrie, 'Rollo Petrie Reminiscences', self-published, 2004.

Hamilton, R.A. and E.T. Fukunaga, *Growing Macadamia Nuts in Hawaii*, Hawaii Agricultural Experiment Station, 1959.

Hamilton, R.A., 'Macadamia Nut Development and Prospects in Brazil', Report to University of Hawaii, 1978.

Hanlon, J., 'Power Without Responsibility: The World Bank and Mozambican Cashew Nuts', *Review of African Political Economy*, 83, 2000, pp. 29–45.

Harari, Yuval Noah, *Sapiens: A Brief History of Humankind*, Random House/ Vintage, 2015.

Hardner, C., 'Squirrelling the Macadamia', *Ecos*, 107, April–June 2001, p. 5.

———— 'Macadamia Domestication in Hawai'i', *Genetic Resources and Crop Evolution*, 63 (8) 2015, pp. 1411–30.

Hardner, Craig, Mark Nikum, Jacqueline Batley and Ainnatul Adawiyah binti Ahmad Termizi, 'Reconstructing the Domestication Pathway of Macadamia from Australia via Hawaii and California', American Society for Horticultural Science (ASHS) Annual Conference paper, 2013.

Herbert, D.A., 'Bidwill, John Carne 1815–1853', *Australian Dictionary of Biography*, Vol. 1, Melbourne: Melbourne University Press, 1966; online https://adb.anu.edu.au/biography/bidwill-john-carne-1778/text1997, accessed 8 March 2024.

Hewitt, N.C., 'Nut Culture', *Daily Mail* (Brisbane), 30 November 1926.

Hill, Robert, 'Origins of the Southeastern Australian

Vegetation', *Philosophical Transactions of the Royal Society*, 2004; online https://royalsocietypublishing.org/doi/10.1098/rstb.2004.1526.

Hill, Walter, Letter to the Editor, *Brisbane Courier*, 5 March 1867.

———— 'Report on Botanic Gardens', *Australasian* (Melbourne), 30 September 1876.

Home, R.W., 'Ferdinand von Mueller, Botanist of Honour', *Australian Heritage*, Winter 2006, pp. 66–71.

Horsburg, C., *Macadamia Nuts: A Marketing Analysis*, QDPI, 1996.

Hoult, Tom, 'Moving in on Macadamias Nuts', *Australian* (newspaper), 26 October 1972, p. 2.

Hu, W., M. Fitzgerald, B. Topp, M. Alam and T.J. O'Hare, 'A Review of Biological Functions, Health Benefits, and Possible de Novo Biosynthetic Pathway of Palmitoleic Acid in Macadamia Nuts', *Journal of Functional Foods*, 2019, pp. 1–12; online: https://doi.org/10.1016/j.jff.2019.103520.

Hurly, R.F., 'A Theoretical Analysis of the 1993 SAMAC Tree Census', *SAMAC Yearbook*, 1994, pp. 11–16.

Iqbal, M. and R. Hampton, 'Macadamia Nut Production in Fiji', *Fiji Agricultural Journal*, 1977, 39 (2) 1977, pp. 97–103.

Ironside, D., 'Macadamia Money Saver', *Nambour Chronicle*, 9 August 1979.

Ito, P., N. Subang and M. Yun, 'Macadamia Acreage and Production in the People's Republic of China', *HMNA Proceedings*, 3 May 1996.

Jackson, Stephanie, 'Going Nuts over Bush Tucker', *Blues Country*, January–February 2003.

James, Clive, 'Cracking Nut for Culture', *Observer Review*, 27 June 1976.

Johansen, R., *Tales of Our Times*, Murwillumbah Print Spot, 1997.

———— 'John Waldron: A Cracking Pioneer', *Tales of Our Times*, 2003, p. 26.

Johnson, Clarence, 'Excerpts from a Letter', *California Macadamia Society Yearbook*, 1968, p. 71.

———— Macadamia Growing in Paraguay', *California Macadamia Society Yearbook*, 1969, pp. 18–21.

———— 'A Report from Paraguay', *California Macadamia Society Yearbook*, 1972.

Johnson, L.A.S., '*Macadamia ternifolia F. Muell* and a Related New Species', *Proceedings of the Linnean Society of New South Wales*, 79 (1–2) 1954, pp. 15–18.

———— '*Macadamia tetraphylla*', *Proceedings of the Linnean Society of New South Wales*, 79 (1–2) 1954.

Jones, Kim, 'Industry Statistics: Trees and Production', 2000.

Jones, K., K. Quinlan and K. Wilson, 'Study Tour of the China Macadamia Industry, February 2014', HAL Report, MC12702.

Jones, Neil, 'Ingenious Farmer with Eye for Ideas', *Australian* (newspaper), 10 January 1994, p. 13.

Kemper, William, 'Distribution of the Macadamia in California', *California Macadamia Society Yearbook*, 1959.

Kerkhove, Ray, *Aboriginal Campsites of Greater Brisbane*, Bowen Hills, Brisbane: Boolarong Press, 2015.

Kiernan, L., 'HAIG Divests Australia's Largest Macadamia Portfolio to Belgian Sugar Producer Finasucre in $60M Deal', *Global AgInvesting*, 22 October 2019.

Kojetin, L., 'Symposium Roundup: IMS 23', *AMS News Bulletin*, Summer 2023.

———— 'Macadamia Production in Brazil', *AMS News Bulletin*, Winter 2024, pp. 14–17.

Lahey, Shirley, *The Laheys: Pioneer Settlers and Sawmillers*, Taringa, Qld: S. Lahey, 2003.

Lake, Bob, 'Nut Market Is Limitless', *Nambour Chronicle*, 14 September 1978.

Lee, Philip, 'South Africa and Africa', *Proceedings Second International Macadamia Symposium*, 2003.

———— *Macadamia 101: A Guide to Macadamia Production*, SAMAC, October 2021.

Loebel, Ross, 'Macadamia Production Expected to Quadruple', NSW Department of Agriculture, 17 September 1985.

Lowndes, A.G. (Arthur), *The World Pattern: A Geography for Secondary Schools*, Sydney: Angus & Robertson, 1947.

———— *South Pacific Enterprise: The Colonial Sugar Refining Company Limited*, Sydney: Angus & Robertson, 1956.

———— 'Some Observations from Australia', *California Macadamia Society Yearbook*, 1964, p. 29.

———— 'Two Historic Macadamia Trees in Australia', *California Macadamia Society Yearbook*, 1966, pp. 29–32.

Maiden, J.H., *The Useful Native Plants of Australia*, Sydney: Turner & Henderson, 1889.

Maiden, J.H. and E. Betche, 'On a New Species of Macadamia together with Notes on Two Plants New to the Colony', *Proceedings of the Linnean Society of New South Wales*, 21, 1897, pp. 624–25.

Mason, R. and I. McConachie, 'A Hard Nut to Crack', *Food Australia*, October 1994

Mast, A. et al., 'A Smaller Macadamia from a More Vagile Tribe', *American Journal of Botany*, 95 (7) 2008, pp. 843–70.

McConachie, Ian, 'Nutta Products Qld Pty Ltd', letter to industry, 18 November 1974.

———— 'Macadamias in Australia', *California Macadamia Society Yearbook*, 1975, pp. 45–6.

———— 'Macadamias in New Zealand', *AMS News Bulletin*, 13 (2) July 1986.

———— 'Twenty-Five Years Old', *AMS News Bulletin*, 1999.

———— 'Who Is This Man?' *AMS News Bulletin*, September 2000, pp. 22–3.

———— 'Greber, Norman Rae (Norm) (1902–1993)', *Australian Dictionary of Biography*, Vol. 19, Canberra: Australian National University Press, 2021; online https://adb.anu.edu.au/biography/greber-norman-rae-norm-25514/text33859, accessed March 2024.

McConchie, Dr C. (CSIRO), 'What Does *Macadamia jansenii* Represent?', Restricted Report to Australian Macadamia Society, 21 August 2012.

McGregor, Andrew, 'A Review of the World Production and Market Environment for Macadamia Nuts', report, 1991.

McKinnon, Ross, 'Hill, Walter (1819–1904)', *Australian Dictionary of Biography*, Supplementary Vol., Melbourne: Melbourne University Publishing, 2005; online https://adb.anu.edu.au/biography/hill-walter-12981/text23461, accessed 8 March 2024.

McMillan-Webster, Fiona, *The Age of Seeds*, Australia: Thames and Hudson, 2022.

McMinn, W.G., *Allan Cunningham: Botanist and Explorer*, Melbourne: Melbourne University Press, 1970.

Mharidzo, I., J. Mwandifura, L. Chikazhe, J. Manyeruke and N. Mashayakure, 'An Insight into the Macadamia Nuts Industry of Zimbabwe: Its History, Current State and Constraints', *IJRISS [International Journal of Research and Innovation in Social Science]*, February 2022, pp. 686–94; online: https://www.rsisinternational.org/journals/ijriss/Digital-Library/volume-6-issue-2/686-694.pdf.

Michall, Naomi, *Food Navigator Newsletter*, 20 March 2019.

Miles, Tallis, 'Flagship Weekend', *Australian AgJournal*, May 2024, pp. 20–5.

Morris, Deidre, 'Mueller, Sir Ferdinand Jakob Heinrich von (1825–1896), *Australian Dictionary of Biography*, Vol. 5, 1974; online https://adb.anu.edu.au/biography/mueller-sir-ferdinand-jakob-heinrich-von-4266/text6893, accessed March 2024.

Nakamoto, S.T., 'Macadamias in Costa Rica, 1992', *Tree Crops Journal* [Hawaii], 1 (3) 1993.

Nock, C.J. et al., 'Wild Origins of Macadamia Domestication Identified Through Intraspecific Chloroplast Genome Sequencing', *Frontiers in Plant Science*, 10 (334) 2019; doi: 10.3389/fpls.2019.00334.

———— 'Genetic Diversity and Population Structure of Wild and Domesticated Macadamia', Final Report MC18004, Horticultural Innovation series, Horticulture Innovation Australia Ltd, 2021

O'Donnell, Dan, 'Ludwig Leichhardt and the Sunshine Coast', *Queensland Agricultural Journal*, September–October 1989.

Ogimori, C., 'Macadamia Specialist Predicts a Big Future for Local Growers', *Northern Star* (Lismore), 30 March 1977.

Ohad, Reuben, 'A Progress Report on Macadamia Culture in Israel', *California Macadamia Society Yearbook*, 1972, p. 51.

Ortega, E. et al., 'Characteristics and Lipid Profiles of Macadamia Nuts', *International Journal of Engineering and Applied Science*, February 2014, pp. 33–9.

Osbourne, Peter, 'The Queensland Acclimatisation Society', *Royal Historical Society of Queensland Journal*, November 2008, pp. 337–47.

Penter, Mark, 'Beaumont Drive to be Paved by Len Hobson', *SubTrop Quarterly Journal*, April–June 2015, pp. 56–8.

———— 'Progress in New Cultivar Trials with the South African Macadamia Industry', SAMAC, *In a Nutshell*, July–September 2016.

'Physical Earth', *National Geographic Magazine*, May 1998.

Pole, M., 'The Proteaceae Record in New Zealand', *Australian Systematic Botany*, 11 (4) 1988, pp. 343–72.

Poole, Dr Mike, 'The Proteaceae Record in New Zealand', *Australian Systemic Botany*, 11 (4), November 1998.

Pope, Willis T., *The Macadamia Nut Industry in Hawaii*, Hawaii Agricultural Experiment Station, 1929.

Powell, M. and L. Gould, *Macadamia Species Recovery Plan 2020–2026*, Macadamia Conservation Trust.

Price, J., 'In-Shell Demand Surge from China Sees Market Dynamics Shift', *AMS News Bulletin*, Summer 2023, pp. 4–5.

Pritchett, John S., 'What Is the Bishop Estate?', 1999; online: https://www.pritchettcartoons.com/bet_essay.htm.

Propagation of Macadamia, NSW Department of Agriculture Bulletin H225, 1972.

'R.B.', in *Sydney Morning Herald*, 30 April 1904.

Read, Stuart, 'Bidwill of Wide Bay', *Queensland Review*, 19 (1) 2012.

Reim, Chris, 'Developments in South Africa', *Proceedings HMNA*, 1976, pp. 116–19.

———— 'Early History of Macadamias in South Africa', SAMAC Mini Symposium, 1991, pp. 1–2.

Reitow, David, 'A Hard Nut to Crack: Macadamia in Hawaii', *HortScience*, 47 (10) October 2012.

Rios, Sonia, Gary Bender and Ben Faber, 'Macadamia Production in California: A Hidden Gem of an Industry?', *Topics in Subtropics* [blog], University of California, Agriculture and Natural Resources, March 2020; online: https://ucanr.edu/blogs/blogcore/postdetail.cfm?postnum=39358.

Roderick, Colin, *Leichhardt: The Dauntless Explorer*, North Ryde, NSW: Angus & Robertson, 1988.

Roehling, Elizabeth E., 'Macadamia Tree Offer Lifeline to Small Growers', *IPS News*, 27 June 2011.

Ross, A., 'The Macadamia Nut', *Queensland Agricultural Journal*, 1952, pp. 21–32.

Rossetto, Maurizio et al., 'From Songlines to Genomes: Prehistoric Assisted Migration of a Rain Forest Tree by Australian Aboriginal People', 2017, online https://pubmed.ncbi.nlm.nih.gov/29117184 and also https://doi.org/10.1371/journal.pone.018666.

Royal Society of Victoria (Melbourne, Vic.), *Proceedings of the Royal Society of Victoria, Vol. 2*, 1858.

Rumsey, Herbert J., *Australian Nuts and Nut Growing in Australia*, Dundas, NSW: Herbert J. Rumsey and Sons, 1927.

Russell, H.S., *The Genesis of Queensland*, Sydney: Turner & Henderson, 1888, p. 134; online https://www.textqueensland.com.au/item/book/b86c5755be-236c74a8fc29e7ae220cb6.

Russell, K.F., 'Macadam, John (1827–1865)', *Australian Dictionary of Biography*, Vol. 5, Melbourne: Melbourne University Press, 1974; online https://adb.anu.edu.au/biography/macadam-john-4054/text6453, accessed 8 March 2024.

Sabate, Joan, 'The Macadamia Nut Effects on Adiposity and Cardiovascular Risk Factors', *Journal of Nutritional Science*, May 2023.

Schlesinger, Larry, 'Canadian Strike Record Deal for Bundaberg Macadamia Orchards', *Financial Review*, 30 January 2023.

———— 'Ryral Funds Nuts Out Macadamia Earnings', *Financial Review*, 24 August 2023.

Schormann, R., 'History and Statistics of the Southern Lowveld', *Mini Macadamia Symposium Proceedings*, 1991, pp. 4–5.

Schroeder, C.A., 'The Macadamia in Malawi and South Africa', *California Macadamia Society Yearbook*, 1970, pp. 20–5.

———— 'Recent Developments in the Macadamia Industry of South Africa', *California Macadamia Society Yearbook*, 1975, p. 77.

Shapcott, A., 'Threatened Plant Translocation Case Study: *Macadamia jansenii* (Bulburin nut), Proteaceae', *Australasian Plant Conservation*, 28 (1) June–August 2019.

Shapcott, A. and M. Powell, 'Demographic Structure, Genetic Diversity and Habitat Distribution of the Endangered, Australian Rainforest Tree *Macadamia jansenii* Help Facilitate an Introduction Program', *Australian Journal of Botany*, 59, 2011, pp. 215–25.

Shepherd, C. and J. McKichnie, 'Mapping the Growth of Macadamia Orchards', Horticulture Innovations, *AMS News Bulletin*, Winter 2021.

Shigeura, G.T. and H. Ooka, *Macadamia Nuts in Hawaii: History and Production*, University of Hawaii, 1984.

Skyring, Z.D.S., 'Hunting with the Wide Bay Blacks', Article 11 in *Gympie Jubilee 1867–1917*, supplement to *Gympie Times*, 16 October 1917.

Smith, Gordon, *Walter Hill of Brisbane's Botanic Gardens*, Greenslopes, Qld: G.D. Smith, 2008.

Smith, L., *Proceedings of the Royal Society of Queensland*, 67 (5) July 1956.

Southern African Development Community, 'Sugar and Macadamias', *SADC Sugar Digest*, 2019, p. 14.

Southern Moreton Bay Islands Heritage Trail, Redland City Council, 2008.

Spurling, A. and D., 'Macadamias in Malawi', *California Macadamia Society Yearbook*, 1971, p. 71.

Stackhouse, Shirley, 'Nuts: A Sensible Crop', *Woman's Day*, 13 June 1977.

Steele, J.G., *The Explorers of the Moreton Bay District, 1770–1830*, St Lucia, Qld: University of Queensland Press, 1972.

———— *Aboriginal Pathways: Southeast Queensland and the Richmond River*, St Lucia, Qld: University of Queensland Press, 1984.

Stephens, E., 'Australia Proud to Lead the World', *Food News*, 7 April 1996, p. 7.

Stephenson, R.A., 'Some Impressions of the Macadamia Industry in South Africa and Malawi', November 1989.

———— 'The Macadamia: From Novelty Crop to New Industry', *Agricultural Science*, November 1990, pp. 38–43.

———— 'Impressions on Macadamias in China', *AMS News Bulletin*, 2012, pp. 32–5.

Stone, Rocky, 'Zimbabwe Macadamias', SAMAC, *In a Nutshell*, 1999, pp. 17–19.

Storey, W.B., 'The Macadamia in Trinidad', *California Macadamia Society Yearbook*, 1967.

———— 'The Macadamia in Venezuela', *California Macadamia Society Yearbook*, 1971, pp. 62–9.

Sullivan, Hon. V. (MLA), 'Two New Macadamia Varieties for the Industry', Queensland Department of Primary Industries, 10 February 1979.

Supamatee, Ito and Jalahan, 'Hawaiian Macadamia Cultivars in Thailand', *HMNA Proceedings*, 1992, p. 107.

Swain, E.H.F., *The Timber and Forest Products of Queensland*, Brisbane, Qld: Queensland Government Printer, 1928.

Tewantin Correspondent, '"Restdown" Orangery and Apiary', *Gympie Times*, 15 August 1895.

Thompson, J. Maldon, Department of Public Lands, *Timber Regulations*, Brisbane, 18 May 1870.

Tonks, Edward, 'A Short Report on the International Macadamia and Pecan Nut Symposium Held in Dunmow [UK], 1971', *California Macadamia Society Yearbook*, 1971, p. 59.

———— 'Macadamias Revisited', *California Macadamia*

Society Yearbook, 1972, p. 45.

Trask, E., 'Macadamias in Mexico', *California Macadamia Society Yearbook*, 1960.

Turner, F., *The Cultivation of the 'Australian Nut'*, Dept of Agriculture (NSW), 1893.

Twentyman, J., *AMS News Bulletin*, May 2000, pp. 30–1.

Vibetnum, 'The Flower Garden', *Western Mail* (Perth, Australia), 8 August 1913, p. 14.

Vievers, Clive, *Walter Hill*, Bulletin 371, Historical Society North Queensland Inc., August 1991.

Volio, A., 'Macadamias in Costa Rica Hawaiian Grown', [Australian] *Tree Crop* [magazine], Summer 1996.

Wagner-Wright, Sandra, *History of the Macadamia Nut Industry in Hawaii 1881–1981*, Lewiston, NY: Edwin Mellen Press, 1985.

Watsford, Penny, *Forest Bountiful*, Murwillumbah, NSW: Nullum Publications, 2010.

Watson, F.J., *Vocabularies of Four Representative Tribes of South Eastern Queensland*, supplement to *Journal of the Royal Geographical Society of Australasia (Queensland)*, 1944.

White, L., 'Change in Ownership for Large Macadamia Farms', *Tweed Daily* (Tweed Head, NSW), 11 May 2016.

Whyte, Jill, 'What's Happening in South Africa', *Proceedings AMS Conference*, 2005.

Wild Plants of Greater Brisbane, Brisbane, Qld: Queensland Museum, 2000.

Wilkinson, Jennifer, 'Macadamias Grown in Australia but Processed Offshore', *Australian Nutgrowers*, September–November 2002.

Williamson, J., *Football's Forgotten Tour: The Story of the British Australian Rules Venture of 1888*, Applecross, WA: J. Williamson, 2003.

Wilson, K.J., 'The Past, Present and Future of the Australian Macadamia Industry', *AMS News Bulletin*, 2008.

Yamaguchi, Alan, Mauna Loa Macadamias Report, 10 March 2005.

———— 'International Research', *AMS News Bulletin*, 2007, pp. 67–70.

Yap, Nelson, 'Zadro Family Hoping Investors Will Go Nuts over Macadamia Offering', *Australian Property Journal*, 27 July 2022.

Zuza, E.J., K. Maseyk, S. Bhagwat, A. Emmott, W. Rawes and Y.N. Araya, 'Review of Macadamia Production in Malawi', *Agriculture*, November 2021.

5) Unpublished manuscripts and theses (by name)

Coleman, Bernard, 'The History of Macadamias in New Zealand', Masters thesis, Massey University, 2003.

Frawley, K.J., 'Forest and Land Management in North-East Queensland: 1859–1960', PhD thesis, Dept of Geography, University of Queensland, 1983.

Lowndes, Arthur, 'Memoir', typescript, 1989.

Mai, Thuy, 'Genomic-Assisted Exploitation of Wild Germplasm for Improvement of Macadamia', PhD thesis, Queensland Alliance for Agriculture and Food Innovation, University of Queensland, 2020.

Ryan, Tom, 'John Carne Bidwill', unpublished manuscript, 2010.

Viola, Ash and Bill, 'Personal Story', manuscript, c. 2008.

6) Talks and Presentations (by name)

Donkin, Derek, 'By the Numbers', SubTrop Marketing Symposium, November 2016.

Husband, Rod and Vanessa Hayes, 'Macadamias in New Zealand', AMS Conference presentation, 2012.

NutPAK, 'The Building Blocks of a Sustainable Macadamia Industry in Kenya', presentation, September 2012.

Osbourne, Myles, AusMac Conference Presentation, Australia, November 2022.

Piza, Pedro Luis Blais de Toledo and Isabel Piza, presentation at Eighth International Macadamia Symposium, China, 2018.

Powell, M., 'A Brief History of Time', Presentation, University of Sunshine Coast (Qld), March 2006, slides 10 & 11.

Tapernoux, Matthias, 'Macadamias in Ecuador', presentation at the International Macadamia Symposium, Brazil, October 2006.

Index

Acknowledgements

I pay my respects to the Traditional Owners of the land where macadamias grow and am grateful to the Elders who have shared their knowledge with me and now you, the reader.

I have been told many stories, and have been corrected and encouraged. Many people have contributed to this book. I am sure to have missed some, but my sincere thanks and gratitude go to one and all. In particular, the following have been most helpful:

The four Australian processors who, in 2005, donated a considerable sum to assist in completing and publishing *The Macadamia Story* (as it was then tentatively titled). They are Marquis Macadamias Ltd, Suncoast Gold Macadamias [Aust.] Limited, Agrimac International Pty Ltd and Macadamias Plantation of Australia Pty Ltd.

The Macadamia: Australia's Gift to the World would probably never have made it into print if not for some incredible, talented women who believed the book was important and 'mothered' both it and the author. Dr Elaine Brown, a professional historian, who over four years researched, disciplined me, and worked through draft after draft. Her skills and dedication to the book converted my early drafts into readable history. Then Denise Bond from the Macadamia Conservation Trust who applied her scientific knowledge, her expanding interest in history, her writing skills and helped progress it to completion. She has gone far, far beyond what was expected and both her own and Elaine's contribution will never be forgotten.

A number of people read, corrected and added to chapters. They were Philip Lee for Africa, Pedro Toledo Piza for Brazil, Ms Chen Yuxiu for China, Carlos Nottebohm for Guatemala, Jolyon Burnett whose perceptive comments and deep understanding have added to many chapters, Nathan Trump for Hawaii, Andrew Burnside for CSR Ltd, Bui Quoc Hoan and Professor Hoang Hoe for Vietnam, Kim Wilson for South East Asia, Mr Maschiro Shiota for Kenya, Matthias Tapernoux for Ecuador, Colleen Wall, a Dauwa-Kabi woman, Vanessa Hayes, a proud Maori leader, Aunty Theresa Williams a Traditional Yugambeh Custodian, Wayne Boldery and Brice Kaddatz.

Many others helped, including Steve Angus, Ted Davenport, Grace Fon, Clare Hamilton-Bate, Dr Craig Hardner who made me a better historian, Ron Johannsen, Kim Jones, Ellie Kaddatz, Bryan Raphael, Geoff Smith, Abbie Grant Taylor, Amanda and Mal Thatcher.

The team from Australian Scholarly Publishing have been enthusiastic and supportive. Publisher Nick Walker, designer David Morgan and copy-editor Diane Carlyle, in particular, have shown patience and professionalism and made the volume more attractive and readable.

Finally, this book would not have been possible without the support and encouragement of my wife Janet over many decades.